Human Senescence
Evolutionary and Biocultural Perspectives

Much current research on the biology of senescence is on cell lines, nematodes, or fruit flies, which may be only of peripheral relevance to the problems encountered in human senescence. *Human Senescence* reviews the evolutionary biology of human senescence and life span, and the evolutionarily recent development of late-life survival. In examining how human patterns of and variability in growth and development have altered later life survival probabilities and competencies, how survival during mid-life contributes to senescent dysfunction and alteration, and the possibilities of further extending human life span, it gives a better understanding of how humans came to senesce as slowly as they do. Bringing together gerontological, anthropological, and biocultural research, it explores human variation in chronic disease, senescence, and life span as outcomes of early life adaptation and the success of humankind's sociocultural evolution. It will be a benchmark publication for all interested in how and why we age.

DOUGLAS E. CREWS is Associate Professor in the Department of Anthropology and School of Public Health at the Ohio State University.

Cambridge Studies in Biological and Evolutionary Anthropology

Series Editors

HUMAN ECOLOGY
C. G. Nicholas Mascie-Taylor, University of Cambridge
Michael A. Little, State University of New York, Binghamton
GENETICS
Kenneth M. Weiss, Pennsylvania State University
HUMAN EVOLUTION
Robert A. Foley, University of Cambridge
Nina G. Jablonski, California Academy of Sciences
PRIMATOLOGY
Karen B. Strier, University of Wisconsin, Madison

Consulting Editor
Emeritus Professor Derek F. Roberts

Cambridge Studies in Biological and Evolutionary Anthropology

Human Senescence

Evolutionary and Biocultural Perspectives

DOUGLAS E. CREWS

Department of Anthropology and School of Public Health,
Ohio State University, Columbus, Ohio, U.S.A.

CAMBRIDGE
UNIVERSITY PRESS

PUBLISHED BY THE PRESS SYNDICATE OF THE UNIVERSITY OF CAMBRIDGE
The Pitt Building, Trumpington Street, Cambridge, United Kingdom

CAMBRIDGE UNIVERSITY PRESS
The Edinburgh Building, Cambridge CB2 2RU, UK
40 West 20th Street, New York, NY 10011–4211, USA
477 Williamstown Road, Port Melbourne, VIC 3207, Australia
Ruiz de Alarcón 13, 28014 Madrid, Spain
Dock House, The Waterfront, Cape Town 8001, South Africa

http://www.cambridge.org

First published 2003

Printed in the United Kingdom at the University Press, Cambridge

Typeface Times 10/12.5 pt   *System* LATEX 2_ε [TB]

A catalog record for this book is available from the British Library

Library of Congress Cataloging in Publication data

Crews, Douglas E.
  Human senescence: evolutionary and biocultural perspectives /
  Douglas E. Crews.
    p.   cm.
  Includes bibliographical references and index.
  ISBN 0 521 57173 1
  1. Aging.   2. Human evolution.   I. Title.
  QP86 .C74   2003
  612.6'7–dc21   2002034951

ISBN 0 521 57173 1 hardback

Contents

Preface

Before writing this volume, I first asked if there really is any need for another volume on the evolutionary biology of senescence. The answer was an emphatic no. The basic evolutionary biology underlying senescence is fairly well described (Rose 1991; Arking 1998), as are evolutionary tradeoffs between the soma and the germline (Kirkwood and Kowald 1997) that push reproductive success over somatic survival in sexually reproducing species. The second question I asked had a positive response, however, and that is whether there was a need for a volume examining the evolutionary biology of human senescence from an evolutionary and biocultural perspective. Prior reports on the bioanthropology of human life span and senescence generally have been in chapters in edited volumes (Weiss 1981; Crews 1990a; Beall 1994; Crews and Garruto 1994) and journal articles (Borkan *et al.* 1982; Weiss 1984, 1989a, b, 1990; Crews 1993a). Although there are numerous volumes addressing life span and senescence from the perspectives of sociocultural anthropologists, none examines these issues from an explicitly bioanthropological perspective. Upon reading Bengtsen and Schaie's edited volume *Handbook of Theories of Aging* (1999) as I was working on this book, I was further convinced of the need for a volume on the biological anthropology of senescence and life span. The Handbook is a quality publication whose contributors represent many current leaders in gerontological research and the latest ideas and theories on life span and senescence. However, two aspects of this representation of current theory building in gerontology bothered me. There is no explicitly evolutionary chapter at the beginning of the volume and the entire field of anthropology is represented by only one chapter in the social science section. This chapter includes none of the contributions of biological anthropology to research on senescence, life span, or longevity. Even fellow anthropologists seem unaware of biological anthropologists' contributions to issues related to human senescence and survival, menopause, chronic degenerative diseases, the evolution of senescence, and variation over the life span.

There are two additional reasons for writing this volume. The first is the lack of information from biological anthropology that characterizes most volumes and publications by mainstream gerontology. While social/cultural anthropologists' contributions to aging are well known to social gerontologists, most

vii

gerontologists have little if any exposure to the methods, data, and analytical results used and generated by biological anthropologists. Conversely, most biological anthropologists are likely to have had too little exposure to mainstream gerontologists through our basic journals and publications. Hopefully, this book will fill in the gap from both directions and will interest biological anthropologists as well as gerontologists who have begun to move away from their traditional emphasis on normal and normative aging and to appreciate areas of long-standing concern to biological anthropologists, i.e. the influences of individual variation and population heterogeneity on senescence and aging. Human variation is a mainstay of bioanthropological research and theory formation and provides the foundation for our studies of life span, chronic degenerative diseases, and adaptability. Thus, biological anthropologists are likely to contribute not only methods and techniques, but also background information and data as gerontology turns its attention toward understanding human variation over the life span.

Biological anthropologists have been leaders in the field of human variation and adaptability over most of the twentieth century, studying worldwide variation in growth and development, fertility, blood pressure, body habitus and physique, genetics, and degenerative processes, along with biocultural, sociocultural, political, and religious inputs to such processes. Biological anthropologists rely on the triangle of biology, environment, and culture when developing explanatory models for empirical observation (Baker 1991). Mainstream gerontology is now recognizing the heterogeneity and variation that characterize the aged of all species; with this recognition has come a call for integrative models that examine senescence as a complex system in whole organisms rather than as isolated cells and tissues (Kirkwood and Kowald 1997). Human beings are unique in the world, and in gerontological research, in their reliance on culture as a major mode of adaptation to the environment. For no other species is there a necessity to invoke culture as an explanatory variable in theory development. Biocultural interactions have structured the life history of humankind such that understanding human senescence will require an integrated examination of environmental, genetic, and cultural factors.

Biological anthropologists, and particularly human biologists, should be keenly interested in how humans have managed to attain their post-reproductive vitality, and why some individuals show great post-reproductive vitality, while others exhibit high frailty with respect to survival. In addition, biological anthropologists are interested in how evolutionary biology and human culture have interacted, and currently interact, to produce current human life history and variable life spans, and how late-life survival influences inclusive fitness in humans. Biological anthropologists not only have the tools and methods to examine such questions, but also background data and theory on human evolution, growth, and development, and a biocultural approach to understand better such aspects of human variation. Few other disciplines have as much experience examining human life history and accumulated data on human variation over the life course from conception to maturity as do biological anthropologists. Data such as those of the Krogman and Fels studies of growth and development provide baselines for assessing how early aspects of life history are associated with later developments during reproduction, maturity, and old age.

Human biologists have also collected extensive data on risk factors and chronic degenerative diseases from a variety of populations over several decades throughout the twentieth century. Today, these data form the baseline for follow-up studies of innumerable person-years of life span and longevity across a wide variety of environmental and cultural settings. Finally, biological anthropologists and human biologists have contributed to understanding population aging. Throughout the twentieth century, population aging has characterized almost every population group on the globe. Only among populations with high fertility are the elderly not an increasing proportion of the total. At the beginning of the twenty-first century, several low fertility populations have exceeded the 20% mark for the proportion of people over age 65 (e.g. Sweden, Japan, Norway) and many are well over 15% (U.S.A., Canada, U.K., France), while some high fertility populations (e.g., India, Ecuador) far exceed 10%. Many individuals survive for decades past their reproductive years, and variabilities in physiological, biological, and molecular factors along with environment and culture influence this survival. As a discipline, biological anthropology is poised to make substantial contributions to assist the understanding of such trends in survival and variations in human senescence, and to assist gerontologists in applying results from laboratory and animal models to humans.

1   *Introduction and background*

Interests in aging and senescence have characterized human thought since the earliest of recorded histories. Ancient Egyptian papyri and Chinese medical treatises, along with the writings of Aristotle and Socrates, describe various aspects of senescence and chronic degenerative conditions. They also detail methods for halting the insidious loss of function that accompanies longevity. Thoughts of mortality and immortality likely characterized the minds of our earliest *Homo* ancestors as well. The search for ways to halt the functional losses associated with growing old continues today. Humans are a long-lived species by any available standard. We are also unusual in that we remember our past and worry about the future: characteristics that we may share with a few other long-lived species or that may set us apart from all other species on earth. Long life provides ample time and opportunity to observe and remark on differences in longevity and vitality among relatives, friends, and acquaintances.

Prior to recent times, it is unlikely that many individuals ever actually survived sufficiently long enough to be considered very old by today's standards. Until recent times, anyone who survived 40 years was likely a grandparent and an elder; those still walking about at ages past 50 years were quite exceptional. Although some small proportion may have survived into their seventh decade of life, few would survive much beyond. Until recent decades, speculation and discourse on why and how particular persons outlived others and why one or another survived all others has outpaced scientific understanding. A major reason for the recency of studies of human senescence is the rapidity with which the aged population has grown. Increasing numbers of elders worldwide and their health care costs have fostered expanded research on the determinants of chronic degenerative conditions (CDCs), senescence, and life span (Smith and Tompkins 1995). These data are generating a greater understanding of both the physiological complexity and evolutionary simplicity of senescence. No simple mechanism(s) of senescence has been found, or ever will be. Instead, a range of phenotypic variability, systemic and local age-related alterations and dysfunctions, and variable genetic influences appear to structure senescence.

Humans represent about 6 million years of hominid and over 65 million years of primate/mammalian evolution. During this period, human life history – including fetal growth and development, neonatal maturation, infant and child

1

growth, ages at menarche and reproductive maturity – life expectancy, and life span have responded to a variety of evolutionary (biological) and sociocultural (biocultural) processes. This biocultural interplay, which does not influence senescence or life spans in cells, worms, insects, or rodents, has structured all aspects of human life history. This biocultural complexity is often slighted or not fully conveyed in both sociocultural and biological studies of human senescence and life span. As gerontologists have turned their attention to individual and population variation in human senescence and to the soma as a complex senescing system, their interests have merged more with biological and biomedical anthropology, human adaptability studies, and biocultural studies on senescence and life history. Anthropologists have helped to document the range of variation in multiple aspects of life history, including reproduction, growth, development, maturation, and adulthood survival. Unlike growth, development, and reproductive adulthood, until recently few humans ever before experienced late-life survival (70+ years). Late life represents a new phase in human and mammalian life history and an emerging area for biocultural, biomedical, and bioanthropological research.

This book explores the biological, cultural, and biocultural processes and environmental stressors through which human senescence, life span, and life history have evolved. The emphasis is on evolutionary, biocultural, and ecological aspects of human aging and senescence, rather than animal and cellular senescence, which are examined extensively elsewhere (Finch 1990; Rose 1991). Human life history evolved as part of the adaptive repertoire of a unique, bipedal, large-brained, large-bodied, gregarious, and polygamous hominid. These specific aspects of hominid evolutionary history necessarily determine to some degree current variation in our species' life history and our individual life spans – minimal/maximum metabolic rates, patterns of reproduction, maximum rates of growth, development and maturation, encephalization, and the DNA content of our cells. Although many such variables show high correlations with observed average and maximum life spans across species, they may provide little information on the determinants of senescence and mortality within species. Many such phenotypic traits simply scale to or are allometric outcomes of antecedent evolutionarily balanced tradeoffs between reproductive investment, environmental stress, and minimum necessary survival times.

In six chapters, this book explores some of the complex interplay of biological, cultural, and environmental forces through which human senescence and life span have evolved. This introductory chapter briefly examines terminological and definitional issues and the genesis and history of studies of human life span, before reviewing demographic trends in human longevity and life span. Chapter 2 examines evolutionary and biological theories of senescence. This is followed by an examination of human variation and the changes in physiological function that appear to be age associated, along with an exploration of how evolutionary biology and biocultural adaptations may help to explain some processes of human senescence. Chapter 4 explores humankind's unique biocultural adaptations to variable environments and biocultural influences on patterns of senescence and life history. This is followed by an examination of the applicability of life extension methods, proven successful in animal models, to humans in Chapter 5. The final chapter discusses current perspectives and future possibilities for advances in our understanding of human senescence from an anthropological and biocultural perspective.

Basic terminology and related concepts

As with any area of scientific pursuit, the study of senescence has its unique vocabulary. A basic division is geriatrics (a branch of medicine that deals with the problems and diseases of old age and aging individuals) and gerontology (a branch of knowledge dealing with aging and problems of the aged) (Webster's Unabridged, 1983, p. 482). Biological or biomedical gerontology is the study of the processes by which individuals within species show post-maturational decline, senesce, and ultimately die. Conversely, geriatrics is a medical specialty concerned with halting and/or retarding the insidious post-maturational changes brought about by the processes of senescence. Both disciplines are predicated on the assumption that there are particular biological processes that underlie changes commonly observed with increasing age. There are two major views as to the genetic bases for these biological processes of senescence: (1) they constitute a specific genetic program for senescence (Clark 1999), or (2) they are an artifact or byproduct of evolutionary forces acting to maximize reproductive success and inclusive fitness in sexually reproducing organisms (Rose 1991). The next chapter will examine evolutionary models of senescence and the molecular and genetic bases of senescence while exploring how these fundamental concepts relate to human senescence and life span.

Senescence and aging

Another fundamental division in gerontology is between aging (to become old: to show the effects or characteristics of increasing age) (Webster's Unabridged, 1983, p. 22) and senescence (the process of becoming old: the phase from full maturity to death characterized by an accumulation of metabolic products and decreased probability of reproduction and survival) (adapted from Webster's Unabridged, 1983, p. 1055; see also Rose 1991) – terms so frequently used incorrectly as synonyms that their individuality is sometimes unclear. All things age, whether living or not. Bottles of wine improve, while rocks and socks weather and wear with age (Harper and Crews 2000). Only the living may senesce. As humans know so well, many physiological phenomena show age-related change, but these are not all senescent changes. Senescence is a biological process of dysfunctional change by which organisms become less capable of maintaining physiological function and homeostasis with increasing survival. This leads to a reduced probability of reproduction and an increased susceptibility to death from both exogenous and endogenous causes. Aging is an elusive term carrying multiple sociocultural and political connotations. Aging best describes social, cultural, biological, and behavioral variability occurring over the life course *that does not directly increase the probability of death.* The areas of social gerontology, death and bereavement, and life course development generally are studies in aging, although some social factors, such as loss of a spouse, are associated with an increased probability of death. Senescence better serves current scientific discussion of mechanisms that preclude continued reproduction and survival in sexually reproducing organisms (Finch 1994; Cristofalo *et al.* 1999).

Researchers and disciplines often define senescence and aging differently (Crews 1993a; Harper and Crews 2000). For example, Comfort (1979) defined senescence as " ... a deteriorating process, with an increasing probability of death with increasing age ... " (p. 8). Fifteen years later, Finch (1994) refined this definition to include " ... age-related changes in an organism that adversely affect its vitality and function ... (Associated with an) increase in mortality rate as a function of time" (p. 5). Rose (1991) faulted earlier definitions for not including any aspect of reproduction, an essential component for an evolutionary definition of senescence, defining aging as " ... a persistent decline in age-specific fitness components of an organism due to internal physiological deterioration" (p. 20). In a recent review of molecular aspects of aging, Kirkwood (1995) defined aging as " ... a progressive, generalized impairment of function resulting in a loss of adaptive response to stress and in a growing risk of age-related disease" that ultimately leads to an increased probability of death, while senescence was defined as "the process of growing old". In the same volume, Johnson *et al.* (1995) provided very different working definitions: "Aging is a naturally occurring, post-developmental process. Senescence is a progressive impairment of function resulting eventually in increased mortality, decreased function, or both." The view of Johnson *et al.* (1995) is that most "but not all, degenerative diseases would thus be manifestations of senescence."

Aging *per se* is simply the fact of existence through time, the phenomenon of becoming older. Senescence is a progressive degeneration following a period of development and attainment of maximum reproductive potential that leads to an increased probability of mortality. Quoting one last definition: "... with the passage of time, organisms undergo progressive physiological deterioration that results in increased vulnerability to stress and an increased probability of death. This phenomenon is commonly referred to as aging, but as aging can refer to any time-related process, a more correct term is senescence" (Cristofalo *et al.* 1999, p. 8). "Aging" and "senescence" are not used interchangeably here. Since animate and inanimate objects alike become older, aging is reserved for such processes and the social, behavioral, cultural, life style, and biological changes that occur as individuals grow older in particular social settings but that do not in and of themselves increase the probability of dying. Biologically, since only certain living forms senesce, senescence is reserved for those detrimental processes that occur secondarily to biological and physiological alterations occurring over the life span that leave individuals less capable of reproducing and more susceptible to extrinsic and intrinsic stresses, and which increase the probability of death.

From a scientific viewpoint, human senescence represents an evolutionary problem to be solved, while, medically, it represents a process to be avoided, halted, or delayed. To do either, senescence must be understood within the context of natural selection. This requires both a better understanding of the evolutionary biology of theories on senescence (reviewed in Chapter 2) and examination of the patterns of life history (changes through which an organism passes in its development from its primary stage of life (gametes) to its natural death) among humans, their closest relatives, and their immediate ancestors. Human life history includes copulation, fertilization, embryogenesis, fetal development, birth, infancy, childhood, adolescence, reproductive adulthood, menopause, post-reproductive survival of women and late-life survival of men, and senescence; each of these is affected by numerous intrinsic (i.e., inborn, biological/genetic) and extrinsic (i.e., not intrinsic) factors. Extrinsic factors include environment, diet, population density, culture, and society (Finch 1994; Wood *et al.* 1994; Finch and Rose 1995). For most natural populations, life history factors are difficult or impossible to measure, thereby limiting the accuracy of available data and their usefulness for comparisons (Finch 1994). Data that are available suggest that rates and patterns of senescence, perhaps even the basic mechanisms of senescence, may differ within and between phylogenic classes and across environmental contexts even within the same species (Finch 1994; Finch and Rose 1995; Johnson *et al.* 1995).

One arguable, but ultimately unfruitful, position is that the processes of senescence are so uniquely individualized and species specific that they are neither interpretable nor understandable. Another is that, as with height, weight, skin color, or blood pressure, human senescence is just another type of phenotypic variation (Johnson *et al.* 1995) and amenable to research. Although its precise method of measurement is unclear, viewing senescence as an individual phenotype is supported by the large amount of interindividual variation in life span (Shock 1984, 1985), the lack of data showing any specific genetic program for senescence (Gavrilov and Gavrilova 1991; Rose 1991; Beall 1994; Wood *et al.* 1994; Arking 1998; Gavrilov and Gavrilova 2001; Mangel 2001), and senescence's multifactorial (where the etiology includes both environmental and genetic factors) and polygenic (an etiology including multiple genetic factors) nature. Common experience tells us that the processes of senescence and death differ between persons. Recognition of this fact is crucial for the diagnosis and treatment of patients. This variation complicates applications of higher order theories to senescence in living individuals. Still, there are consistent patterns within and across populations, suggesting that, as with other complex phenotypes, although there is a wide range of variation, senescence can be measured and experimentally manipulated. Wide variation also suggests that neither life span nor senescence may be subject to strong selective pressures in wild (natural) populations. In this book, senescence is viewed as a multifactorial and detrimental physiological process affecting all organs and bodily systems that, although accelerating with increasing age, is itself time independent and increases individual risk of death.

Although senescence is an individual phenomenon, different in its details across somas, certain generalizations are true. Senescent changes are encountered in most organisms (Finch 1994) and apparently are universal in sexually reproducing species (Rose 1991). No non-senescing sexually reproducing species has been reported. A broad range of organisms show mortality (or survival) curves that indicate an increasing vulnerability to death with increasing time of survival – the hallmark of senescence (Comfort 1979) – many also display similar and specific changes in proteins and DNA along with accumulations of lipofuscin and mitochondrial DNA (mtDNA) mutations with increasing survival time (Reff 1985; Wallace 1992b). Such broad similarities across species suggest that at least some common biological processes and genetic factors underlie individual and species manifestations of senescence. Current research is directed to finding such root causes of senescence and physiological dysfunction and to determining their relevance for each species.

Longevity and life span

In addition to aging/senescence, inconsistency characterizes many additional terms found in the gerontological literature (Crews 1990a; Olshansky *et al.* 1990; Finch 1994; Olshansky and Carnes 1994; Harper and Crews 2000). Terms such as longevity, life span, average and maximum life span, life expectancy,

Table 1.1 *Male and female life expectancy (in years) at birth, age 40, and age 85 in the U.S.A.*

	Men			Women		
Year	Birth	Age 40	Age 85	Birth	Age 40	Age 85
1900	46.6	68.0	88.8	48.7	69.1	89.1
1910	48.6	67.7	88.8	52.0	69.2	89.1
1920	54.4	69.1	89.0	55.6	69.9	89.1
1930	59.7	69.1	89.0	63.5	71.6	89.8
1940	62.1	69.9	89.0	66.6	73.0	89.3
1950	66.5	71.2	89.4	72.2	75.7	89.8
1960	67.4	71.6	89.3	74.1	77.1	89.7
1970	68.0	71.9	89.6	75.6	78.3	90.5
1980	70.7	74.0	90.0	78.1	80.1	91.3
1990	72.7	75.6	90.2	79.4	81.0	91.4
2000	74.3	76.9	90.4	80.9	82.0	91.7

Data from Wright, 1997.

maximum achievable life span (MALS), mortality rate doubling time (MRDT), and maximum life span potential (MLSP/MLP) all have very specific meanings, but like aging and senescence are not always used appropriately. Expectation of life at birth or life expectancy at birth (e_o) is a demographic measure of average life span resulting from the all-cause mortality of a cohort (a group of individuals born in the same year). Expectation of life (e_x) at any age (x) is a well-defined basic life table (an actuarial table based on mortality statistics that follows an entire cohort from birth to death) function. Although well defined, life expectancy data may be used misleadingly in aging research because they are based on both child and adult mortality rates and are influenced by prevailing sociocultural, political, economic, and environmental factors (Olshansky *et al.* 1990; Olshansky and Carnes 1994). For example, comparing e_o of populations in very different cultural or ecological settings, where one group experiences high and the other low infant and child mortality, reflects sociocultural and environmental factors associated with preventable diseases and illnesses, rather than processes of senescence. However, e_x calculated for ages other than birth may provide more meaningful comparisons between populations and time periods (see Table 1.1 to examine e_0, e_{40}, and e_{85} for the U.S. population between 1900 and 2000).

MLSP and MALS are closely related theoretical concepts commonly defined as the longest known life span or the oldest living individual of a species or the maximum predicted life span (Weiss 1981; Hoffman 1984; Harper and

Table 1.2 *Estimated average and maximum life spans
and ages at puberty for selected mammalian species*

| Name | Life span | | Age at puberty (months) |
	Average (months)	Maximum (months)	
Human	849	1380	144
Gorilla	–	472	–
Chimpanzee	210	534	120
Rhesus	–	348	36
Cow	276	360	6
Swine	192	324	4
Horse	300	744	11
Elephant	480	840	21
Cat	180	336	2
Dog	180	408	2
Whale	–	960	12
Mouse	18	42	1.5
Rat	30	56	2
Guinea pig	24	90	2

From Table 2, Finch and Hayflick (1997), p. 9.

Crews 2000). Maximum life span is commonly estimated based on captive and domestic samples. Some researchers have suggested that the MALS represents the genetic capacity of a species for long-term survival (Cutler 1980; Fries 1983; Hoffman 1984; Susser *et al.* 1985). However, both life expectancy and current maximum life span are sensitive to environmental influences, vary widely between different populations of the same species, and are easily modulated in controlled laboratory settings (e.g., dietary restriction, temperature variation) (Finch 1994). Among extant lineages, MALS is thought to have increased over evolutionary time and to have changed over the course of evolution of multiple species. Unfortunately, documentation of such change cannot be obtained directly from the fossil record. There is no direct measure of either e_o or MALS for extinct species such as dinosauria, dryopithecines, australopithecines, or erectines. Rather, allometric relationships between life spans and either body or brain size established for extant, often domestic, species are used to estimate MALS for fossil specimens (see Table 1.2 for estimates of average and maximum life spans and age at puberty for some modern species).

Longevity (long-lived, a long duration of individual life) is an individual phenomenon, identical to life span. The individual with the greatest longevity (maximum life span) in any particular environment is an outlier, a unique individual.

The maximum verified age for any human is over 122 years (Jean Calment of France), which is 7 years above the maximum life span reported in Table 1.2 for humans, and 2 years beyond the MALS for humans predicted by proponents of a limited life span model (Fries 1980; Fries and Crapo 1981; Fries 1983, 1984, 1988). Available data on maximum life span from zoo specimens or capture–recapture studies in the wild (such as are presented in Table 1.2) do not provide sufficient information to assert anything regarding either patterns or rates of senescence in natural populations. What they do illustrate is that maximum life span is often much greater than the average. Paraphrasing Finch (1994, pp. 12–13), little evidence about the role of senescence in limiting life span is garnered from such comparisons. However, similar comparisons of the same species in different environmental settings do show that average and maximum life spans of most lengthen in response to simple environmental modulations that include improved nutrition, reduced disease, and lack of predation. These data illustrate that most wild species have a potential for long life not often expressed in their natural ecological setting. The domestic cat (*Felis catus*) provides a clear example. When kept as a house cat without access to the outdoors, the life expectancy of *F. catus* is about 15 years. Conversely, a feral cat's life expectancy is only about 18–36 months. Extended life expectancy among domestic house cats results without change in genes or biology. Rather, improved nutrition, negligible predation, and reduced disease (an altered environmental setting produced by human culture) lead to improved survival, and, if not surgically controlled, greatly enhanced reproductive success.

Evolutionary biology

Fundamental to grappling with the complex biology of senescence is a basic understanding of the terminology and principles of evolutionary biology. The basic hereditary unit, DNA, is composed of four nucleotides – thymine, adenine, guanine, and cytosine. In humans, DNA molecules form 46 linkage groups (chromosomes) sequenced into about 30 000 coded subunits called genes (a segment of DNA that can be translated into RNA, a locus). Loci provide RNA templates for proteins and differ in DNA sequence across chromosomes. Each DNA variant at a specific locus is a unique allele. Such coding loci are separated by intervening nucleotides (perhaps 90% of all human DNA, but only about 10% in flying mammals and birds); these are not known to code for RNA. Each allele codes for a specific RNA molecule, but the same RNA molecule and thus protein may be coded for by a variety of possible alleles. For most loci and segments of intervening DNA (iDNA), many different sequences of DNA nucleotides (alleles) are available to occupy the locus.

Genetic traits and conditions (e.g., the ABO blood group, the enzyme pheny-lalanine hydoxylase, albinism, Huntington's disease, sickle cell anemia) are due to the inheritance of different alleles at a specific locus. Any alleles that differ from the wild type (the most common allele in the wild population) represent mutations (change in DNA sequence) of the supposed original allele in the founding population. Loci with but a single common allele are monomorphic, which is an uncommon situation. Loci generally show two or more common alleles. These are polymorphic (many types) when the second most common allele occurs more frequently than its mutation rate, or is above 1%. Alleles oc-curring at low frequencies (1/1000 or 1/10 000) cause a variety of detrimental phenotypes (e.g., Duchenne muscular dystrophy, hemophilia, cystinuria, cystic fibrosis, phenlyketonuria). These are frequently termed mutants compared with alleles predisposing to what are considered 'normal' phenotypic outcomes. In such cases, normal and mutant may include a variety of specific alleles produc-ing either phenoype.

DNA alleles are the raw material acted on by the forces of evolution (natural selection, mutation, gene flow, and genetic drift). Mutation creates entirely new DNA sequences by small (base pair (bp) substitutions that change one nucleotide, e.g., A → T) and large steps (insertions and deletions covering a few or a few hundred of bases, e.g., a 9 bp deletion of mtDNA or a 240 bp deletion of the angiotensin converting enzyme (ACE) locus). Natural selection, flow, and drift only shape this variability. Natural selection limits the reproductive success and inclusive fitness of individuals carrying mutations less viable in the current environment. In a constant environment, natural selection may lead to organisms remarkably well adapted to a specific ecological niche (e.g., koala bears in eucalyptus forests, giant pandas in bamboo forests). Most environments are not so stable nor are most organisms so highly specialized. Eating almost anything, surviving in a range of habitats, and using culture to manipulate the environment, humans may be included among the most generalized of species, along with, for example, other primates, rodents, and insects.

Gene flow and drift act to spread/mix and eliminate genetic variation. Flow is simply the exchange of gametes (DNA) between populations, such that variants arising in one area may migrate throughout an entire species if not eliminated by natural selection or genetic drift. In highly mobile organisms such as humans, the spread of novel alleles with reproductive or survival benefits may be very rapid (Lasker and Crews 1996). Alleles with no (or very little) effect on fitness and reproductive success are selectively neutral. These may be lost or become fixed through chance alone as their frequencies change from one generation to the next in relatively small populations through, random genetic drift. High fre-quencies of conditions such as Huntington's disease, pseudohermaphroditism, xeroderma pigmentosum, polydactally, and diabetes in human isolates illustrate the power of random genetic drift, through Founder's effect, to alter allele frequencies. During phases of hominid evolution when local populations (demes) were smaller and more dispersed, drift was more influential on human variation and likely contributed to the variety of human types found around the globe.

Genetic influences developed under the influence of evolutionary forces structure multiple quantitative aspects of human phenotypic variation. To the degree genetic variability influences human life history, these same evolutionary forces have structured human senescence and life span. Mutation continually alters the DNA of all organisms. This provides new alleles that potentially enhance or retard senescent processes. Clearly, many alleles contributing to rapid loss of function and lower fitness have already been culled from the human gene pool, or we would not survive twice as long as other large-bodied primates. Such culling continues today. The pace at which detrimental alleles are eliminated depends on their penetrance (degree to which genotype is expressed in the phenotype), and any established dominant–recessive, pleiotropic, and epistatic interactions. Alleles neutral with respect to fitness (including early life survival, growth and development, reproductive success, and the fledging of offspring) but carrying senescence-enhancing or -reducing effects have, over most of hominid and human existence, been most affected by random drift. Alleles that retard or enhance senescence are likely to be widely dispersed and represent a genetic reservoir of senescence-delaying allelic propensities available to not only humans, but also most wild species.

Several additional terms related to evolutionary biology are frequent in discussions of senescence. Phenotype or phenotypic traits are the observed manifestations of human form and function. Phenotypes result from the interplay of genotype (the sum of nuclear and mitochondrial alleles carried by or the specific allele(s) at one (or more) locus in an individual) with the cellular and external environment, and, in humans, the sociocultural environment. Some phenotypes are quantifiable characteristics – height, weight, blood pressure, glycemia, visual acuity, enzyme activity, skin reflectance, number of children produced, and observed life span – while others are qualitative – amino acid sequences of proteins, eye color, and sex. Many are difficult to measure – for example, rate of senescence, psychological inclinations, culture/ethnicity, stress response, and personality type. Throughout life, from the cytoplasm of the ova that becomes our zygote, to our mother's uterus and internal physiology, birth, and ultimately death, the genotype constantly interacts with the environment. Natural selection and numerous random factors determine how well phenotypes produced through this process survive and reproduce. If the phenotype does not reproduce, the entire genotype is lost without representation in future generations.

Today, in more cosmopolitan (cultures that look outward and are not bounded by local customs and beliefs, the antithesis of traditional) settings, genotypes producing phenotypes that survive through birth generally also survive to re-productive age. The force of natural selection falls less heavily on them during infancy and childhood than on members of populations living more traditional (cultures that look inward and are bounded by local customs and beliefs) life styles or those who were conceived and survived to birth during earlier epochs of human evolution. Throughout hominid evolution, natural selection, through high fetal, infant, and childhood mortality rates, likely eliminated many alleles and genotypes that today allow their carriers to survive to maturity. Today, in privileged settings, differential pre-reproductive survival is very low ($<5\%$) and likely to be secondary to differential reproduction in changing the gene pool (the sum total of alleles represented in a breeding population).

History of research on senescence and longevity

Alfred Russel Wallace (1823–1913), who along with Charles R. Darwin (1800–1882) first developed the concept of evolution by natural selection among vari-able phenotypes, was also the first naturalist to propose an evolutionary theory of senescence. His original formulation appeared in an unpublished note written between 1865 and 1870, and was quoted in a footnote to an 1889 translation of an 1881 essay by August Weismann (1834–1914) entitled "The Duration of Life" (for a review of Wallace's comments see Rose 1991, pp. 4–5). It was Weismann (1891; for a review see Kirkwood and Cremer 1982), however, who articulated the most substantive evolutionary theory of senescence prior to the twentieth century. Weismann (1891) proposed that death was an adaptation which arose when an infinite life span became disadvantageous to the species, such as when old and decrepit, yet immortal, individuals take the place or re-sources of those who are healthy. The notion that death would be the direct result of natural selection and, thus, an adaptation is not supported by available data and few today accept Weismann's explanation for senescence as an out-come of group selection. However, Weismann was among the first to examine senescence from the viewpoint of evolutionary biology and natural selection, to see the separation between the mortal soma and the potentially immortal germ cells, to observe that there was no fixed relationship between size of the soma and length of life, and to propose that the "rate at which animal lives, influences duration of life" (Weismann 1891, p. 8).

Weismann (1891) also observed that larger animals tended to have greater difficulty obtaining sufficient food and need additional calories to reproduce than smaller ones. Thus, larger animals exhibit slower rates of reproduction. He further noted that the rate of living influenced longevity "not because of rapid consumption of the body" but rather because the "...increased rate at which vital processes take place permits more rapid achievement of the aim and purpose of life, *viz.*, the attainment of maturity and reproduction of the species" (p. 8). Thus, Weismann proposed that growth and development of the soma, reproduction, and life span shared common evolutionary origins. Weismann did not believe that physiological factors alone could determine length of life; rather, he found that "duration of life is really dependent upon adaptations to external conditions . . . determined by precisely the same mechanical process of regulation as that by which the structures and functions of an organism are adapted to its environment" (p. 10). Based on these postulates, Weismann suggested " . . . that, as a rule, life does not greatly outlast the period of reproduction except in those species which tend their young" (p. 11) and that evolutionarily there is a "necessity for as short a life as possible" (p. 23). Furthermore, he refers "the question as to the means by which the lengthening or shortening of life is brought about . . . to the process of natural selection" (p. 20). Weismann saw long life as a "luxury without an advantage" (p. 25), but knew that death was not "an absolute necessity, essentially inherent in life itself" (p. 26). Rather, death " . . . is an adaptation that first appeared when, in consequence of a certain complexity of structure, an unending life became disadvantageous to the species" (p. 111). He based this assertion on phenomena that first Medawar (1952) and later Kirkwood (1977) would use to develop further the evolutionary theory of senescence. Weismann proposed that the " . . . limited duration of Metazoan life may be attributed to the worthlessness (of individuals) . . . liable to wear and tear . . . (and that the) perishable and vulnerable nature of the soma was the reason why nature made no effort to endow this part of the individual with a life of unlimited length" (p. 156). Finally, Weismann clearly understood that: "As the soma becomes larger and more highly organized, it is able to withstand more injuries, and its average duration of life will extend . . . (Such) lengthening of life is connected with an increase in the duration of reproduction . . . (thus) there is no reason to expect life to be prolonged beyond the reproductive period; so that the end of this period is usually more or less coincident with death."

In a series of writings early in the twentieth century, Raymond Pearl (Pearl 1922, 1928, 1931; Pearl and Pearl 1934), following closely Weismann's work, suggested that longevity resulted as an epiphenomenon of other aspects of life history (e.g., metabolic rate, body size, and reproduction; see also Comfort 1979; Finch 1994). Among mammals, smaller forms tend to have higher basal metabolic rates and shorter life spans than larger animals and metabolic rates tend to co-vary with brain weight and body size (Harvey and Bennet 1983; Hoffman 1983), while both body and brain size show positive associations with life span (Cutler 1975, 1980; Sacher 1980; Hoffman 1983; Finch 1994). The basic premise for metabolic control of senescence or rate-of-living theories is that smaller animals have more rapid metabolic rates, expend their life's allotment of energy more rapidly, and die more quickly than larger animals with slower metabolic rates. However, not all species conform to this model. For instance, bats live longer than other rodents of comparable body size and metabolic rates, while birds live longer than comparably sized mammals or reptiles. Weismann noted this latter phenomenon and suggested that the "long life of birds ... (was) compensation for their feeble fertility and great mortality of their young ... ". However, it may also be related to lower rates of extrinsic mortality from predation in adults of species that can fly, which then allows such feeble fertility to be a successful evolutionary strategy.

About a quarter of a century after Pearls' writings, Medawar (1952) elaborated an evolutionarily based theory of senescence. He demonstrated that the force of natural selection decreases with age and that this decline holds not only for organisms that senesce, it also decreases in theoretical populations of immortal individuals with unlimited reproductive capability. Medawar showed that, given real, life-threatening hazards (e.g., illness, predation, accidents) that act to limit reproductive success, the force of natural selection always decreases with increasing age. Weismann (1891) previously had recognized that the susceptibility of the soma to accidents would diminish an organism's reproductive ability; however, he did not develop a model that showed why this must ultimately be true. Eventually, Hamilton (1966) and Charlesworth (1994) both provided rigorous mathematical proofs of Medawar's intuitive results. Between them, they proposed and mathematically proved what today is regarded as the ultimate evolutionary cause of senescence and death: regardless of the physical capabilities of the organism to prevent internal degeneration of the soma, or its continued capability to reproduce, the force of natural selection decreases with age.

Population genetics shows that the force of natural selection decreases throughout most of one's adult life. This was first parameterized by R.A. Fisher (1930) as: "reproductive value", or $v(x)$, representing the relative reproductive contribution individuals aged x can expect to make to the next generation. Fisher hypothesized that this variable would be proportional to the force of natural selection at age x. Of primary interest for gerontology is the maximum value of $v(x)$, the point at which the force of natural selection begins to decline, for this is when senescent changes may begin to emerge. The age at which $v(x)$ is maximum is expected to vary across populations; however, for the Australian women Fisher used as an example, maximum $v(x)$ was reached at about 19 years of age. Similar results are cited for Taiwanese men and women by Hamilton (1966), Chilean, German, and American women by Keyfitz (1968), and American women by Goodman (1969). Thus, among humans, the process of senescence likely begins at a relatively early age (one-sixth of the maximum known life span) as the force of natural selection and reproductive value reach their maximum and begin to decline after about two decades of life. Rather than occurring exclusively in the elderly, senescence is a long-term process occurring over many decades of life (Harper and Crews 2000).

In addition to remarking on the role of declining natural selection with age, Medawar (1952) proposed that the age-specific actions of genes and mutation accumulation were basic components of the senescent process. Because natural selection diminishes with age, mutant alleles producing deleterious effects at ages beyond that of maximum reproductive potential, but with little or no effect at younger ages, may remain in the gene pool unaffected by natural selection. When sufficient mutations with late-acting, detrimental effects accumulate, broad senescent changes may result. Recently, Martin *et al.* (1995, 1996) suggested that any alleles that accumulated in this fashion should be "private" (alleles that segregate only in specific families/kindreds), as opposed to "public" (alleles found in broad segments of the population) alleles. Since natural selection is negligible for any such neutral mutations, their ultimate frequencies are presumed to be dictated by genetic drift alone (Martin *et al.* 1995, 1996). Evolutionary theory suggests that gene flow, mutation rates, and, specifically for alleles with post-reproductive influences on senescent processes, linkage (when loci are found close together on the same chromosome) to non-selectively neutral alleles – a specific form of genetic drift – would determine the fate of these late-acting alleles. Martin *et al.* (1995, 1996) suggest that such late-acting genes should aggregate in certain lineages, but they do not anticipate that they will be responsible for population-level senescent processes. However, given linkage, gene flow, variable mutation rates across populations and environments, and local selective forces, such alleles, particularly if they are linked to other alleles with positive influences on reproduction, are likely to spread widely across the population and ultimately to be represented throughout the species (Lasker and Crews 1996).

Medawar (1952) saw that both pleiotrophy (when the product of one locus has multiple effects on physiology or phenotype) and linkage to beneficial early-acting alleles could lead to an increased representation of alleles with late-acting detrimental effects on survival at late ages. However, G.C. Williams (1957) was the first to develop fully the theory that pleiotrophy might be an important determinant of senescence. Williams proposed that genes exhibiting pleiotrophy could have early-life beneficial effects but later-life detrimental effects. Such late-occurring detrimental effects could then account for multiple senescent changes observed in various species. Clearly, natural selection favors alleles that increase vigor and vitality at young ages, including any alleles that enhance acquisition, utilization, or retention of the resources needed for growth and development or that lead to the optimal use of resources whose scarcity may have limited reproductive success during earlier phases of human evolution

(Crews and Gerber 1994; see Chapter 2). Natural selection also favors any mutations that postpone undesirable pleiotrophic effects of valuable early-acting alleles to ages beyond the point of maximum reproductive potential. Dubbed "antagonistic pleiotropy" because of the counteracting pleiotropic effects such alleles have during different phases of the life course, natural selection is the major factor in this model. In Martin and colleagues' (1995, 1996) jargon, antagonistic pleiotropy is a "public" mechanism and, as such, should explain senescent processes which are common across a wide variety of life forms.

Fisher's (1930) model showing reproductive value as proportional to the force of natural selection at any given age was refined by Hamilton (1966) who demonstrated that under very specific conditions $v(x)$ may increase at the same time as the force of natural selection decreases. Hamilton proposed a slightly different concept, "expected reproduction beyond age x", where $w(x)$, is an explicit formulation of "reproductive probability" as defined by Williams (1957). Hamilton's equations more precisely estimate the changing force of natural selection with age (see Rose 1991, p. 14). However, the relationship between reproductive potential and natural selection in humans may be approached more intuitively. Maximum reproductive potential (MRP) may be defined as "the point in life at which an organism is sufficiently mature to not only bear/sire offspring, but also best able to rear and fledge any offspring produced with maximum efficiency" (Crews and Gerber 1994; Gerber and Crews 1999; Harper and Crews 2000). In species with large investments of parental care in offspring, the fledging process is as important as the ability to reproduce. Natural selection will have less influence on alleles carried by individuals who have passed their MRP as their expected genetic contributions to future generations decrease with time (both $v(x)$ and $w(x)$ may have declined prior to MRP and generally will continue to decline). MRP differs from either $v(x)$ or $w(x)$ because sexual maturity and thus reproduction may occur before future parents are physically, emotionally, or behaviorally prepared to fledge successfully any offspring produced. Neither the onset of reproduction nor the maxima of either $v(x)$ or $w(x)$ signal the point of MRP. Total "reproductive probability" (Williams' terminology) and reproductive value (Fisher's terminology) generally decline following sexual maturity, favoring maximization of reproductive potential at about the same time. In humans, the effectiveness of natural selection decreases slowly after the attainment of MRP. Incipient senescent changes are slower while reproductive potential remains high, increasing only slightly through ages needed for continued reproductive effort (e.g., mating, gestation, birth, rearing, and fledging multiple offspring). Thereafter, senescent processes proceed in a random and time-independent, but seemingly synchronized, fashion. Senescent dysfunction increases rapidly in those with enhancing, and more slowly in those with retarding, propensities. This leads to some having short (45 years) and others long (>100 years) life spans, while the majority falls in the mid-range (70–75 years).

Williams (1957), Charlesworth (1994), Rose (1991), and others have described how synchronicity may have come to characterize multiple aspects of senescence. Over evolutionary time, human life history, reproductive processes, patterns of growth and development, and senescence have responded to multiple evolutionary pressures in a similar way to all other existing species. Most populations that have ever existed, including hominids, probably experienced relatively high mortality rates throughout most of their life spans (recall Malthus and Darwin). Such high mortality would necessarily have limited the effectiveness of natural selection beyond early adulthood. This suggests that multiple senescent changes should coalesce just beyond the average life expectancy of a species in the wild (Williams 1957). Hominid evolutionary history may provide one example. Australopithecines and early *Homo* species are estimated to have averaged only about 15–20 years of life over the 280 000–350 000 generations they prowled the earth (Cutler 1976; Weiss 1981, 1984, 1989a), while over 400 generations of early agriculturists and nomadic pastoralists could expect to live only about 25 years (Weiss 1981, 1984, 1989a). Only over the last ten or so generations has life expectancy increased from 43 to 75 years (Deevey 1950; Weiss 1981, 1984, 1989a). Given such brief life expectancies among our ancestors, it is likely that multiple adult-onset, age-related declines in function result from the actions of alleles that, due to their infrequent expression and exposure to natural selection in earlier hominid and human ecological settings, have been retained in humanity's gene pool. Increases in life expectancy among humans in recent generations have followed cultural elaborations and novel environmental adaptations among our recent ancestors, rather than any major biological innovation, also supporting this model. For humankind, and our domesticated animals, culture has added an extra dimension of variation to senescence.

Originally introduced by Weismann (1891), the disposable soma theory of senescence was later further elaborated by Kirkwood (1977, 1995). Both Weismann and Kirkwood observed that the decreasing force of natural selection and the corresponding loss of reproductive potential with increasing age ultimately cause senescence. Although the disposable soma theory is fundamentally related to both antagonistic pleiotrophy and the age-specific action of alleles, Kirkwood (1990) proposes an additional direct relationship between evolutionary theories of aging and physiological descriptions of senescence in organisms with a distinct soma and germ line. In sexually reproducing species, available resources – most importantly energy – may be used either to preserve the soma or reproduce the germ line. Because death due to environmental mishaps, accidents, and hazards remains inevitable, once sexual maturation is complete natural selection favors investment in reproductive success over somatic maintenance. Attainment of MRP demarcates the timing of this change to investment in reproduction. Kirkwood (1977, 1981) suggested that senescent changes result from insufficient development of maintenance and repair mechanisms in somatic tissues, leading to their ultimate decline and failure. Decline and failure are closely linked to the number and fidelity (accuracy of function) of redundant sub-units (cells, sub-units) in somatic tissues (Gavrilov and Gavrilova 1986, 2001). Based on the disposable soma theory of aging, multiple candidate alleles for antagonistic pleiotropy will directly or indirectly mediate processes of cellular, organ, and organism generation, maintenance, repair, and protection. The disposable soma theory fits well with observations that the physiological manifestations of senescence often appear to result from "wear and tear", manifesting as physical damage to cells, organs, and organisms (Kirkwood, 1990).

The disposable soma theory is a broad evolutionary model that places natural selection in the position of a prime mover and details how tradeoffs between reproduction and maintenance of the soma ultimately result in somatic instability, dysfunction, and death. Numerous additional theories describe the mechanisms by which somatic damage, or a lack of repair thereof, may occur and accumulate. Prevalent among these is the oxidative damage theory of aging, in which naturally occurring oxidants impart functional damage to proteins, nucleic acids, lipids, and carbohydrates (Mead 1976; Sevanian 1985; Davies 1987; Richter *et al.* 1988; Stadtman 1992; Fridovich 1995; Newcomb and Leob 1998; Luciani *et al.* 2001; Sozou and Kirkwood 2001). Although the human organism has several enzymatic and non-enzymatic antioxidant defenses, the age-related increase of markers of oxidized protein and oxidized DNA suggests a long-term imbalance in oxidant production and defense (Martin *et al.* 1995, 1996). "Oxidative damage" subsumes Harman's (1956, 1981, 1984, 1999) free radical theory of aging, since only a portion of damaging oxidative processes result from free radicals *per se* (Swartz and Mader 1995). The more inclusive term for such molecules is "reactive oxygen species" (ROS) (Swartz and Mader 1995).

Not only oxidative processes cause somatic damage. Senescence also may result from long-term, low-intensity stressors, including, in addition to oxidative damage, temperature, physical trauma, radiation, diet, and toxic agents, which cannot be perpetually counterbalanced by protective and repair processes without some damage accumulating (Masoro 1996). Other proximate theories of senescence include intrinsic cellular aging, protein error catastrophe, and somatic DNA mutations (Vijg and Wei 1995). These all assume gradually accumulating damage to DNA, proteins, or cells and declining somatic function with increasing survival times. In a general way, senescence results from accumulating somatic damage and impaired function as the body's maintenance, protective, and repair processes gradually become overwhelmed. Damage results from a variety of basic mechanisms – ROS, wear and tear, toxic agents, error accumulation, temperature extremes, trauma, loss of repair capability, metabolism, and mtDNA mutations – and is cumulative. The common outcome of these multiple processes is somatic damage, which is the key connector between senescence and chronic degenerative conditions (CDCs) (Esser and Martin 1995).

A modern revival of Pearl's (1928) rate-of-living and Kirkwood's (1977) somatic damage theories followed observations of cellular damage due to the byproducts of metabolic processes, in particular ROS (highly reactive charged molecules such as O–, COH–, H–) (Harman 1987; Adelman *et al.* 1988; Cutler 1991). Although rate-of-living theories have been popular in gerontology and age-related changes in physiology are closely correlated with life span, experimental data refute their applicability to senescence (Hart and Tuturro 1983; Rose 1991; Austad 1992; Finch 1994). Comparative life spans in birds and mammals illustrate these problems (Austad 1992). Rate-of-living theories predict that in both birds and mammals metabolic rate should decrease with increasing size. At any given body size, birds show about twice the metabolic rate of mammals. Therefore, one would expect that, at any particular body size, mammals should live about twice as long as birds. This is not true: birds generally live much longer than same-sized mammals. Based on exhaustive review of many body systems, from whole body metabolism to cellular and biochemical levels, Finch (1994) demonstrated that there are neither strong nor consistent relationships among body size, metabolism, and life span. Weismann had suggested this based on his review of avian life spans in 1891. Primates, our nearest phylogenetic relatives, provide another example. Folivorous primates have lower metabolic rates than their body sizes would predict. This could be because they have relatively smaller brains than frugivorous species, or perhaps because the leaves they eat are energy poor and require greater processing (Harvey and Bennet 1983). Relationships between body sizes, metabolic rates, and life span are affected by multiple aspects of diet, ecology, habitual activity, and phylogeny. In addition, there is no reason why the rate of senescence must be related to length of life. Selection pressures are likely to vary for these traits (e.g., Pacific salmon). Why we senesce and why we live a certain life span are two related, but separate, questions. Rate-of-living theories address the questions of how, not why, we senesce, and may apply to senescence in specific organisms or organ systems composed of variable numbers of redundant sub-units.

Age-related, age-determined, and senescent

"Age-related" changes are changes in physiology, structure, or function that show an increase in the probability of occurrence with increasing age. These need not manifest in all individuals at the same age or even occur in all prior to death. "Age-determined" changes include events that are invariable and universal aspects of a species' life history, but which may occur across a range of ages with an accepted mean. These would include such human universals as loss of juvenile, and eruption of adult, dentition, closing of cranial sutures, attainment of puberty and reproductive function, decreased hormone production in later life, accumulation of lipofuscin, menopause and the lessening of reproductive potential in men, graying of hair, loss of skin elasticity, enlargement of the prostate, and decreased visual acuity with increasing age. Although the precise age at which such changes will occur is variable, they will occur in all individuals who survive sufficiently long. Age-related changes include those alterations that occur more frequently with the passage of time – loss of teeth and hair, hyperglycemia, decreased ability to complete activities of daily living (ADLs) and instrumental activities of daily living (IADLs), decreased bone density, hypertension, and hypercholesterolemia – but which may not affect all individuals. These increase in frequency with age, but show variable age patterns of onset. As with the distinction between disease and senescence, the distinction between "age related" and "age determined" is arbitrary and to a great degree related to our current lack of knowledge of the multiple causal pathways linking a chronological variable, age, to a physiological process, senescence, in a complex adaptive system.

Many physiological systems show age-related alterations; not all of these alterations are senescent. Attempts have been made to enumerate characteristics required for an age-related change to be an aspect of senescence (i.e., it increases the hazard of mortality). Arking (1998, following Strehler 1982) suggests some combination of cumulative, progressive, intrinsic, and deleterious. Additional criteria might include irreversible or degenerative. Criteria such as Strahler/Arking's provide a baseline for evaluating age-related change as senescent or not. Although such criteria may change with additional research (Arking (1998) dropped universal from his 1991 list (Arking *et al.* 1991)), they provide a guide for determining if a specific age-related change may be a senescent alteration. Currently, it is difficult to assess processes or delimit early alterations representing senescence. For many CDCs/senescence, early stages are insidious and progressive – such as the build-up of plaque in arteriosclerosis or the uncontrolled cell proliferation in neoplastic disease – but almost undetectable with current technologies. Many age-determined changes seem to have little impact on the survival probabilities of individuals; conversely, many age-related changes significantly alter survival probabilities with increasing age.

Given exceptions to every generalization, numerous age-related changes in humans appear to be senescent alterations – e.g., progressive dementia, lowered

immune function, decreased reproductive capability, declines in hormone production, loss of reserve capacity in respiratory and circulatory systems, atrophy
of muscle, loss of bone, accumulation of lipofuscin, reduced repair of damaged DNA, decreased permeability of cell membranes, increased density of the
internal cellular matrix, and loss of reproductive function in women. Others appear to be age-related disease processes (e.g., Alzheimer's disease, non-insulin-
dependent diabetes mellitus, atherosclerosis, neoplasia disease, cross-linking of
lens crystallin, osteoporosis) that afflict large proportions of elders in some societies, but few in others. The lone factor differentiating these sets of changes is
that the former describe systemic alterations and the latter specific clinically defined degenerative processes. These system-wide senescent alterations underlie
manifestations of CDCs. In elders, attempts to separate pathology and dysfunction into senescence versus disease may be misleading and counterproductive
(Miller 1995).

Not all aspects of physiology show a decline with increasing age; some show
stability while others show age-related enhancement (Finch 1994 appendices;
Mayer 1994). After attainment of MRP, many aspects of physiological function
show an increased variation in resiliency and responsiveness to perturbations.
Age-related increases in variation (e.g., standard deviations) are observed at
both the individual and population level. Changes in physiological function
(e.g., blood pressure, glycemia) between MRP and late adulthood (ages 50–
64) vary more widely between individuals than do height, weight, or fatness.
Reproductive adulthood is commonly defined as the period between menarche
and menopause for woman. It is arbitrarily set between MRP and age 50 in
men, who may reproduce in their ninth decade. Reproductive adulthood may
be further divided into early reproductive (ages 20–34) and late reproductive
(ages 35–49) stages, based on variable reproductive efforts, mate acquisition,
reproduction, infant care, late child-rearing and fledging, and grandparenting.
During late adulthood (ages 50–64) variation may decline as the extremes show
higher mortality hazards. After age 75+, a more homogenous representation
of alleles (e.g., increased frequencies of specific alleles at the apolipoprotein
E (APOE), angiotensinogen converting enzyme (ACE), atrial naturitic peptide
(ANP), and human leucocyte antigen loci), physiological function (e.g., blood
pressures, lipid levels, and body fat), and morphology (e.g., lower body mass
index (BMI), height, and trunkal fat) is observed.

All somatic systems are susceptible to disease and breakdown at about the
same age because senescent changes that precipitate dysfunction are similar and must accumulate over time before causing problems (Williams 1957;
Hamilton 1966; Rose 1991). As expected, multiple diseases commonly afflict
humans of the most advanced ages (Zeman 1962; Howell 1963). This is so

common that the assignment of death to a single cause for most elders must disregard multiple contributing senescent and disease processes (Manton and Stallard 1984; Crews 1988, 1990b; Crews and James 1991). Additionally, alleles predisposing to CDCs may underlie multiple processes of senescence (Martin and Turker 1994). Osteoporosis and atherosclerosis/arteriosclerosis illustrate how disease and age-related change occur in concert, proceed throughout adult life, and increase risk of death. They also show how multiple cultural, genetic, and environmental factors structure senescence and risks for CDCs. Multiple degenerative processes acting in concert produce a range of age-related declines in function; these limit an organism's ability to maintain homeostasis and survive in a hostile environment (Crews 1993a). Separation of age-related pathological processes and senescent changes does not seem to be a profitable approach to understanding the multiple degenerative processes that constitute senescence (Martin and Turker 1994; Miller 1995; Cristofalo *et al.* 1998, 1999). Although some may disagree with this suggestion (Fozard *et al.* 1990), distinguishing between the processes of senescence and age-related diseases seems impossible.

Normal and normative

Among commonly measured phenotypes – height (inches/cm), weight (pounds/kg), skin color (wavelength), glycemia (mg/dl or mM) – senescence is likely to be the most poorly operationalized. Complex, life-long interactions among genes, environment, and culture produce an array of senescent phenotypes. As yet, no accurate method of measuring this phenotype has been developed. Age at death is the most common variable used to study senescence. In such designs, life span is presumed to measure resistance to, or be the inverse of, senescence. Another view is that senescence is age-independent because the biological processes, physical forces, and extrinsic stresses that produce it are not themselves time dependent; however, they are cumulative, providing the impression of time dependence.

Given no specific scale for senescence, researchers have defined concepts such as "normal" or "normative" aging (Shock 1984, 1985; Andres 1971). What generally passes for "normative" and "normal" are average levels, differences, or changes in measurable factors (e.g., hormone activity, visual and audio acuity, muscle strength, and skin elasticity) of a cohort followed longitudinally, or averages at specific ages (e.g., 50–59, 60–69) from cross-sectional samples. Estimates of "normal" ignore fluctuating coefficients of variation with age for many physiological measures. Loss to follow-up due to illness, death, lack of interest, and/or ability to participate also produces non-representative samples in such designs. Loss to follow-up is particularly problematic in studies of senescence, since those lost may represent the extreme phenotypes. "Normative aging" defined as cohort averages is illusory, representing only the mean of a group of highly variable individuals, but few real persons. Looking at the means of physiological measures by age measures no process of senescence; rather, it ignores the structure of human variation.

To reduce problems inherent in "normal aging", "biological age" (BA) was proposed as an alternative to "chronological age" in studies of senescence. Unfortunately, BA is also estimated as the mean value of a trait at a specific chronological age (Borkan *et al.* 1982). As with "normal" and "normative" aging, estimates of BA are based on chronological age and averages. To estimate a person's "biological age", one determines the mean chronological age (CA) at which their specific measures place them. BA represents the average CA of a large set of physiological traits (e.g., blood pressure, weight, BMI, serum cholesterol, glucose) measured in a standard cohort similar to that used to determine "normal aging". Clinically, BA assesses whether someone is "biologically" younger than their chronological age relative to age-matched peers. A BA greater than the CA is thought to reflect more rapid senescence than the norm; a younger BA than CA represents slower senescence. Probability of death is associated positively with older BA profiles than CA (Borkan *et al.* 1982; Borkan and Norris 1982; Borkan 1986). This association is similar to that reported between CA estimated by a medical researcher on first observing study participants (Borkan *et al.* 1982; Borkan and Norris 1982). Although BA appears to provide information about age-related change and disease risk, it says little about senescence *per se*. A standard physician's examination likely provides as much information on future life span as does the calculation of BA.

Senescence is an individual phenomenon and most physiological functions increase in variance with increasing age. Analyses of "normal" and "normative" provide few insights into these patterns, nor do they inform us of how individuals are experiencing old age and senescence. Furthermore, current norms are based on historical trends and patterns and are unlikely to apply to elders today – or to tomorrow's either. Examining variation from norms and between and within populations may show why some progress more rapidly through the processes of senescence and others more slowly (Crews 1993b; Crews and Garruto 1994; Crews and Gerber 1994; Gerber and Crews 1999; Harper and Crews 2000). Heterogeneity and individual variation have also recently re-emerged as issues of concern to gerontologists (Bengtson and Schaie 1999). Defining variable senescent phenotypes (e.g., rapid, gradual, negligible; accelerated/rapid, slowed/retarded) reflects this conceptualization of variation. Specification of variable phenotypes will aid in identifying genetic and physiological factors underlying individual manifestations of senescence.

Demographic perspectives on human longevity and life span

Life expectancy and life span

Life expectancy (e_o, the expectation of life at birth is a basic life table parameter) is calculated as the sum of all years lived by a birth cohort divided by the total number born into the cohort ($\Sigma l_x / N_x = e_o$). e_o varies widely across genera and species and often across populations within species. e_o is often compared between and among populations and species, but it is not clear what is being compared, since e_o is responsive to biological, environmental, and cultural factors, and does not measure senescence. Life span is the length of time an individual has lived. In population genetics and demography (and previous equations), its average is e_o and its maximum is ω. Some view life span as being limited by species life history to a certain maximum (Fries 1984). Evolutionary models suggest that the current maximum life span is the outcome of competing risks representing tradeoffs between the soma and germ line (see Kirkwood 1990; Finch and Rose 1995; Harper and Crews 2000). The former view sees life span as intrinsically fixed; the latter sees life span as probabilistic (the ultimate outcome balances survival mechanisms against the need to reproduce and fledge offspring). Even similar species within single genera often show significantly different values for e_o and ω. These differences are likely to represent a variety of factors in addition to responses to prevailing environmental pressures. Among all species, natural selection and random genetic isolating mechanisms have likely exerted strong patterning effects on the timing and rates of acquisition of all life stages; among humans, biocultural factors have further shaped life history. Differences in life history represent heterochrony (differences between species in rates of growth and differentiation; timing of life history events within a species) and different survival and fitness strategies (see Finch and Rose 1995, p. 8).

As late as 1850, e_o in the U.S. (Massachusetts) approximated only 38.3 years for "white" males and 40.2 years for "white" females; in 1900–02, among "non-whites," e_o was 32.5 years and 35.0 years for males and females, respectively, while e_o for "white" males was 48.2 years and, for "white" females, it was 51.1 years (see Erhardt and Berlin 1974 for a review of trends in mortality and longevity in the U.S.A.; note that, herein when race/ethnicity and population terms are used, they are used as they were in the cited references they are enclosed in quotations). In the U.S.A., expectation of life only exceeded 50 years in 1900 (among "white" women; Erhardt and Berlin 1974). In the tenth century, only the very lucky survived their fiftieth birthday, and as late as 1959–61 only about 88% of persons living in the U.S.A. saw it. At the close of the twentieth century, it was an unlucky man or women who did not celebrate their seventieth birthday – an age that almost 70% of men and 80% of women might expect to reach today (Anderson 2001).

Life expectancy at both birth and age 40 increased steadily for men and women in the U.S.A. after the start of the twentieth century (Table 1.2). Expectation of life at age 85 did not increase as steadily or as dramatically as that for younger persons; rising by only 2.6 years for women and 1.6 years for men over the same period. Until recently, life expectancy increased most at young ages with little change at ages over 75 years. However, since about the 1980s, life expectancy at ages over 75 years has been steadily increasing, resulting in a large and increasing percentage of the U.S. population aged 65 and over. At the beginning of the twenty-first century in the U.S.A., 65-year-old men may expect to live an additional 16 years, and women 20 years, compared with only 13 and 15 years as recently as 1940. Worldwide, there are over 500 million people aged 55+, and they are increasing rapidly, by about 15 million per year. More people surviving has led to increases in chronic degenerative diseases, long-term morbidity, functional declines, and a greater representation of slow-senescing phenotypes in human populations.

Through the twentieth century, life expectancy has increased steadily for residents of most societies. Today, life expectancy for U.S. residents is about 76 years, with women living about 5–7 more years than men (about 10% more). Currently, the highest known life expectancies are among Scandinavian and Japanese populations (Table 1.3) – societies characterized by little genetic, environmental, or cultural heterogeneity as compared with the U.S.A., where average e_o represents life expectancy among an amalgam of different ethnic/cultural sub-groups. The lowest e_o values are among populations with generally lower socioeconomic status (SES), whether across nations – e.g., India, Kenya, Somalia – or within one nation such as the U.S.A. – e.g., Native Americans, African-Americans, Hispanic Americans. In 1997, the oldest person with a verified birthdate died at 122 years; she lived in France. Today, over 13% of the U.S. population is aged 65 and older. Proportions are higher in several other populations – notably, Sweden, Norway, and Japan (Table 1.3).

Because some families show familial disease syndromes (e.g., cancer, hypertension, cardiovascular diseases), it is likely that others show a reduced susceptibility to CDCs and slower senescence. Based on males born between 1750 and 1910 in the Connecticut Valley, Swedlund and colleagues have explored both family patterns of longevity and the persistence of families over time. In general, given obvious caveats, low correlations of life spans between sib pairs, twin pairs, and parent/offspring pairs indicate some degree of either social or genetic familial influence on life span (Swedlund *et al.* 1976, 1980, 1983). Persistence of specific families (family longevity) over time was more influenced by wealth than by patterns of fertility, mortality, or individual

Table 1.3 *Expectation of life, percentage of population over age 65, and proportion of population living to age 70 for various countries of the world in the late twentieth century*

Country	Expectation of life				% of population aged >65		% alive at age 70	
	Birth		Age 40					
	Male	Female	Male	Female	Male	Female	Male	Female
U.S.A. (1994)	72.4	79.0	75.5	80.7	10.7	14.8	65.3	79.0
Great Britain (1995)	74.1	79.3	75.9	80.5	13.0	18.4	70.0	81.0
Japan (1990–95)	76.4	82.9	77.9	83.9	12.2□	16.8	75.1	87.7
Sweden (1994)	76.1	81.4	77.6	82.4	15.0□	20.0	75.0	85.2
Norway (1993)	74.2	80.2	76.1	81.4	13.7	18.8	71.0	83.9
Ecuador (1990–95)	67.3	72.4	75.1	78.7	3.9++	4.5	59.4	68.9
Brazil (1996)	64.12	70.6	71.5	75.6	4.4+	5.2	50.6	65.8
Samoa (1990–1995)	65.9	69.2	n/a	n/a	3.3+++	3.6	n/a	n/a
Somalia (1990–1995)	45.4	48.6	n/a	n/a	n/a	n/a	n/a	n/a
Kenya (1990–95)	52.7	55.4	n/a	n/a	3.2**	3.4	45.7*	54.1
India (1986–90)	57.7	58.1	69.8	72.3	4.1	4.4***	41.4	47.1

*1979–1989   **1989   ***1993   +1991   ++1992   +++1990   □1995
From *Demographic Yearbook* (1996, 1998) United Nations Department of Economic and Social Affairs, New York.

longevity (Swedlund *et al.* 1983). A recent evaluation of health in the U.S.A. (Pamuk *et al.* 1998) reports that education and income remain major influences on health. Those with the highest income and education live the longest – an average life span of 76.1 years – compared with those with the lowest income, who average only 70 years (Pamuk *et al.* 1998). Although genes contribute significantly to the chances of long life, disparities in longevity in the U.S.A. over the past few centuries have been more attributable to differential access to resources than to biology. Still, members of some families are freer of disease-associated alleles, have more efficient immune and repair mechanisms, and/or are endowed with more stable neurological and physiological functioning than are others. Members of other families experience vascular and neoplastic disease morbidity and mortality significantly earlier in life than do most. Given human mating patterns and interdemic mate migration, any traits predisposing to improved fitness will spread relatively rapidly through all human populations whether they also predispose to retarded or accelerated senescence (see Lasker and Crews 1996).

Life expectancy shows variable degrees of familial aggregation and heritability. Mayer (1991) reported a heritability of 10–30% for length of life from 1650–1874 in the U.S.A. This level of heritability apparently persisted over a 225-year span of genealogical data, including 13 656 deaths, from a socially elite sample of six New England families. This suggests that the variation in longevity during this period was not due to genetic differences and that the influence of genes on longevity did not increase through the nineteenth century as life expectancy increased. Although longevity may be familial in nature, factors other than genes tend to have a large influence on longevity in humans. Over the past several centuries, gains in longevity have followed from changes in environments and culture, not in alleles. Still, rapid environmental changes in recent decades, centuries, and millennia have likely altered multiple gene–environment, epistatic, and selective interactions among systems that were established over many thousands and millions of years. To some degree, these have likely led to alterations in the survival and reproduction of specific alleles in various human populations. Slowed senescence likely results from multiple alterations in pleiotropic, epistatic, and coordinated relationships among genes/alleles as various thresholds and transitions across life stages are passed. These life stages have themselves been altered by increasing degrees of bioculturally derived selective pressures in a culture-using hominid.

Mortality and survival

In most species, including humans, survivorship is a negative function of time. This is because mortality rates accelerate with age in populations that show senescence. The Gompertz equation, an exponential function of time, has been used extensively to model human survivorship. The log of the age-specific mortality rate (the fraction of survivors that die in a specified time interval (t)) increases linearly with age over most of the adult life of humans and many other species. This mortality rate, graphed on a semi-log scale by age, fits a straight line from the age of sexual maturation through the estimated average life expectancy for the population (see Harper and Crews 2000 for more details). The slope of this line is known as the Gompertz coefficient (G), which measures

Table 1.4 *Estimates of initial mortality rate and mortality rate doubling time (MRDT) for various species*

Species	Initial mortality rate	MRDT
Human Female U.S.A. 1980	0.0002	8.9
Dutch Civilians 1945	0.0014	7.8
Rhesus	0.02	8.0
Mouse	0.01	0.3
Rat	0.002	0.3
Hippo	0.01	7.0
Dog	0.02	3.0

From Finch, 1994.

acceleration in the rate of mortality with increasing age (Figure 1.1A). This rate is considered by many to be a simple estimate of the rate of senescence for a species (Finch 1994; see also Table 1.4). Others suggest that this model is too simplistic to model human mortality and that it is based on several misleading assumptions (Harper and Crews 2000; Ossa and Crews in press). Conversely, survivorship (Figure 1.1B) is graphed as the percentage of a cohort still alive at age x. In high mortality populations, this curve approximates a right triangle by dropping sharply in early life and flattering out over the remaining life span (Figure 1.1B, B1), while in low mortality populations it becomes rectangular-ized (Figure 1.1B, B4). Interestingly, in populations that show no age-related increase in mortality with age – the same percentage of the remaining popu-lation dies in each age class – the curve appears to be concave (Figure 1.1B, B2), whereas, in populations where mortality remains at a constant number of deaths in each age class, the survivorship curve is a straight line from 100% to zero over the life span (Figure 1.1B, B3). Type B1 applies to most tradi-tional living human populations with high infant and childhood mortality and relatively high mortality rates throughout the life span; examples include the Yanomami, !Kung, and Turkana. Type B4 applies to many modern day popula-tions with low infant, childhood, and young adult mortality, with the majority of deaths occurring after reproduction has ceased for all women and most men; the U.K., Japan, Sweden, and the U.S.A. provide examples. Pattern B2 is found among some animal populations (e.g., hydra, songbirds, and oysters) and the non-living (water glasses in a restaurant are the best example). Pattern 1B4 includes toughened water glasses in a restaurant and sheep in a zoo (see Wilson and Bossert 1971, p. 113; Arking 1991, pp. 27–8).

An additional term used to characterize population senescence is MRDT (the amount of time that it takes a species' mortality rate to double), which is approximately every 9 years after age 30 in modern humans (see Table 1.4).

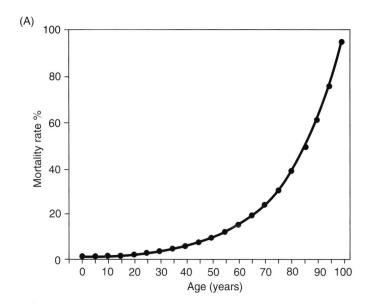

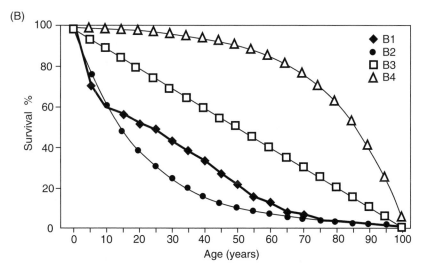

Figure 1.1 (A) Gompertz model and (B) hypothetical survivorship curves for a variety of populations.

MRDT is represented graphically as the natural log of the mortality rate against age; recall the slope of this line is G, the rate of acceleration of mortality with age. Using G as a measure of senescence has been critiqued. First, MRDT is based on an exponential relationship using curve fitting parameters; thus, deviations from expected values decrease as a function of increasing time (Hart and Tuturro 1983). As a result, G becomes less sensitive to deviations in survival with increasing age (Curtsinger *et al.* 1992). Second, accurate estimation of G is dependent on there being few survivors beyond a specific MAL (Hart and Tuturro 1983); therefore, possible values for G are restricted by the estimate of MALS used in the model. Although Gompertz coefficients of human populations in drastically different environments are remarkably similar (Finch 1994), few reports have examined the magnitude of variation between populations or across different cohorts within populations (Hart and Tuturro 1983). In addition, the conformity of mortality parameters to Gompertz equations may be related to the intensity of selection acting on age-specific mortality rates, a phenomenon that changes as patterns of pleiotropy change within a population (Rose 1991). This is likely to happen when gene–environment and gene–biocultural interactions are altered in human populations. Finally, empirical data to complete such analyses do not exist for most living, prehistoric, or fossil populations (Finch 1994). Researchers still use Gompertz models to suggest a human species-specific maximum life span of about 120 years (Fries 1980, 1983, 1984; Fries *et al.* 2001) and to suggest that the increasing rectangularization of survival curves for the more cosmopolitan societies during the twentieth century, including a sharp downward slope at about age 85, illustrates a specific limit to human longevity. As Fries and Carpo (1981, p. xiii) state: " . . . Increasing life expectancy is thus on a converging course with fixed life span represented by the downslope of the rectangular curve." However, there are multiple problems with suggestions that the human survivorship curve has become rectangularized or that any intrinsically programmed limit to life span exists. Survival curves only look rectangular if truncated to 85 or 100 years; past these ages, the slope is far from perpendicular (Figure 1.2; see also Myers and Manton 1984). After age 85, the mortality curve for 1980 in Figure 1.2 is reminiscent of the curve illustrated in Figure 1.1, B3, with a long tail out to the oldest survivor, suggesting little association of late life survivorship with age.

The three survival curves in Figure 1.2 are based on the best available data from the U.S. Social Security Administration and show several important demographic trends. There were overall gains in survival from 1900 to 1980. Infants, children, and young and middle-aged adults all showed larger increases than those over 80. The largest proportional increases in survival between 1900 and 1960 were for those aged around 60 years; these improvements were greater than in the younger groups from 1960 to 1980. Improved survival stretched survival

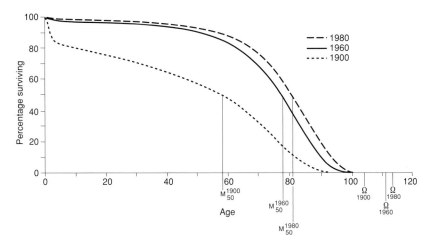

Figure 1.2 Survival curve based on 1980 social security administration data. (Redrawn with permission from Myers and Manton 1984, Fig. 1, p. 347).

curves out over time and caused a shift to older ages of all variables along the abscissa (e.g., life expectancy (ω), Myers and Manton 1984, pp. 347–8). Rectangularization of survival as articulated by Fries (1980, 1983, 1984, 2001) and coworkers (Fries and Carpo 1981) is poorly illustrated in these curves (Myers and Manton 1984). Based on their review of available data, Myers and Manton (1984) find little demographic evidence that humans are near any biological limit to life span that would constrain future increases in life expectancy.

Another question that can be addressed with demographic data is whether mortality is being compressed into later ages as any model based on a maximum life span and rectangularization of survival suggests. Myers and Manton (1984) used the distribution of ages at death to examine the possibility of a compression of mortality. Although the proportion of all deaths occurring after age 60 increased from 1962 to 1979, the mean age at death for these deaths also increased, as did the standard deviation of age at death, showing that mortality compression did not occur at older ages over the period and that elders showed more variable life spans (Myers and Manton 1984). In fact, Myers and Manton (1984) find support for an increased dispersal of ages of death at older ages rather than the reverse, and "no evidence that rectangularization has had a significant effect on the population and mortality dynamics of the elderly, at least through 1979."

It is now well established that the acceleration of mortality rates, which characterizes most species over their life span, slows at advanced ages in humans – currently at about the age of 75–85 years (Myers and Manton 1984; Perls 1995; Carey 1997; Perls 2001). Mortality patterns of neither humans nor most species fit assumptions of the Gompertz model or Fries' rectangularization of survival (Myers and Manton 1984; Carey 1997). Populations are not composed of identical individuals; rather, some members are more likely to succumb to both exogenous and endogenous factors and die in greater numbers at younger ages (frailty) than other individuals who are endowed with a greater resistance to death (vitality) (see Vaupel *et al.* 1979; Vaupel 1988; Weiss 1990; Carey 1997). Given such demographic heterogeneity, i.e., sub-groups within the population with different inherited levels of frailty, mortality of a population is expected to depart from the Gompertz model as age increases (Carey 1997).

Population growth and change

Currently, "elders" (arbitrarily set at ages 65+) are the fastest growing segment of the U.S. population and of populations in many developed countries. This is partly because over the past century, and more rapidly over the last 40 years, longevity has increased at all ages. For example, from 1950 to 1980, the U.S. population aged 65 plus years increased at an average of 2.5% a year (range 0.2% in 1951 to 3.6% in 1974) (Guralnick *et al.* 1988; Verbrugge 1989; Brock *et al.* 1990). Although life expectancy is significantly lower in developing countries and among some inner city and rural populations in developed countries (Susser *et al.* 1985; Grigsby 1991), the elderly population is also increasing in these areas (Kinsella and Suzman 1992). Worldwide, there is a net increase of about 1.2 million people aged 55 plus per month; of these, four-fifths reside in developing countries and their annual growth rate is 3% compared with only 1% on average in developed countries (Kinsella and Suzman 1992). Currently, about 60% of all persons aged 55 and over reside in the developing countries of the world (about 400 million persons); by 2020, this proportion will be 72%, representing 1 billion persons (Kinsella and Suzman 1992). Population aging occurs as fertility declines and mortality shifts to older ages. Today, the oldest old, those aged 85 or more years, are the fastest growing population sub-group in many countries due to reduced mortality at these ages (Rosenwaike 1985; Grisby 1991; Kinsella and Suzman 1992).

Although older cohorts are increasing partially due to a reduction in mortality at older ages, the primary factor producing increases in the population aged 55 and over has been sustained decreases in childhood mortality throughout the twentieth century and relatively high birth rates in developing nations (Olshansky *et al.* 1990; Kinsella and Suzman 1992; Olshansky and Carnes 1994). A striking example of how fertility affects population aging is the baby-boom generation born between 1946 and 1964. This cohort is currently aging toward elderhood in the U.S.A., U.K., and other nations that saw high birth rates during the decades following World War II. As these individuals pass their sixty-fifth birthdays over a period of about 20 years (2011–2029 A.D.), the proportion of elderly in places like the U.S.A. will increase rapidly from its current level of about 13% to over 20%. A decrease in fertility, except for a few populations, is also contributing to population aging worldwide. In several Latin American and Caribbean countries, the proportion over 55 years currently exceeds 10% (Argentina 17.5%, Cuba 15.9%, Jamaica 11.8%, Uruguay 21.7%), as it does in places like South Korea (11.1%) and Sri Lanka (11.0%).

Chapter synopsis

The purpose of this introductory chapter was to present a background for discussion of evolutionary and biocultural aspects of human senescence. To do so, terminology and critical concepts in senescent biology along with a framework of historical thought and research were reviewed. This was followed by a brief exposition on demographic aspects of senescing populations and changes in life history, life expectancy, and life span that have occurred among humans. Chapter 2 will expand this discussion to current evolutionary theories and proposed biological mechanisms of senescence that affect human life span.

2 Evolutionary and biological theories of senescence

Background

Developments in general evolutionary theory and quantitative population genetics during the early twentieth century set the stage for theorizing on the evolution of senescence. Medawar (1946) developed the "wear-and-tear" theory based on observations that with age organisms have an increasing likelihood of dying due to intrinsic (belonging to the real nature of a thing; not dependent on external circumstances; inherent) factors (reviewed in Chapter 1). He illustrated this point using the now classic example of test tubes in a laboratory (Austad 1992). Test tubes do not senesce; still, through time, they tend to break due to laboratory mishap and intrinsic fragility. If most test tubes break (die) within 2 years, a change (mutation) that affects test tubes only after 2 years will have little effect on longevity in the overall population. However, a change that improves a 1-year-old test tube's survival, or, better yet, that of a 6-month-old tube, will have a substantial effect on both average and maximum time until loss in that test tube population. Medawar further noted that with age selection pressure declines in concert with decreasing fertility in natural populations, ultimately becoming minimal or non-existent.

Two conflicting views typify evolutionary theories on senescence. Most early theorists saw senescence as the outcome of a specific genetic program designed by natural selection to eliminate unneeded post-reproductive individuals. A minority of gerontologists now subscribe to various permutations of this genetic programming theory (see Clark 1999 for a recent review). Most current evolutionary theories see senescence as an artifact of forces acting to maximize reproductive success in sexually reproducing organisms with mortal somas (Rose 1991; Austad 1997). This has been described as an "epidemiological model"(Wood et al. 1994). In general, processes of senescence are set into action by the development of an inherently imperfect somatic system, susceptible to degenerative alterations with increasing survival times, for reproducing the germ line. Over time, physiological wear-and-tear, loss of redundant sub-units, and damage from internal and external insults lead to chronic dysfunction, degenerative conditions, and an increased probability of death.

No known single or simple biological mechanism is responsible for the multiple manifestations of senescence – they arise secondary to multiple environmental, biological, and, in humans, cultural variables. Genetic and biological contributors to senescence are based in the inherent fragility of DNA, proteins, organelles, cells, and organs. These are not programmed to cause senescence *per se*; rather, they reflect multiple compromises between efficiency, survivability, and reproduction intrinsic to all multicellular life forms. Natural selection acts to maximize total fitness of an organism relative to its competitors. Promoters of the programmed theory have never clarified how evolutionary process could both promote individual senescence and improve lifetime fitness. Observations of multiple physiological alterations, selective replication of mutant mitochondria, DNA damage from reactive oxygen species (ROS), loss of terminally differentiated cells, wear-and-tear, telomere shortening, unregulated apoptosis, and autoimmunity do not suggest genetic programming. If somatic senescence were programmed, there would be no need for it to involve complex multiple system failure or for the process to take decades to complete. Rather than suggesting a programmed process, individual variability, along with the pattern and complexity of somatic dysfunction seen during senescence, points to loss of multiple systems of regulation with increasing survival past the prime reproductive years.

Another problem with programmed senescence is that, at different levels of biological organization, the process is quite variable; this would not be expected if senescence were due to a genetic program that had been conserved through evolutionary time (i.e., plesiomorphic). To the contrary, homologs of proteins that alter life history in less complex multicellular models, such as roundworms, generally have new uses and functions and interact with different proteins and macromolecules in fruit flies, rodents, and humans. In addition, senescence varies across species, within species, and even within the different organs and systems of a single organism. Evolution is conservative: if there were a genetic program needed to promote senescence, it should be observable across all sexually reproducing species. In the case of senescence, a mechanistic probability model based on the inherent frailty of cells, their internal components, and the organs/systems they produce appears most appropriate. Each physiological sub-system is composed of various organs (e.g., lungs, kidneys, heart, blood vessels) that are made up of numerous redundant sub-units (e.g., alveoli, nephrons, myocytes, capillaries). These sub-units randomly fail because of variably distributed insults to the cytoplasmic and nucleic components of their cells. During growth and development, sufficient numbers of redundant sub-units generally are laid down to attain the maximum reproductive potential (MRP) possible for the specific genotype (in some cases, MRP may be zero). However, available redundancy is not sufficient to last indefinitely (see Gavrilov and Gavrilova, 1991, 2001 for reviews of reliability theory and senescence). The realization that no single biochemical switch is responsible for senescent changes has led to a greater emphasis on the soma as a complex adaptive system striving for survival and reproduction in a hazardous world (Kowald and Kirkwood 1994; Kirkwood 1995; Mangel 2001). Senescence is a complex systematic, multifactorial, and multilevel phenomenon. Ascribing it to the declining force of natural selection with increasing age is neither practical nor is it medically, psychologically, or biologically helpful. To intervene in the process, numerous proximate mechanisms that cumulatively and progressively increase the risk of death with continued survival must be understood.

Williams (1957) enumerated four conditions that any valid evolutionary theory of senescence will meet. First, "Senescence should be found where ever the conditions specified are met (a soma) and should not be found where these conditions are absent" (p. 403). Clonal species for which the soma (the entire body of an organism with the exception of the germ cells) and germ plasm (reproductive cells: in sexually reproducing species, sperm and ovum) are the same should not senesce; senescence should only characterize sexually reproducing species. All available data on senescence support this deduction (Finch 1994). Second, "Low adult death rates should be associated with low rates of senescence, and high death rates with high rates of senescence" (p. 404). The continuum of senescence patterns outlined by Finch (1994) fits this prediction. Rapid senescence occurs in annual plants, marsupial mice, and Pacific salmon; these populations all experience exponential increases in adult mortality after maturation and reproduction (Finch 1994). At the other extreme, negligible senescence occurs in populations with little adult mortality, such as anemones, clams, sharks, tortoises, and some species of trees (Finch 1994). Third, "Senescence should be more rapid in those organisms that do not increase markedly in fecundity after maturity than those that do show such an increase" (p. 405). Mice do not show increased fecundity (the quality or ability to produce offspring) with age and senesce rapidly, whereas sharks and tortoises do show increased fecundity and fertility (the production of offspring, or process of having reproduced a new individual) with age and senesce slowly. Lastly, "Where there is a sex difference, the sex with the higher mortality rate and less rate of increase in fecundity should undergo the more rapid senescence" (p. 406). Among most mammals in protected settings, females outlive males (Williams 1957; Hazzard 1986; Holden 1987). In "natural" or "wild settings" and in most human populations without modern health care and sanitation, women do not generally outlive men (Gavrilov and Gavrilov 1991). This may be explained partially by the variable reproductive strategies used by, and different evolutionary pressures on, men and women. Among women, fecundity appears to increase until the mid-twenties and then decline slowly until the mid-thirties, after which it declines rapidly through the forties till menopause at about age 50. Cultural developments have had little influence on this pattern. For men, cultural factors may have greatly altered reproductive patterns, allowing older men to father children well into their sixth and seventh decades of life, and even later.

The remainder of this chapter will examine evolutionary theories and biological mechanisms of human senescence. This division separates ultimate causes of senescence (*why*) from proximate models of *how* dysfunctional alterations decrease the probability of survival with increasing age after the attainment of MRP (Austad 1992; Crews and Gerber 1994; Gerber and Crews 1999). Ultimate causes are the outcome of evolutionary pressures; these lay a framework of life history possibilities for all species. Proximate causes are specific mechanisms that pace and pattern life history within the context of specific phylogenetically determined evolutionary frameworks. Evolutionary theories answer the fundamental question of: "Why do all sexually reproducing species senesce?" Biological models provide answers to questions of "Through what mechanisms do species senesce?" This distinction benchmarks the explanatory nature and applicability of theories and models of senescence. Understanding why we senescence is a basic question in evolutionary biology, but this contributes little to geriatrics or to the care of aging individuals. Although contributing little to current treatment and preventive strategies, understanding the evolutionary basis of senescence helps to guide much of the biomedical, genetic, and epidemiological research on chronic degenerative conditions (CDCs). Understanding the proximate biological mechanisms that lead to senescent physiological dysfunction aids in developing interventions and treatments for CDCs and halting detrimental processes. A recent estimate is that over 300 proximate explanations for senescence have been proposed, but only two theories on ultimate causes (Austad 1992).

Evolutionary (ultimate) theories: the basis of senescence

Antagonistic pleiotropy

Following Medawar's intuition that the force of natural selection decreased with age, G.C. Williams (1957) elaborated the concept of senescence secondary to the pleiotropic actions of genes. He saw that any alleles predisposing to better survival and/or fitness at young ages are favored by natural selection, even if they pleiotropically predispose to short life after reproduction is complete. Later, this model was labeled "antagonistic pleiotropy" (neither I, nor several of my colleagues, have been able to find who originally coined the term "antagonistic pleiotropy", – it did not appear in Williams' original 1957 article, nor does he take credit for it elsewhere or recall who first used this terminology; Linda Gerber, personal communication). At any locus, alleles that enhance survival, MRP, or fitness during early and reproductive phases of life history, but produce detrimental ones that decrease the probability of later survival, show antagonistic pleiotropy. The relative fitness values of such alleles early in life outweigh any loss of fitness at older ages when fitness tends to zero. Even in a population of immortal individuals, alleles with antagonistic pleiotropy will become fixed, driving all alleles not conferring early life or fitness advantages, even those predisposing to immortality, to extinction (Albin 1988). Such alleles should rapidly increase in a population, where their presence increases the likelihood that other alleles with similar countervailing effects might also increase and spread (Albin 1988). Later, Hamilton (1966) proved mathematically what Medawar (1952) and Williams (1957) had intuited – that the force of natural selection declines with increasing adult age as fecundity declines and the probability of death increases (Rose 1991; Finch and Rose 1995). Alleles with antagonistic pleiotropy generally only produce dysfunction when individual life history extends beyond the time needed for maturation, mating, reproduction, and the fledging of offspring. Most species' genomes are likely to include many alleles with early life benefits and later life detriments.

Specific examples of antagonistic pleiotropy have proved difficult to identify. As late as 1988, Albin lamented that no-one had yet identified any antagonistic pleiotropic senescent alleles. Today, alleles at several human loci are thought to show some antagonistic pleiotropy. Angiotensin-converting enzyme (ACE), apolipoprotein E (ApoE), Huntington's disease, any number of oncogenes (e.g., *ras*, *myc*), molecular mechanisms associated with corticosteriod-dependent neuron damage and plasma sodium balance, testosterone levels/activity in men, alleles contributing to the glycation of long-lived molecules, and protein-mediated dopamine-related damage to neural tissue all appear to show some degree of antagonistic pleiotropy in humans (Albin 1988; Schächter *et al.* 1994; Finch and Rose 1995; Crews and Harper 1998). A strong case has been made for the $\varepsilon*4$ allele at the apolipoprotein E locus (Schächter *et al.* 1994), although some suggest this may be an example of mutation accumulation rather than antagonistic pleiotropy (Austad 1997). In traditional-living populations, this allele may have been at a higher frequency and selectively advantaged due to its cholesterol-elevating effects in fat-poor environments (Crews *et al.* 1993). In natural settings with low fat and cholesterol diets, carriers of $\varepsilon*4$ may have experienced better survival, maturation, or reproduction than those with the $\varepsilon*3$ or the $\varepsilon*2$ allele; the latter is generally absent in Native American samples (Crews *et al.* 1993). In low-fat and -cholesterol dietary environments, the $\varepsilon*2$

allele may not predispose to adequate fitness. In modern populations with a surfeit of dietary fat and cholesterol, the $\varepsilon*4$ allele predisposes to increased lipidemia, cardiovascular disease (CVD) mortality, late-onset Alzheimer's disease, and to shorter survival than either of the other two common alleles.

Alleles at loci regulating testosterone formation or its circulating levels, binding, and activity are also candidates for antagonistic pleiotropy in men. Dihydroepiandrosterone (DHEA), produced by both the fetus and placenta during embryonic development, regulates differentiation of the fetal sex organs, closure of the vaginal opening, and, in conjunction with the SRY gene product, formation of the testes from hermaphroditic gonads *in utero* (Crews 1993c). Testosterone is the major androgen regulator of puberty. It actively promotes masculinization of secondary sexual characteristics and the pubertal growth spurt. After puberty, testosterone continues to promote cellular mitosis and cellular growth in prostate and muscle tissues. With increasing time of survival, men with the highest testosterone activities apparently are at a greater risk of prostate hypertrophy and carcinoma (Finch and Rose 1995); they also tend to show higher blood pressure and an increased risk of CVD. These outcomes suggest antagonistic pleiotropy in the levels and/or activity of testosterone. Inter-individual variations in testosterone activity are likely to reflect allelic variability in the proteins determining the manufacture of receptor proteins for, and/or intracellular protein transmitters and secondary carriers of, testosterone signaling.

Lifetime patterns of estrogen activity in women also suggest testosterone-mediated antagonistic pleiotropy. Prior to menopause, normal estrogen activity is associated with lower blood pressure and lipids and fewer cardiovascular events than seen in same-age men. High estrogen activity may protect women from testosterone. At menopause, ovaries produce estrogen at significantly lower levels than seen at young and middle ages. As estrogens decline, relatively high testosterone levels, compared with earlier decades, are seen in women. Increased blood pressure, lipidemia, CVD, and reproductive tissue neoplasms accompany this alteration in relative estrogen/testosterone levels. High estrogen levels during the reproductive period can be presumed to be necessary for women to achieve MRP and fitness. Unfortunately, these are also associated with depletion of primary oocytes from the ovary within six decades of life and loss of the ability to respond with increased estrogen to hypothalamic follicle-stimulating hormone (FSH), which also increases to well above pre-menopause levels. Production of high estrogen levels appears to be necessary for women to reproduce, even though the outcome of this reproductive investment is loss of hormonal homeostasis and an increased risk of death post-menopause.

A variety of single locus human disorders also seem to conform to the predictions of antagonistic pleiotropy – e.g., Huntington's disease (HD), sickle-cell anemia, and peptic ulcers (Albin 1988). HD, an autosomal dominant condition whose gene is located on chromosome 4, shows no obvious clinical signs prior to onset, usually during the late fourth to fifth decades of life. Thereafter, HD shows as progressive loss of cognitive function and voluntary movement, and increased tremor and loss of personality, leading to death usually within 10–20 years of onset (Albin 1988). Pedigree analyses of affected kindreds show larger than average sibship sizes; those with HD also tend to have more offspring than their unaffected sibs, suggesting that the HD allele may promote higher fertility (Albin 1988). It is unclear whether increased fertility of HD patients and kindreds is due to behavioral or biological influences. Higher fertility seems to characterize women with HD, who tend toward higher gonadotropin-releasing hormone (GRH) levels in their hypothalami, while men show no increased production of GRH (Albin 1988).

The polymorphism for sickle-cell anemia is a classic example of heterozygote advantage; it also is a candidate for antagonistic pleiotropy. Located on the long arm of chromosome 4, the β chain of hemoglobin comes in multiple forms; a mutation at the sixth base pair (A $\rightarrow$ T) causes an amino acid substitution (val$^+$ $\rightarrow$ lys^0), resulting in a change in protein charge from positive to neutral and a shift in protein conformation. Homozygotes suffer an extreme form of anemia termed sickle-cell disease. Historically, these patients have succumbed before reproducing. Heterozygotes show a milder anemia, which is most obvious during periods of stress or heavy work; they also are more resistant to malarial parasites, providing them with an early-life and reproductive advantage in malarial areas. In non-malarial settings, non-sickle-cell allele homozygotes are advantaged: they have neither malaria nor anemia. In malarial environments, heterozygotes are advantaged: they survive and reproduce better than either homozygote. Ultimately, however, heterozygotes do not survive as long as homozygous non-carriers in non-malarial settings, suggesting that the sickle-cell allele contributes to decreased survival.

Another candidate locus for antagonistic pleiotropy may be associated with peptic ulcer disease (PUD). Persons who overproduce pepsinogen 1, the precursor of pepsin, due to an autosomal dominant allele, often show PUD (Albin 1988). Overproduction of pepsinogen 1 produces gastric hyperacidity, which is reported in 50% of PUD patients. This hyperacidity may provide a non-specific barrier to enteric infections, tuberculosis, and cholera. Before the advent of pharmacological interventions to control PUD, overproduction likely led to higher mortality during mid and later life. However, hyperacidity likely also protected carriers from a variety of food/waterborne and crowd diseases during their formative and reproductive years.

Age-specific gene action/mutation accumulation

Alleles showing deleterious effects only beyond the ages associated with the majority of reproductive effort (age-specific gene action) were first proposed by Medawar (1952). Williams (1957) further elaborated this model, suggesting that alleles with harmful effects should accumulate around ages when reproductive potential is low (mutation accumulation). In humans, this age cannot occur before allowing an adequate time for maturation, attainment of MRP, mating, multiple births, and the necessary investment in fledging offspring (a minimum necessary life span), when natural selection is most potent. The same selective pressures that structure a species' life history determine the minimum necessary life span before which such ill effects should not accumulate. Physiological changes in modern humans with age, along with data from historical human populations, place this in the fifth or sixth decade of life (ages 40 and over). Multiple alleles with such age-specific effects could produce broad and synchronized changes in physiology. When their ill effects all appear at about the same age, this could be misinterpreted as a single switch. Martin *et al.* (1996) suggest that most late-acting alleles should be private polymorphisms, aggregating in specific lineages and kindreds. However, it seems likely that, through the effects of genetic drift, linkage, and gene flow, which in humans include culturally determined patterns of mate exchange (see Lasker and Crews 1996), such alleles will be widespread across human populations and will continually arise. Alleles with age-specific action are likely to influence differences in longevity across families and kindreds and between individuals.

As with antagonistic pleiotropy, few alleles are candidates for increasing the age-specific risks for senescence and disease. The most often cited candidate is HD, described earlier as a possible case for antagonistic pleiotropy; HD may also reflect age-specific gene action and mutation accumulation. Although there is evidence for increased fertility in HD women, this may be a behavioral (e.g., make-up fertility by couples trying to produce disease-free offspring) rather than a biological response. Before onset, most Huntington's victims have had opportunities (20+ years) to mate, reproduce, and fledge offspring. Until life expectancy exceeded four decades, little or no selection against the HD allele is likely to have occurred. The HD allele appears to show aspects expected from both antagonistic pleiotropy and age-specific gene action, providing an example of how difficult it often is to separate these mechanisms in any practical, rather than theoretical, fashion. Both predict that multiple "senescence-enhancing" alleles become more influential in promoting dysfunction with increasing survival. Numerous degenerative conditions are likely to be influenced by subtle differences in alleles with such age-specific effects. For many late-life CDCs (e.g., CVD, neoplasms, diabetes, hypertension, dementias), multiple predisposing alleles are likely to exist; several have even been identified. They include private alleles, as might be expected for recently emerged variants, and others that are widespread among and across today's populations. Other allelic variants with "senescence-slowing" effects predispose to longer life spans. Candidate senescence-slowing alleles include ApoE Malino, some Y-chromosome variants among a kindred of Mennonites, and the *ApoE*2* allele.

CVD and elevated blood pressure occur more frequently in individuals carrying either the *Apoε*4*, ACE deletion, or aldosterone 287A→T allele than in those with two wild-type alleles. The *Apoε*4* allele is also associated with late-onset Alzheimer's disease. Those carrying one or two copies of *Apoε*4* tend to have elevated serum cholesterol, low high-density lipoprotein (HDL) cholesterol, and a higher mortality from CVD. Homozygotes for *Apoε*2* have the lowest cholesterol and CVD mortality: an "anti-senescent" effect. *Apoε*4* homozygotes also show the highest frequencies of late-onset Alzheimer's disease. Most effects occur late in life, suggesting mutation accumulation. However, these effects may reflect antagonistic pleiotropy. Both the ACE deletion (D) and the aldosterone 287A→T alleles are associated marginally with blood pressure within populations. The ACE DD genotype is also associated with myocardial infarctions (MIs) and stroke and is under-represented in elderly samples (Crews and Harper 1998), but the D allele is over-represented in others, relative to frequencies in young and middle-aged adults. These data suggest age-specific selection pressure at the ACE locus. The ACE and aldosterone loci may also exhibit gene–gene interaction such that individuals homozygous for both the ACE D allele and the aldosterone 287A→T variant show the highest blood pressures and most frequent hypertension (reviewed by Crews and Williams 1999). As with the *Apoε*4* allele, it is unclear whether these are age-specific gene actions or antagonistic pleiotropy. For the ACE D allele, one advantage may be its length; 270 base pairs shorter than the insertion (I) allele, it produces a protein with 90 fewer amino acids. This lowers the metabolic cost of manufacturing DNA, messenger RNA, and protein, decreases wear-and-tear on cytoplasmic and nuclear components, and reduces opportunities for error accumulation. Over human life history, conserved resources may offset predispositions to MI and stroke in late middle-aged adults, by enhancing maturation and fitness in earlier life.

Both the *BRCA1* and *BRCA2* loci harbor alleles that increase the risk of mid-life disease and death with age-specific actions that lead to breast cancer at about the fourth decade of life – an age that likely was at about the upper limit of survival among human females through much of hominid/human evolution. Women carrying risk alleles are more likely to die before completing their fifth decade of life. Non-insulin-dependent diabetes mellitus (NIDDM, adult-onset diabetes, type II diabetes) is another CDC showing mid-life onset that may reflect mutation accumulation and age-specific gene action. Multiple loci, including insulin and glucagon, along with their promoters, receptors, and intracellular cytokines, influence the onset and progression of NIDDM. Before the advent of a surfeit of calories and socioculturally determined opportunities to limit physical activity, most such predispositions remained unexpressed. In settings with overabundant calories and low physical activity, they manifest age-specific effects that increase obesity, hyperglycemia, insulin resistance, and diabetic pathologies that limit life span. The human genome probably has many loci with various alleles that exhibit age-specific gene action and/or antagonistic pleiotropy, enhance or slow senescence, and limit individual life spans. Suggesting that alleles predisposing to increased risks for CDCs are senescence enhancing conforms to current trends viewing CDCs as the visible manifestations of senescence (Miller 1995; Cristofalo *et al.* 1998, 1999; Harper and Crews 2000).

Thrifty/pleiotrophic gene model

Physiologically, humans are the product of over 65 million years (MY) of primate evolution and about 5–6 MY of hominoid evolution. Humans evolved as a highly mobile bipedal hominid, exploiting an omnivorous ecological niche, and consuming a diet of mostly high-fiber, low-fat, plant products. Seeds, grains, nuts, fruits, roots, and vegetables, along with protein from insects, carrion, and prey, made up the bulk of hominid diets over the past 5 MY (Cohen 1989). Such diets are low in cholesterol, fats, and salt, contain no refined sugars or grains, and do not expose humans to industrial and crowd-related pollution, nitrates, or overnutrition. When our fully modern human form (Cro-Magnon) appeared about 100–200 000 years ago, hominid life expectancy was probably less than 20 years, few individuals (probably less than 5%) survived past 40 years, and adaptive mechanisms for the retention and conservation of scarce resources (e.g., calories, cholesterol, salt, fats) were already fully developed. Conversely, adaptive mechanisms for the retention of plentiful resources (e.g., calcium, vitamin C, fiber) or the elimination of non-existing chemicals (e.g., PCBs, nitrates, n-butyl acetate) never needed to develop. Today, well-developed genetic-based adaptive mechanisms may underlie a variety of risk factors for age-related CDCs, including salt-sensitive hypertension, arteriosclerosis, obesity, hyperglycemia, and hyperlipidemia; conversely underdeveloped genetic mechanisms are likely to underlie conditions such as osteoporosis, some cancers, multiple chemical sensitivity syndromes, and deficiency diseases (Crews and Gerber 1994; Gerber and Crews 1999).

In 1962, J.V. Neel (1962) defined "thrifty" genes – alleles that predispose carriers to more efficient extraction and storage of energy from dietary sources. Those with thrifty genes/genotypes are predisposed to store more energy as fat during times of plenty. When energy availability is low, thrifty individuals may use stored energy to maintain homeostasis. Conversely, non-thrifty individuals are disadvantaged, show poorer survival, and reproduce fewer copies of their alleles. The generalizability of a thrifty alleles model to multiple dietary and metabolic components (e.g., cholesterol, salt, essential fatty acids) is obvious (Crews and Gerber 1994). Innumerable thrifty alleles may have been incorporated into the genome during human evolution (Crews and Gerber 1994; Gerber and Crews 1999).

Combining the concepts of thrifty alleles and antagonistic pleiotropy produces a "thrifty/pleiotropic gene" theory for explaining multiple aspects of human CDCs and senescence (Crews and Gerber 1994). The thrifty/pleiotropic theory suggests that some CDCs are secondary to excessive accumulations of once scarce, but now abundant, resources, including fats/lipids, protein, cholesterol, iron, and salt, in addition to energy/calories as originally described by Neel (1962). Different CDCs arise following decreased dietary and metabolic availability of resources previously plentiful, but now rare in some human diets, including calcium, vitamins (e.g., C, B_{12}), micronutrients (e.g., selenium, cadmium), iodine, fiber, and antioxidants. There never were selective pressures to increase the extraction or retention of resources previously abundant in the diets of evolving hominids and humans (e.g., vitamin C). During youth and reproductive periods, thrifty predispositions improve metabolic extraction, endogenous production, or retention/conservation of what were once scarce resources. As survival increases to late middle age and later life, continued efficiency produces detrimental alterations in the soma (e.g., hyperglycemia and obesity from energy storage, and hyperlipidemia and hypertension through retention of lipids and sodium). Sociocultural changes (e.g., processing of grains, preserving of foods, social desirability) have led to physiological scarcity of many once abundant dietary nutrients in modern diets. This lack has exposed multiple non-thrifty predispositions related to the conservation, retention, and extraction of nutrients previously widely available, leading to vitamin deficiencies, osteoporosis, bowel and gastric pathologies, and cancer in many societies.

Most, if not all, genetic loci have variant alleles, although some that code for only three amino acid proteins must have very few, if any. As a general rule, all alleles predisposing to pre-reproductive mortality, infertility, and low relative fitness are selected against, while those without such predispositions are more frequently passed to the next generation. Pre-reproductive selection removes alleles regardless of any beneficial or detrimental survival effects they may confer at later ages. Thus, the fate of alleles with survival benefits at later ages is determined partly by whether or not they, or the other loci they were in linkage disequillibrium with, conferred an advantage earlier in life. Alleles predisposing to enhanced early-life survival, growth, development, maturation, and reproduction are not selected against regardless of any influence (positive or negative) they exert on disease resistance later in life. To the degree that they enhance early life history factors and relative fitness, alleles with antagonistic pleiotropy are retained. Alleles exhibiting both early-life and late-life benefits are also retained. Alleles that confer late-life benefits but have no influence on earlier life history and fitness differentials respond only to random genetic drift and may be retained or lost. Given the variety of possibilities, multiple alleles affecting survival probabilities in later life are likely to exist. Frequencies of alleles that show neither antagonistic pleiotropy nor influences on relative fitness change as natural selection affects alleles at linked loci. When reproductive potential or fitness-enhancing behaviors are extended to ages beyond which such detrimental predispositions reduce relative fitness, alleles that enhance fitness will outcompete them as natural selection eliminates the earliest-acting detrimental variants. Selection against senescent-enhancing alleles has likely characterized hominid and human evolution since their separation from other hominoids. This selection pressure continues today, although at an altered pace.

The thrifty–pleiotropic theory was developed to help explain the current worldwide epidemic of CDCs (Crews and Gerber 1994). A necessary component for its development was knowledge that at most loci there are numerous alleles. These produce variable proteins with differences in activities, functions, gene–gene, gene–environment, and pleiotropy interactions in metabolic pathways and tissues allowing one mutation to alter multiple physiological relationships. Another component was knowledge of how extensively humans have altered their own environment and ecological circumstances producing biocultural selective pressures. This process was slow during most of hominid evolution, occurred more rapidly during human evolution, but culminated in the past 100–200 000 years with global colonization and a population of over 6 billion (Fenner 1970; Stini 1971; Harrison 1973; Baker and Baker 1977; Eaton and Konner 1985). The thrifty–pleiotropic theory simply expands the proposal that changes in human life history in recent periods have exposed numerous predispositions that, although advantageous earlier in human evolution, today lead to more rapid senescence and reduced life span. Because the ill effects of these thrifty-pleiotropic alleles occur after the attainment of MRP and after most reproductive effort is complete, they do not alter the relative fitness of their carriers and are passed to succeeding generations (Williams 1957; Levi and Anderson 1975; Eaton and Konner 1985; James *et al.* 1989; Crews 1990a; Crews and James 1991; Crews and Gerber 1994; Gerber and Crews 1999). As discussed elsewhere (see Eaton and Konner 1985), a major contributor to CDCs is a lack of fit between humankind's "Stone Age genes" and the ecological settings of technological and cosmopolitan life styles.

Alterations in life style have affected both our social and physical environments (Baker 1984); included among these alterations are new nutritional and dietary patterns (Stini 1971; Eaton and Konner 1985; Eaton and Nelson 1991), increased public health activities, along with improved medical technologies (e.g., vaccinations, chlorinated water, chemotherapy, pre-natal surgery), altered exposures to new and old pathogens (e.g., multidrug-resistant tuberculosis, sexually transmitted diseases, ebola virus, human immuno-deficiency virus, other multidrug-resistant bacteria), increases in exposure to environmental and human-made toxins (e.g., PCBs, nitrates), and changes in physical activity. These changes are played out against a background of coadapted gene complexes that show multiple epistatic (i.e., when one gene or genetic locus influences the activity of another gene or locus), multifactorial (i.e., affected by both genetic and environmental factors), and pleiotropic relationships. Our current gene pool – the product of millions of years of evolution in a mobile, primarily plant-eating, but omnivorous, ecological niche with short or at least shorter than present life expectancy – did not develop in response to today's sedentary life styles, fat-rich/fiber-poor diets, and extended life spans. Until recently, most humans who have ever lived probably did not live long enough for the late-acting effects of pleiotropic, age-specific, or thrifty–pleiotropic genes to be expressed. Our current burden of CDCs is the product of evolutionary "lag time", the recency of improved life expectancies, and the declining force of natural selection with increasing age (Rose 1991; Crews 1993a; Crews and Gerber 1994; Gerber and Crews 1999).

Type II diabetes (NIDDM) provides one model for a thrifty/pleiotropic CDC. If any thrifty alleles predisposing to NIDDM via the efficient or rapid storage of excess energy (glucose) exist (Neel 1962, 1982), it is not likely that their detrimental outcomes – impaired glucose tolerance and NIDDM, blindness, amputation, and death (Andres 1971; Kreisberg 1987; Shimokata *et al.* 1991) – or non-enzymatic glycation (Cerami 1985, 1986), were observed often among nomadic gatherer/hunters. Conversely, multiple early-life advantages likely accompanied efficient storage of excess in ecological settings of scarce resources. Thrifty alleles also might provide enhanced opportunities for growth and development, leading to early maturation, reproduction, or improved survival. In marginal low-calorie environments, even relatively small gestational increases in blood glucose may better nourish developing fetuses. This provides thrifty mothers with larger, healthier, and more appropriately developed and proportioned neonates than conspecifics. Thrifty mothers do not experience the hyperglycemia of gestational diabetes associated with macrosomatic infants and birth complications in high-calorie settings. Ancestors of modern humans would only have realized the early benefits of such thrifty/pleiotropic alleles, improved reproductive success and survival, in times of scarcity. Their ecological settings did not allow hyperalimentation, high-calorie diets, and low physical activity. Neither did they commonly survive to experience the deleterious CDCs seen today. In contemporary societies, the early benefits of such alleles are minimized, while their detriments are often fully realized. Mutations in the nematode's insulin-signaling pathway locus produce a thrifty genotype with extended life span; mutations in the human homolog are associated with obesity and diabetes (Barzilai and Shuldiner 2001), also suggesting a thrifty genotype. However, among humans with abundant nutrients, these mutants may accelerate senescence and shorten life span.

Numerous CDCs may result from early-acting thrifty or non-thrifty processes that, remaining active through life, contribute to risk. Multiple conditions may result including coronary artery disease, hypertension, osteoporosis, anemia, and goiter (Crews and Gerber 1994; Gerber and Crews 1999). Humans are likely to share some thrifty–pleiotropic mechanisms with multiple organisms, including fish, reptiles, worms, or insects (plesiomorphies); others are likely to be exclusive to mammals, primates (symplesiomorphies), or only hominids (apomorphies). Although multiple diseases and senescent processes appear to be secondary to evolutionary adaptations in previous environments, returning to prehistoric diets is an unlikely panacea for CDCs. CDCs and senescence are inevitable outcomes of long life. Everyone who lives sufficiently long suffers from at least one, if not multiple, CDCs, which typify the oldest-old.

Biological (proximate) theories: mechanisms/models of senescence

A variety of senescent processes are observed in different cells, organs, organ systems, and species (Arking 1991; Gavrilov and Gavrilova 1991; Rose 1991; Clark 1999). This has led to a proliferation of mechanistic theories of senescence (e.g., wear-and-tear, damage due to ROS, error catastrophes, rate-of-living, loss of cell proliferation capacity, telomere shortening, and glycation). Several proposed mechanisms, including damaged molecules secondary to ROS (Adelman *et al.* 1988; Cutler 1991), defective byproducts of non-enzymatic glycation (Cerami 1985; Brownlee *et al.* 1988), increases in faulty proteins and DNA as a result of mutation (Armstrong 1984; Pryor 1984; Richter 1995; Sohal and Orr 1995; Swartz and Mäder 1995), and increases in inert waste materials (lipofuscin) (Garvilov and Garvilova 1991; Harrington and Wischik 1995), suggest the accumulation of metabolic wastes with age. These have been labeled "garbage can" models (Gibbons 1990), reflecting the fact that cells do not have a place to dump all of their wastes and byproducts.

Data suggest that various mechanistic processes act at all levels of biological organization from intracellular organelles, to cells and organs, and across a variety of organisms. For example, a limited ability to repair DNA damage and low levels of antioxidants are associated with shorter life spans within and across species (Adelman *et al.* 1988; Cutler 1991; Rose 1991). DNA repair and antioxidant defenses are mechanisms through which innate variability in abilities to repair and retard degenerative processes lead to individual differences in senescence and life span. This is not programmed, which would involve specific genetic instructions to halt or reduce DNA repair. Progressive loss of DNA repair ability is an inherent aspect of intrinsic chromosomal senescence, including progressive heterochromatinization, information loss, and mutation (Lezhava 2001). Multiple senescent mechanisms are likely to affect all cells, organs, and systems. Thus, the presence of any one does not exclude the possibility that others also limit or enhance senescence and life span in the same system. Multiple physiological processes are integrated into complex somatic adaptive systems (Luciani *et al.* 2001; Mangel 2001; Olofsson *et al.* 2001; Sozou and Kirkwood 2001); there exist multiple intrinsic and extrinsic stressors that may interfere with their continued performance, maintenance, and repair. This produces variable patterns of senescence across species, between individuals within a species, between different organs and cells within the same individual, and between the same types of cells and organs in different individuals.

Proximate mechanisms inducing senescent changes (cumulative, progressive, irreversible, degenerative) are sub-divided in multiple ways: cellular/ organ/organism, intrinsic/extrinsic, dividing/non-dividing cells, and random/ programmed. For many, the most useful division is between loss of functional capacities in non-dividing, long-lived (e.g., neurons, myocytes, nephrons) and dividing (e.g., fibroblasts, osteoblasts, hemopoietic cells) somatic cells. Earlier theories and models of cell senescence were premised on the assumption that the replicative capacity of dividing cells (their ability to produce a monolayer of cells covering a culture flask within a specific time frame) was infinite. This is so with cellular organisms that do not differentiate their somatic and reproductive functions – the basis for early models of dividing cells. Thus, continually dividing cells were thought to be unimportant in senescence. Following reports that dividing cells progressively lose replicative capacity in culture (Hayflick and Moorhead 1961), they were given a greater role in models of senescence (Hayflick 1988).

Studies of senescence in non-dividing, long-lived cells revealed loss of integration and function due to oxidative damage from ROS, mitochondrial DNA dysfunction and mutations, wear-and-tear, DNA and protein alterations, and loss of cellular and structural redundancy in organs composed of such cells. They did not show any specific program for senescence. Studies of dividing cells have examined how loss of replicative ability (the Hayflick limit) may limit organismal life span. In addition to the Hayflick limit, mitotic misregulation and telomere shortening over subsequent cell divisions are likely to affect dividing cells, while their terminally differentiated forms often senescence over a very short period of time (i.e., red blood cells without DNA, about 60 days, skin cells, about 3 weeks, and white blood cells, about 7–14 days). Processes such as inappropriate apoptosis (either premature or lack thereof), heterochromatinization, and loss of DNA repair capabilities are likely to apply equally to loss and uncontrolled growth of both non-dividing and dividing cells.

The Hayflick limit

Until 1961, when Hayflick published research showing that the cell-doubling capability of fibroblasts is limited *in vitro*, prevailing theory was that cells possessed an unlimited capacity for replication (Hayflick and Moorhead 1961; Hayflick 1965). However, upon introducing controlled and standardized culturing techniques and eliminating cross-contamination, Hayflick and colleagues found that cultured fibroblasts from young donors only double about 50 times *in vitro*. Previous culturing methods were found to either promote cellular aneuploidy (transformation to an odd number of chromosomes, which is common in neoplastic cells) or allow contamination with extraneous fibroblasts. Once contamination and transformation were limited, fibroblasts from infants to centenarians showed a linear decline in replicative capacity (Hayflick 1987). Cells obtained from infants doubled about 50 times on average, while those from centenarians doubled only 5–10 times. Cells obtained from pre-teen progeria patients show about the same number of doublings as those from centenarians. Cells from long-lived animals also show more doublings than do those from short-lived ones (e.g., Galapagos turtle, 120 doublings, 180-year life span; humans, 50 and 122; and rodents, 12 and four, respectively; Clark 1999). Eventually, the Hayflick limit was found not to apply to all dividing cell populations (Weiss 1981). Factors other than intrinsic doubling capability, halving the pleuipotent fibroblast cell population with each subsequent culture (generation), setting a time limit for the process of confluence to be completed, or other artifacts of culturing may limit apparent cell doublings *in vitro*. Still Hayflick's findings, along with replications and refinements thereof, prove a general slowing of replication with increasing number of culture cycles (passages) and may reflect a senescent process. This possibility influences some theoretical models and has led to dividing cells being viewed with greater suspicion in senescent processes (Finch 1994; Clark 1999); others see the Hayflick limit as having little influence on general mammalian aging (Kirkwood 1995).

Animals do not "run out" of cells over their life spans due to limited cell proliferations *in vivo* nor do doublings in culture correlate as strongly with age as originally suggested (Hart and Tuturro 1983). For example, the range of variation in total doublings for cells cultured from the same individual (average = 50, range = 10–90), different individuals of the same age (infant: average = 50, range = 25–120; aged 65, average = 20, range = 5–50), and individuals of all ages (average = 35, range = 0–120) is great. Furthermore, in a small sample of healthy Baltimore Longitudinal Study of Aging (BLSA) participants ($n = 42$), no significant correlation of fibroblast doublings and age of cell donors was observed (Cristofalo *et al.* 1998, 1999). Also problematic is finding that fibroblasts from some very old individuals show no doubling potential at all, yet the person continues to survive. Relationships between *in vitro* doublings and senescence are far from clear.

Some analyses suggest dividing cells may come close to their *in vitro* doubling limit when they produce a mature reproductive adult human. Human zygotes use about 42 cell divisions to produce a mature viable infant at birth, but only about five additional cycles are needed to develop and maintain the reproductive adult, leaving only a few divisions for error and reserve capacity (Dennison *et al.* 1997). Also, during terminal passages *in vitro*, cells show many alterations similar to senescent cells *in vivo* (Clark 1999). Data on telomere shortening (reviewed later) suggest that skin, thyroid, immune, and digestive cells may fail to replace themselves adequately with increasing age (Clark 1999). To some, replicative failure is a model for cellular senescence *in vivo*, predicting where deterioration in cellular functions should be observed in whole organisms (Clark 1999). In general, loss of proliferative capacity *in vitro* is not thought to produce or reflect processes of organismal senescence. Rather, loss of replicative capacity provides but one model of how complex molecular systems fail to maintain themselves infinitely due to intrinsic fragility and random losses of function. Although the average numbers of doublings *in vitro* are correlated loosely with the age of cell donors, they are not highly predictive of individual life span nor does loss of replicative capacity appear to underlie senescence in humans or mammals.

Programmed senescence

Lack of understanding of how molecular factors trigger senescence has led to proposals that there exist specific genetic programs or a "molecular clock" for senescence (Driscoll 1995; Miller 1995; Clark 1999). The basic premise is that a simple and parsimonious master gene or genes exist that, when switched on, cause senescence. Proponents of the programmed model believe that if these senescent-initiating loci were switched off, organisms would then naturally express their innate, but blocked, immortality. Limited replicative capacity of dividing cells (The Hayflick limit), thymic involution and subsequent immune suppression, shortening of chromosomal telomere ends during repeated cell divisions, and uncontrolled programmed cell death (apoptosis) in non-dividing cells have all been proposed as possible "master regulators". Many of these processes do correlate closely with life span within and across species. However, it is not likely that any genetic program specific for senescence or any single system controls or promotes all aspects of senescent change. Any investment in master genes or senescent-promoting loci would waste finite resources in somas that are by their molecular design inherently predisposed to dysfunction and disintegration. Chronological correlation of multiple aspects of senescence and life span results because of evolutionarily imposed limits on the continued development of and investment in structures that need only last while their somas are engaged in reproductive investment (see Kirkwood 1995).

Through millions of years of selective pressure, cells, tissues, and organs in sexually reproducing species have been designed to last only sufficiently long to assure the evolutionary success of the germ line that produced the soma as a reproductive organ to reproduce itself (Weismann 1891; Kirkwood 1995). Senescence is not "nature's backup plan", designed to kick in and produce senescence "as a programmed outcome" if we do not die from natural causes (Clark 1999, p. 8). Senescence is the random and uncontrolled degeneration of a complex adaptive system designed to achieve a single end: reproduction of the germ line. Once reproductive effort is complete, the need for additional programming or control of the machinery built for that single purpose is obviated. Among humans, which component fails first or fastest largely depends on luck in having alleles viable in the current environment and culture. Among animals, without human intervention, alleles, luck, and environment determine failure. As with many non-human animals today, during earlier stages of human evolution, reproduction of the germ line and reproductive effort likely occurred at younger ages (due to extrinsic pressures) than the ages at which many somas currently reproduce and survive. It is only in recent generations that selective pressures for human cells and organs to survive beyond the ages previously associated with maximum reproductive effort (the minimal necessary species life span, including mating, birthing, and fledging of offspring) are likely to have occurred. Cells never were programmed to senesce. To the contrary, mammalian cells lack programs to ensure survival beyond their species' minimal necessary life span for reproductive effort.

Unless one postulates that synchronicity of dysfunction in physiological processes itself represents programmed senescence, the case for programmed senescence is weak. Programmed senescence reflects the philosophy that at some "... point the "soma" is better disposed of than maintained ..." (Clark 1999, p. 51). However, there is no evolutionary return on investments in DNA systems to eliminate an already useless soma. Furthermore, if the soma were not already useless, continued reproductive effort would better enhance fitness than programmed destruction. A non-programmed, stochastic model, positing that somas are designed to last only sufficiently long to maximize reproductive success/effort and relative fitness, constrained only by phylogenetic history and current ecological conditions, better fits observations. In this model, no resources are invested either to maintain or shut down the system once its primary objective is achieved. The analogy of an old barn left to itself and slowly falling in on itself comes to mind. Today, the search for a master mechanism of senescence focuses on loci associated with programmed cell death (Driscoll 1995; Miller 1995; Clark 1999). Few experimental data support programmed senescence (Rose 1991; Stadtman 1995).

Although there is no specific genetic program for senescence, as older adults became more numerous in human populations they likely experienced more opportunities to invest in their own and their kindred's total fitness. To the extent that elders' reproductive investments benefit the survival and reproduction of subsequent generations of their kindred, their inclusive fitness increases and natural selection falls more heavily on those without senescence-slowing propensities. As elders continue to reproduce (particularly men) and invest in kin, and cultural circumstances change to provide opportunities for longer life span, selection will fall more heavily on variants that limit life span. The greater the opportunities for parental and grandparental investment in offspring and grandchildren, the more rapidly natural selection will lead to increased frequencies of the alleles that retard senescence. Although, as Weissmann (1891) suggested, longevity generally is a "luxury without an advantage", among humankind long life may provide reproductive and fitness advantages.

Humankind's evolutionary history suggests that natural selection has eliminated, and continues to eliminate, alleles associated with early onset and progression of senescent processes and age-related disease. This trend has probably been occurring over at least the past 2 million years (see Hawkes *et al.* 1997), although it is likely to have become more effective in more recent millennia. Among recent cohorts, cemetery and family records from colonial Massachusetts, U.S.A., suggest that kindreds in which more women survived to later ages (grandmotherhood) were more prolific (Mayer 1991). Data available from several present-day populations living relatively traditional lives show that families with living grandparents tend to be larger and to care better for infant and children than families without grandparents (Hill and Hurtado 1991; Hawkes, *et al.* 1997, 1998; Hill and Hurtado 1999). Once grandparents and great-grandparents have sufficient opportunities to either reproduce (men) or contribute to their descendants' well-being and reproductive success, this reinforces selection for senescence-slowing alleles. To the degree long life improves total fitness (reproductive success/Darwinian fitness plus inclusive fitness), alleles promoting senescence slowing will increase. Fitness differentials of only 1–2% per generation or even lower will over the long term increase representation of senescence-slowing alleles. Finally, patterns of senescence and life span have been altered in wild and laboratory species, contrary to suggestions that senescence is programmed. Both senescence and life span respond to Darwinian selective pressures and multigenerational influences on total fitness – neither is static as a programmed model suggests.

Rate of living/metabolic rate

Rate-of-living and metabolic limits on life span have been central tenets of theory development in senescence since Weismann's (1891) and Pearl's (1928) observations that metabolism and body mass were associated with species life span. Modeling species' life spans as a dependent variable has produced significant regressions with basal metabolic rate (BMR), brain weight, body size, the encephalization index (brain weight/body weight × 100), and total lifetime metabolism. In general, large-bodied and/or -brained species show lower BMRs and longer life spans than do smaller-brained and/or -bodied species with higher BMRs and more rapid energy expenditure. One refuted suggestion was that all species experience similar total lifetime energy expenditure per unit somatic mass. Conversely, cellular metabolism generates ROS, toxic byproducts, and intermediate compounds that alter DNA, RNA, proteins, and macromolecules. By merely maintaining cellular function, molecular processes deteriorate the integrity of cells, promote dysfunction and tissue breakdown, and inflict some of the somatic damage defined as senescence (antagonistic pleiotropy on a system-wide scale?). These produce cumulative, progressive, irreversible, and degenerative alterations in cell function and lead to correlations between metabolic rates, body mass, or brain size and life span.

Many organisms, including primates, birds, and rodents, do not fit predictions from regression models of life span based on size and metabolic rate. Foliverous primates have low metabolic rates for their body sizes and relatively smaller brains than frugivorous species, perhaps because leaves are low in energy content, require high energy expenditures to process (Harvey and Bennet 1983), and do not require complex thought or behavioral patterns to locate and consume. Fruit eaters must forage more and across wider areas using more energy to locate concentrated food sources. This likely requires more complex behavior and neural integration to recall and locate fruiting trees. Still, both primate groups experience about the same expectation of life. Among primates, associations of body size, brain size, and metabolic rate with life span apparently vary in concert with dietary preferences, ecology, and foraging strategies.

Bats are flying rodents generally about the same size as a house mouse. BMR in bats generally exceeds that of similar-sized rodents (which already have fairly high BMRs). However, a bat's life expectancy is about twice that of similar-sized non-flying rodents. Obviously, bats differ from ground-living rodents in that they fly, making them less susceptible to predation, while increasing their metabolic demands. Bats also feed on concentrated energy sources – insects, fruit, blood – rather than the more eclectic diets of grounded rodents. Variability in locomotion (flying–walking), predation rates, and diet alter associations of metabolism and body size with life span in rodents just as they do in primates. Birds also occur in a range of sizes, include flightless and flying varieties, and have variable behavioral patterns. Flying birds are less susceptible to predation, have higher metabolic rates, and are smaller, but generally live longer than flightless birds. Neither aspects of body size nor metabolic rates consistently predict life span or expectancy even within a single order of animals, making it unlikely that rate-of-living or energy expenditure per unit mass are major determinants of life span.

Another problem for rate-of-living models is that the rate of senescence in organisms need not be directly related to life span. Selection pressures molding metabolic rates, senescent processes, and life span are likely to vary. Life may be long, but senescence rapid, or any other combination (short–rapid; long–slow; short–slow), and metabolic rates may be slow or fast and all variations between. Pacific salmon show little senescence before they migrate, swimming upstream to spawn. During migration, they cease to feed. After laying eggs or expelling sperm, salmon show rapid senescent alterations in metabolic processes secondary to the induction of hormones involved in reproductive functions. Increased cell death leads to rapid senescence and somatic death within a short period of time. There is little correlation between rate of senescence and age in salmon; rather, endocrine alterations related to spawning itself seem to trigger senescence. This may appear to some to validate the hypothesis that a single molecular switch causes senescence. However, this model validates the close connections between reproductive effort, somatic maintenance, and survival. Salmon have but a single opportunity to reproduce, so all reproductive effort must go into one attempt. During this process, a hormonal cascade is initiated that causes salmon to migrate, cease feeding, and spawn, ultimately leading to death of the salmon soma. The soma (a reproductive organ) is exhausted during this single reproductive effort. Once this primary objective is achieved, insufficient resources remain to repair the now useless somatic reproductive machinery. Why organisms senesce and why they live a specific life span are different questions. In the wild, individuals do not outlive their innate capacity to use energy nor do they wear out. They usually succumb to intrinsic and extrinsic factors well before any limits on metabolism are reached.

Redundancy, optimality, wear-and-tear, and error accumulation

Most organs (e.g., lungs, kidneys, heart, and brain) are composed of multiple sub-units, which are composed of many terminally differentiated cells (e.g., alveoli, nephrons, myocytes, and neurons). These provide reproductive-age organisms with great redundancy. This redundancy usually allows for sufficient reserve capacity (RC) for most humans to survive through reproductive adulthood. As reproductive potential wanes after the attainment of MRP, these redundant units fail at an accelerating rate, but are neither replaced nor repaired (see Gavrilov and Gavrilova 1991; Gavrilov and Gavrilova 2001 for reviews). After the seventh or eighth decade of life, humans tend to have little RC left compared with that at the age of 25, although some retain enough to be reckoned among the oldest-old (Perls 1995). During childhood and reproductive adulthood, redundancy and RC (along with plasticity and adaptability) allow wide ranges of responses to external and internal stimuli and stressors. In response to hypoxic stress, we may upregulate heart and ventilatory rates; in response to changing blood pH or fluid volume, we may alter kidney function. Elders generally show less flexibility in response to stressors and a reduced ability to maintain and/or return to internal homeostasis. What amounts to a minor hypoxic or heat stress for a healthy 25-year-old may be a fatal stress for a frail 75-year-old. This occurs because as we age our redundant sub-units, in many cases made up of non-dividing terminally differentiated cells, are lost from multiple organs. Disease, intrinsic fragility, and wear-and-tear reduce our systemic responses to stressors such that much greater perturbations are needed to initiate a response, and, once initiated, responses are more sluggish and less likely to return the system to homeostasis (see Beall 1994). Circulatory, ventilatory, and immunological capabilities all decline with age, due in part to intrinsic loss and in part to recurring environmental stresses. These losses occur in concert and affect multiple physiological systems. As with other processes, such synchronicity gives the appearance that a single master switch controls senescence.

Human ovaries exemplify redundancy and optimality. Early in embryonic development, several million primary oocytes are laid down in the developing ovary. Even before birth, these begin to disappear and they continue to be lost throughout infancy and childhood. At reproductive maturity, only 10–25% of the original number remain. At menopause, so few primary oocytes remain that they no longer produce sufficient estrogen to promote ovulation in response to pituitary FSH (see Wood *et al.* 1994; Leidy 1999; Leidy-Sievert 2000 for more complete reviews). Through atresia, ovaries come to a point where they retain too few oocytes to promote ovulation. Whether this is due to factors intrinsic to the ovaries, hypothalamic–pituitary alterations, or both remains unclear. If reproduction after about 45–50 years were a viable reproductive strategy for women, natural selection should have favored women who retained oocytes into late life, had later or no menopause, produced more fledged offspring, and did not show menopause in their late forties and early fifties. However, late-life reproduction does not occur in women, their ovaries run out of oocytes, they fail to produce offspring after age 50, and survive decades beyond natural menopause. Either late-life reproduction carries high fitness costs, or was so rare during hominid evolution that cessation of reproductive function did not decrease relative fitness, or mammalian oocytes have an intrinsic limit to their viability at about age 50 (a plesiomorphic trait). The ovaries are just one of many organs in mammalian somas. Investment in oocytes is just one aspect of somatic investment and must be balanced against other systems; too few oocytes and there are not enough to bring one to maturity every month or so, too many and they are a wasted investment due to finite life spans. The optimal solution is to provide as many redundant oocytes as will allow most women sufficient numbers to reproduce throughout their prime reproductive years.

Human kidneys are responsible, in conjunction with the adrenal glands and the hypothalamic–pituitary–adrenal axis, for maintaining sodium balance, and thereby systemic blood pressure and blood pH, and, along with the liver, for cleansing and detoxifying the blood. The working structure of the kidney is composed of millions of redundant functional subunits, nephrons, which complete these complex tasks. During growth and development, kidneys increase in cell numbers and cell size. They attain a huge RC of nephrons, sufficient to allow life to continue with one kidney removed. Throughout reproductive adulthood, these numbers are maintained at high levels and optimal function ensues. However, after about age 40, nephrons decline by about 10% per decade. After age 70, kidney failure is common and individuals no longer have sufficient RC to donate a kidney. Although most organs appear to overshoot their necessary capacity during maturation (redundancy), this apparent surplus is continually reduced with increasing age as function decreases from its optimum. The RC of key organs is critical for late-life survival (Hales 1997). Fetal and infant development periods are pivotal times for attaining adequate RC. When stresses such as undernutrition occur, growth retardation of strategic organs (e.g., brain, kidneys, pancreas) may lead to restricted RC. Loss of function in these organs likely plays a key role in individual life span (Hales 1997).

Genetic and molecular alterations

Non-dividing cells: oxidative damage
Non-dividing, terminally differentiated cells accumulate altered proteins, DNA, and RNA as they age (Harmon 1956; Orgel 1970; Harmon 1981, 1999). Damage to DNA and proteins follows exposures to environmental toxins (e.g., radiation, chemical agents, dietary oxidants), internally generated ROS (e.g., oxide (O_2^-), hydrogen peroxide (HO_2), hydroxyl (OH^-), singlet oxygen (O^{2-}), all byproducts of normal mitochondrial metabolism; Richter 1995), and a variety of other mutagens (Richter 1995; Sohal and Orr 1995; Stadtman 1995; Swartz and Mäder 1995; Harmon 1999). Some ROS – such as free radicals – tend to react rapidly with local (often scavenger) molecules. Others may travel far from their origin before reacting with another molecule. When ROS react with non-scavenger molecules, additional ROS are created. Resulting chain reactions continue until the final products are scavenged by cellular ion receptors (e.g., glutathione peroxidase, superoxide dismutase (SOD), catalase, glutathione (GSH), ascorbate (vitamin C), tocopherols (vitamin E), carotenoids (vitamin A), ubiquinone (co-enzyme Q), or urate, among others), halting their progression. Oxidative damage is observed in proteins, nucleic acids, and lipids (Richter *et al.* 1988; Stadtman 1992; Newcomb and Leob 1996). Proponents of damage and subsequent error models of senescence view ROS as the smoking gun of senescence in terminally differentiated cells (Swartz and Mäder 1995). Membranes damaged by ROS are hypothesized to reduce the permeability of cells, leading to lost and/or altered cellular functions, reduced cellular and organ function, and stimulation of cellular apoptosis. Damaged DNA is thought to produce faulty RNA and proteins. These errors and functional losses accumulate over time, leading to altered and/or diminished function of cells, organs, and systems.

Harmon (1956) originally proposed that free radicals (highly reactive molecular species such as OH^- and O^{2-} produced during normal cellular metabolism) would react "near the area where they were produced with the more easily oxidized substances . . . including nucleoproteins and nucleic acids. The organic radicals formed in this manner (by removal of a hydrogen atom) could then undergo further reactions . . . In this manner the functional efficiency and reproductive ability (of organisms) would be expected to be attacked . . . occasionally it would be anticipated that mutations and cancer would result every now and then" (pp. 298–9). Orgel (1970, 1973) anticipated the "slow oxidation of connective tissue" and that "cellular protection via increased cellular concentrations of an easily reduced compound such as cysteine . . . would be expected to slow down the aging process", both of which have been documented since his original paper was published. Harmon's insight that normal cellular respiration and processes involved in the utilization of molecular oxygen would be important in generating such highly reactive molecular species has also proven quite accurate. The postulate that free radicals/ROS are the principle cause of senescence (Harmon 1956; Richter 1995; Sohal and Orr 1995; Swartz and Mäder 1995; Harmon 1999) is, however, not well supported (Rose 1991; Austad 1997). ROS do damage cellular components, contributing to senescent alterations that inhibit cellular function (Swartz and Mäder 1995). These, along with multiple other processes jointly produce senescence.

Over the human life span, mitochondrial DNA (mtDNA), located within organelles that are responsible for cellular respiration, incurs more oxidative damage than does nuclear DNA (nDNA) and experiences mutation about 16 times more rapidly (Wallace 1992b). Proximity is the first cause; another is fragility – mtDNA is not protected by a histone coat as is nDNA. Also, mtDNA is single stranded, lacking a template for repair, and mitochondria do not have efficient repair systems compared with the nucleus. The finding that mitochondria not only produce ROS but that their DNA suffers more damage has led to refinements in the oxidative damage theory of senescence (Wallace *et al.* 1995; de Grey 1997; Wei 1998). Caloric restriction of rodents is associated with longer mean and maximum life spans and reduced oxidative damage to mtDNA and nDNA. Senescing rodents also show lower antioxidant levels, more ROS production, and decreased transcription and translation of mtDNA, suggesting oxidative damage (Richter 1995). mtDNA is fragmented by ROS; these fragments occasionally migrate to the cell nucleus and insert themselves into nDNA, where they accumulate in a time-dependent manner, progressively altering nDNA information (Shay and Werbin 1992; Richter 1995). Transpositions of mtDNA into nDNA have a long evolutionary history and are not confined only to somatic cells. A variety of transposed mtDNA sequences occur at frequencies of between 10–130 copies in human nDNA. Some transpositions occurred as recently as 5 MY ago (MYA), about the time of the earliest hominids, post-dating or coinciding with the human–chimpanzee phyletic divergence (Shay *et al.* 1995). Others are as ancient as 43 MYA, prior to the phyletic divergence of the anthropoids (~40 MYA) or the catarrhine–platyrrhine split (~35 MYA) (Shay *et al.* 1995). Somatic cell mtDNA transpositions may contribute to senescent changes through the loss of information from, and disruptions in, the nDNA of coding or control loci (Shay *et al.* 1995).

The elaborate defensive network that cells marshal to stop ROS also suggests that oxidative damage hinders cell and somatic survival (Sohal and Orr 1995). This network includes catalases and glutathione peroxidase that remove hydrogen peroxide and superoxide dismutase (SOD) which removes superoxide, along with the low molecular weight antioxidants, glutathione, co-enzyme Q, and the vitamins E, C, and A (Cristofalo *et al.* 1999). Transgenic rodents overexpressing the SOD locus show extended life spans and reduced oxidative damage to their mtDNA (Sohal and Orr 1995), while fruitflies that overexpress both SOD and catalyse show a 34% greater maximum life span. Oxidized proteins, along with inactive, less active, and less stable forms of many enzymes, increase exponentially with age in animal models, while measures associated with longer life span inversely correlate with oxidized proteins in the intercellular matrix (Stadtman 1995).

Alterations in physiological function following exposure to ROS (e.g., mtDNA, nDNA, and mitochondrial, cellular, and tissue degradation) differ across somas and determine, in part, individual variation in CDCs, senescence, and life span. Oxidation of amino acids, DNA, RNA, and lipids of cellular membranes (lipoxidation) and the glycation of proteins are primary promoters of functional loss to organs and senescence (Paz *et al.* 1995). This is particularly so for non-dividing cells, but may also apply to dividing ones (Sozou and Kirkwood 2001). As organisms age, three related problems occur: generation of ROS increases, antioxidant defenses decrease, and the innate capacity to degrade oxidized proteins declines (Stadtman 1995). If senescence results in any large part from oxidation of cellular components, numerous variable alleles will produce these same endpoints (e.g., dysfunction, CDCs, senescence, life span) in different individuals. With this variety of possible promoters and outcomes, no specific genetic trait can underlie all oxidative processes or senescent phenomena (Stadtman 1995). Dysfunctions induced by ROS are cumulative over the life span, affecting all tissues and organs composed of differentiated nondividing cells. Some oxidative damage appears to be irreversible once incurred. Progressively accumulating over the life span, such oxidative damage is one basis for the functional losses of senescence (Sohal and Orr 1995).

ROS, antioxidant defenses, and oxidative stress do not, however, explain all aspects of senescence. Total measures of plasma antioxidants do not show strong correlations with life span nor do there appear to be sufficient amounts of oxidized proteins, amino acids, DNA, and RNA in cells for cellular or organ malfunction to result (Cristofalo *et al.* 1999). Such results might be expected if those with low antioxidant levels die early and only those with high levels survive past middle age. Antioxidant (e.g., SOD, catalase) activity and age-related changes in activity are also tissue and organ specific; this is in addition to being species and gender specific. Variance in SOD activity suggests that it may be induced as a response to superoxide exposure rather than being a constant (Warner 1994; Okabe *et al.* 1996). Contrary to popular culture, ingestion of exogenous antioxidants does not correlate with higher antioxidant levels or defenses; quite the opposite, they may suppress endogenous antioxidant production and defenses. Data also suggest that, among the major intracellular antioxidants (i.e., SOD, catalase, and glutathione), a biological optimum exists such that it is the ratio of SOD to the sum of the other two (SOD/catalase + glutathione peroxidase) that relates to oxidative stress rather than the absolute value of any one (de Haan *et al.* 1995).

Non-dividing cells: mitochondrial alterations

As discussed in the preceding section, mitochondrial alterations are closely related to oxidative damage theories of senescence. Rates of generation of ROS, particularly SO and HO_2, increase with age, correlate with age in the tissues of mammals and insects, and are inversely correlated with life span (Richter 1988; Sohal and Weindruch 1996). ROS also mutate and fragment mtDNA. In addition, bioenergetic patterns of mitochondria vary over the life span in ways unrelated to oxidative processes. Mutations and alterations of mtDNA occur for a variety of reasons and accumulate exponentially with age. Mitochondria with damaged mtDNA increase in frequency with age, suggesting that mutant mitochondria, and the cells in which they reside, are selectively replicated (Richter 1988; Wallace 1992b). Loss of wild-type mtDNA may lead to chronic heart disease, late-onset diabetes, Parkinson's and Alzheimer's diseases, and senescence (Wallace 1992b).

Cellular energy production depends on mitochondrial oxidative phosphorylation (OXPHOS), which depends on a variety of nDNA- and mtDNA-derived proteins (66 and 13, respectively) and RNAs. In both cohort and cross-sectional samples, OXPHOS declines with increasing age after MRP. Aged muscle shows lowered mitochondrial activity than younger muscle, suggesting loss of functional mitochondria with age. In cells that use OXPHOS as their primary source of energy (e.g., brain, muscle, heart, kidney, liver, and pancreas), such losses are likely to promote degenerative diseases (Wallace 1992b). One suggestion is that individuals who "...start with a high initial energy capacity do not fall below tissue-specific thresholds before old age ... (while others who) inherit mildly deleterious mtDNA mutations manifest late-onset degenerative diseases ... (while those) who inherit severe mutations get childhood diseases ... " (Wallace *et al.* 1995). Damage to mtDNA and loss of wild-type mitochondria are cumulative over the life span, leading to irreversible and progressive functional decline in cells and organs, suggesting that they may represent a basic senescent process.

mtDNA is inherited maternally in the ovum's cytoplasm; less than one in 1000 mitochondria are contributed by the sperm. Each somatic cell contains hundreds of mitochondria, each with multiple copies of the circular 6569 base pair (bp) mtDNA genome (Wallace 1992b). These copies are not identical. The cytoplasm of all ova carry a variety of mtDNA – a condition described as heteroplasmy. These various mtDNA genomes may not alter function (normal) or promote various dysfunctions (mutants). Once the fertilized ovum divides, mitochondrial variants are randomly distributed to daughter cells. This is similar to the chromosomes at mitosis, except that division of mitochondria is unequal. This results in the predominant mtDNA genotype sometimes differing between daughter cells. Over multiple divisions, clones of cells come to include either mainly mutant or mainly normal mtDNA (homoplasmy) (Wallace 1992b, p. 628). When oogenesis begins, primary oocytes in the developing ovaries already differ in the proportions of mutant and normal mtDNA genomes. Later, when fertilized, they will produce offspring with variable risks for, and penetrances of, mtDNA-related disease. During life, disease and its severity are related to the amount of defective mtDNA and the specific cells and tissues in which the mutants are most prominent. Diseases related to heteroplasmic and homoplasmic mtDNA mutations are relatively common in humans (Wallace 1992b). As discussed earlier, mtDNA also suffers greater damage by ROS and has higher mutation and lower repair rates than nDNA. Thus, OXPHOS and energy production within cells decreases with increasing age. Loss of cells through apoptosis due to the loss of OXPHOS ability and reduced energy output may precipitate organ failure, CDCs, and senescence (Richter 1988).

CDCs due to specific mtDNA mutants prove their potential for harm. Myoclonic epilepsy and ragged-red fiber (MERRF) disease results from a mutation at mtDNA nucleotide 8344, Leber's hereditary optic neuropathy (LHON) is related to at least four different mtDNA mutations, and neurogenic muscle weakness, ataxia, and retinitis pigmentosa (NARP) is caused by a mutation at mtDNA nucleotide 8993 (Wallace 2001). In each condition, mutations are heteroplasmic in families and the proportion of mutant mtDNAs (a measure of the quantitative reductions in OXPHOS) correlates with severity and age of onset, which are highly variable. In MERRF, individuals may have 15% normal mtDNAs and mild symptoms at age 20, but severe symptoms and few remaining normal mtDNA at age 60 (Wallace 1992b). Age of onset of LHON ranges from 8 to 60 years, and patients may show mild phenotypes at young ages and severe ones at older ages. Some mitochondrial-related conditions are spontaneous rather than inherited, suggesting that they arise secondary to somatic deletions of mtDNA; examples include Kearns–Sayre syndrome, chronic external ophthalmoplegia plus, and Pearson's marrow/pancreas syndrome (Wallace 1992b).

Individuals inherit different OXPHOS capacities; these capacities then decline with age. When energy production falls below the levels needed for continued function, cells may die, thereby promoting CDCs (Wallace 1992b). Those who inherit high OXPHOS capacity may live sufficiently long to die of non-mtDNA-related causes before the loss of functional mitochondria leads to disease; others carrying highly deleterious or many mutations may cross loss-of-function thresholds for multiple organ system before death (Wallace 1992b). Chronic cardiac ischemia is associated with an 8- to 2200-fold increase in a 5 kilobase (Kb) deletion of mtDNA, hearts of cardiomyopathy patients more frequently have a 7.4 Kb mtDNA deletion, and patients with non-ischemic heart failure have increased levels of the 5 Kb deletion (Wallace 1992b). In late-onset diabetes mellitus (a heterogeneous group of disorders included within a single clinical category), a chronic deficiency in OXPHOS due to a 10.4 Kb mtDNA deletion inhibits insulin production and produces a mitochondrial protein synthesis defect (Wallace 1992b). Among elders, a significant proportion of mtDNA carries alterations and deletions, and mutant mtDNA may be associated with Parkinson's, Alzheimer's, and Huntington's diseases (Wallace 1992b). Although Wallace and colleagues (1995, 2001) suggest data are inadequate to conclude that damaged mitochondria and mutant mtDNA are promoters of age-related CDCs and senescence, few data contradict this conclusion. Contrary to nuclear genes, which appear unlikely to accumulate mutations that lead to senescence, degenerative disease processes parallel the kinetics associated with slow accumulations of random mutations in post-mitotic tissues predicted by a mtDNA model (Wallace *et al.* 2001).

Additional lines of evidence suggest involvement of mitochondria in cellular senescence (deGrey 1997, p. 161). Mitochondrial generation times are very short relative to their host cells, some of which never divide. This rapid replication of mitochondria does not lead to overpopulation of cells with mitochondria; rather, lipofuscin, believed to be composed of indigestible cellular debris mainly from mitochondrial membranes, accumulates in these cells with age. Mitochondrial membranes of long-lived species incorporate fewer highly unsaturated fatty acids and show less lipid peroxidation than those of short-lived species relative to their metabolic rates (deGrey 1997). Mitochondrial-coded proteins are essential for metabolism, but are not needed for the replication of mitochondria or mtDNA. This allows mtDNA mutations to become homoplasmic in a mitochondrion, thereby creating a clone of mutant mitochondria. As mutant mitochondria clones continue replicating, cellular respiration may be completely eliminated due to the loss of functional mitochondria (deGrey 1997). Rates of mtDNA replication are directly proportional to the length of the molecule; molecules with deletions have an advantage and may increase in frequency more rapidly than wild types (Wallace 1992b). Increasing proportions of deletion mtDNA in tissues suggest a mechanism whereby disease and disease severity may increase with age. Mitochondria within any cell constitute a single replicating population. Preferential replication of mutants or lysosomal digestion of non-mutants increases the proportion of mutant forms and decreases cellular respiration. Mutant mitochondria provide less ATP (adenosine triphosphate), show lower respiration, produce fewer damaging radicals, accumulate less membrane damage, and remain viable after most normal mitochondria have succumbed to oxidative damage (deGrey 1997). When cells copy their mitochondria, these long-lived mutants are preferentially replicated. Eventually, such cell lineages may have no properly functioning mitochondria. This is a model of "survival of the slowest" (deGrey 1997). In dividing cells, mitosis is rapid enough that no significant loss of properly functioning, actively respiring mitochondria ever occurs. In non-dividing or rarely dividing cells, the story is very different: mitochondria with mutant DNA replace wild types and these ATP-deficient cells then accumulate in the tissue leading to senescent alterations (deGrey 1997). Oxidation of mtDNA and selective reproduction of mitochondria carrying deletions are cumulative, irreversible, progressive, and degenerative, producing senescent change in non-dividing cells.

Non-dividing cells: non-enzymatic glycation

Although ROS and mutation are best known, non-enzymatic glycation (NEG) by glucose circulating in the blood also alters DNA and proteins. Age-related increases in glycemia and the increased NEG of macromolecules have been suggested as universals of human senescence (Raeven and Raeven 1985; Cerami 1986), although some recent results do not support this suggestion (McLorg 2000). Neither insulin nor glucagon levels tend to vary greatly with increasing age in humans. However, glycemia does increase in many. Some proteins (e.g., the various forms of collagen) undergo NEG as part of their normal physiological development. NEG may hamper the function of other molecules, making them less permeable, less elastic, stiffer, or harder. It may also alter the functions of nucleic acids. Continued NEG of collagen over its life span decreases its elasticity and increases age-related brittleness. NEG is proposed to contribute to multiple age-related alterations in proteins and cells. These include the decreased elasticity of cutaneous tissues, ligaments, and tendons, permeability of cell membranes, and contractility of the heart (Cerami 1986).

NEG is a reversible process; cells have specific enzymes to remove glycation. NEG is, however, only the first step in a process that leads to advanced glycation endproducts (AGEs) which are not reversible by normal physiological processes. AGEs are thought to build up in cells and organs, where they decrease the perfusion of blood to cells, leading to local necrosis and tissue death. By reducing the elasticity of tissues, they may decrease ventilatory capacity, arterial compliance and recoil, increase joint stiffness, reduce mobility, and contribute to the senescent changes seen in cells and organs. NEGs of macromolecules may not occur as frequently or extensively in all populations (McLorg 2000). NEG and AGEs may contribute to loss of function with increasing age in some populations, but may not be a species-wide manifestation of senescence in all environments. NEG and AGEs may be of recent genesis in populations with a surfeit of calories, representing a multifactorially determined promoter of senescent alteration.

Dividing cells: telomere shortening

A molecular alteration proposed to explain senescence in dividing cells is the shortening of chromosomal telomeres that occurs during each round of cellular mitosis (De Lange *et al.* 1998). The ends of chromosomes are made up of a repeated sequence of DNA nucleotides that has been highly conserved over evolutionary time; in humans, it generally is TTAGGG. Telomeres appear to be essential for chromosomal and genomic stability. A specific RNA-containing enzyme, telomerase is needed for the replication of chromosomes. Without telomerase, telomeres become shorter, until ultimately the cell fails to replicate its chromosomes and dies (Shippen *et al.* 1993; Allsopp 1996). In cultured human fibroblasts, mean telomere length decreases by about 50 bp per doubling and by 2 Kb over cumulative population doublings through the terminal passage as the proportion of total DNA composed of telomeric DNA decreases (Harley *et al.* 1990). Fibroblast donor age is also significantly negatively associated with telomere length (Harley *et al.* 1990). Telomerase is active in all single cell organisms, in germ line cells, and in 90% of malignant cells from adult multicellular organisms, but not in their normal somatic cells. Sperm telomeres are over two times longer, about 9 Kb, than telomeres of normal somatic cells, which are about 4 Kb in length (Harley *et al.* 1990).

One hypothesis is that shortening of telomeres is related to the Hayflick limit in dividing cells and, consequently, cellular life spans (Olovnikov *et al.* 1996; Shay 1997; De Lange *et al.* 1998). This suggests that chromosome shortening represents natural cell aging, i.e., an internal clock of sorts. A number of variables – cell type, donor age, and heredity – influence the correlations between telomeric shortening and age, both across and within individuals. Telomere length varies across peripheral lymphocytes from the same individuals, between similar cells from different individuals of the same age, and across cell cultures of fibroblasts from the same donor (Harley *et al.* 1990; Hastie *et al.* 1990; Slagboom *et al.* 1994). Studies of human peripheral lymphocytes suggest an average loss of 31–33 bp/year (Hastie *et al.* 1990; Slagboon *et al.* 1994). The average, across various types of human cells, ranges from 40 to 200 bp/division (Allsop 1996), compared with an average of 50 bp/doubling in cultured cells.

A Dutch study of both monozygotic (MZ) and dizygotic twins suggests that heritability for the length of terminal restriction fragments (TRF) of telomeric DNA is high – about 78% (Slagboon *et al.* 1994). Variation in TRF length is highest between unrelated pairs and smallest in MZ twin pairs (Slagboon *et al.* 1994). Human lymphocyte telomerase production also shows high heritability (about 81%) (Kosciolek and Rowley 1998). Variations in telomere size and production across individuals appear for the most part to be heritable. This makes sense since individual variability in cell turnover rates are also genetically determined (Slagboon *et al.* 1994). This also suggests that either *in vivo* cell replication or turnover do not reflect *in vitro* cell culture models or that cellular mitosis follows an intrinsic genetic program that is differentially expressed *in vivo* and *in vitro*. With such wide variation, some have suggested that telomere shortening may not be a major contributor to longevity differences among individuals (Barzilai and Shuldiner 2001). In general, quantitative traits with high heritability show little association with differential fitness; that is, they do not have strong selective pressures forcing them to a specific level as do, for example, blood pH or body temperature. Traits with high heritability and low selective pressure, such as TRF length, are exactly the type that should underlie senescence (life span itself is highly heritable and variable). During previous epochs of human evolution, TRF length was likely to be unrelated to relative fitness/inclusive fitness; any effects on life shortening, if they existed would not be expressed in short-lived hominids. Only once large proportions of people survived to their sixth and later decades of life would such effects (if there are any) be observed as differential senescence and life span.

The degree to which the loss of dividing cells promotes senescence is unclear. If loss of pluripotent cells contributes to senescence, which has not been proven, telomere shortening may be important in senescence. As with ROS in non-dividing cells, telomere shortening in cultured cells is sometimes viewed as the molecular mechanism producing cellular senescence and the Hayflick limit (Harley *et al.* 1990; Allsop 1992), and as the link between failing replicative capacity *in vitro* and senescence of whole organisms (Shay *et al.* 1995; Olonikov *et al.* 1996; De Lange *et al.* 1998). This suggests that reactivation of the telomerase locus may halt progression of cellular senescence *in vitro* and *in vivo*. Introduction of exogenous telomerase does induce additional division cycles (+40 doublings) of human cells *in vitro* (Bodner *et al.* 1998), suggesting that telomere length influences replicative capacity. However, simple reactivation of the telomerase locus or introduction of exogenous telomerase *in vivo* may produce additional complications, since this is the normal state of cancerous and germ line cells. Furthermore, a variety of enzymes show variable activity, mRNA concentrations, and DNA expression in cell cultures attaining the Hayflick limit (e.g., collagenase, stromelysin) (Cristofalo *et al.* 1999). In concert rather than independently, alterations in telomerase activity and multiple additional factors are likely to influence cellular senescence *in vitro* and overall somatic senescence *in vivo*; altering only one factor *in vivo* could have deleterious consequences for the organism. Telomere shortening may be linked to other aspects of senescence; for example, thymic involution is thought to result from the loss of cellular replicative capacity, suggesting TRF shortening may influence immune system senescence and/or that thymus cells may have shorter TRFs to begin with.

Dividing cells: mitotic misregulation

A recent and compelling addition to theories of senescence in dividing cells is mitotic misregulation (Ly *et al.* 2000; Marx 2000). Examining over 6000 loci for active mRNA transcription with the use of DNA microarrays, Ly *et al.* (2000) reported that only 61 loci showed consistent changes in expression in cell cultures from young, middle-aged, and older donors. Expression patterns between the older donor cultures and those of Hutchinson–Gilford progeria patients also showed great consistency. About 25% of these 61 senescense candidate loci control cell cycle progression (mitosis), and are downregulated 2.6–12.5-fold over the age span examined (Ly *et al.* 2000). A high proportion of these 61 loci are also involved in maintaining integrity of the extracellular matrix and cytoskeleton (31%) (Ly *et al.* 2000). Cells cultured from older donors also show a higher percentage with 4N DNA content, indicating cells with multiple nuclei, similar to progeria-derived cultures. In both progeria and older donor cultures, the loci involved in cell division, synthesis, and processing of DNA and RNA, and mitochondrial genes are downregulated; those controlling stress responses and heat shock proteins are upregulated (Ly *et al.* 2000). Based on these findings, Ly *et al.* (2000) suggested that alterations in the mitotic machinery of dividing cells in the post-reproductive period are a general mechanism of senescence leading to chromosomal pathologies – particularly errors in the normal separation of genes during mitosis (Marx 2000). Observed alterations in gene expression are not the same as those observed during cellular senescence of non-dividing cells, among quiescent or G_1-arrested cells, or those reported for cells from aged muscle or hypothalamus cells. This suggests that altered expression patterns are specific to mitotic (dividing) as opposed to post-mitotic (non-dividing) cells (Ly *et al.* 2000). Misregulation of genes controlling cell division results in increased rates of somatic mutation which manifest as senescent processes. In this model, the original misregulation may be due to any of several mechanisms, including intrinsic lack of fidelity, oxidative damage, mutation, radiation, or other factors. Following the original damage, misregulation multiplies with each succeeding round of cell division, producing gradual and mosaic senescence (Ly *et al.* 2000).

Apoptosis

Apoptosis (programmed cell death) is an organized process of cell destruction which is important during growth and development for sculpting the soma, eliminating extra cells produced during embryogenesis, and during life for eliminating old, damaged, and defective cells (Pearson and Crews 1998). Apoptosis is a housekeeping function designed to enhance somatic survival by eliminating defective and unneeded cells. During apoptotic death, cells undergo a programmed sequence of events that includes shrinkage and fragmenting of the cytoplasm, cleavage of cellular DNA into 180 bp pieces, and digestion of these remnants by macrophages without an inflammatory response. Specific proteins mark cells for apoptosis, while others halt or prevent the program for cell destruction. Continued cell survival represents a constant balance between proteins designed to halt or implement apoptosis. Apopotosis is a highly conserved (plesiomorphic) aspect of multicellular organism. Cancerous cells with aneuploidy, a condition that promotes apoptosis, and other markers have tipped the balance to allow their continued survival. As a counterpoint to cells escaping programmed cell death when needed, functional cells may be inappropriately marked for apoptosis and eliminated. This leads to loss of both dividing and non-dividing cell populations. Conversely, non-functional terminally differentiated cells – e.g., due to homoplasmy for a mtDNA mutant – may escape apoptosis, remain viable, and lower organ function. Numerous molecular alterations could initiate inappropriate apoptosis or halt appropriate apoptosis. Faults either in the apoptosis process or inappropriate apoptosis secondary to other mechanisms may promote senescent change. Cumulative apoptotic death of cells could produce the gradual loss of organ function and RC seen in senescence and these results would be irreversible, progressive, and degenerative.

Chronological synchronicity

As previously mentioned, many senescent processes manifest at about the same point in life. Often, this has been taken as support for the programmed senescence theory (see Clark 1999). However, chronological synchronicity may result from multiple processes. Wide variability is likely to have characterized the timing of allelicly determined senescent processes. However, over time, natural selection is likely to have acted sequentially against the earliest occurring processes. Eventually, through such selective pressure, most will be held in abeyance until some common age beyond which reproductive potential has dropped sufficiently that natural selection is no longer capable of eliminating or masking them (Williams 1957). Organisms passing on alleles promoting somatic survival at the expense of current fitness will generally be outcompeted by their faster reproducing conspecifics. However, alleles that slow senescent alterations by remanding them to later ages, while not decreasing or even enhancing overall inclusive fitness, could or would increase in the population, contributing to synchronicity. This is the mutation-accumulation model of senescence (Medawar 1952), wherein multiple mutant alleles with adverse affects on life span may be passed on generation to generation because they occur well beyond the ages of MRP in the current environmental/cultural setting. When populations experience new environments that allow for longer-term survival, these adverse outcomes are exposed and may then be subject to natural selection.

In short-lived, small-bodied, small-brained, rapidly reproducing species, somatic design is generally constrained evolutionarily by the need to reproduce as quickly and as often as possible to be competitive. In humans, somatic design is constrained by low birth rates, long-term infant and childhood development, and the need to invest significant parental care into each offspring. This life history necessitates individual survival over at least several decades: the species' minimum necessary life span. Among humans and their near relatives, senescence need be halted only long enough that individual survival is sufficiently long for offspring from multiple consecutive births to be reproduced and fledged, a process that likely took no more than three or four decades throughout most of hominid evolution. Therein lies the dilemma for modern humans wishing to survive beyond current life spans – our evolutionarily based adaptations produced a minimum necessary life span (MNLS) of only about five decades. During human evolution, the capacity to achieve this MNLS was attained as early as the *Homo erectus* stage (including *H. eqaster* and *H. heiedborgensis*). Since then, genetic variation for senescence has accrued through both random drift and, once culture provided opportunities for survival beyond the fourth or fifth decades of life, natural selection, as competition between late middle-aged adults investing in their own relative fitness or the inclusive fitness of their kindred became more commonplace.

Until recently, there has been little evolutionary pressure to build organisms with life spans extending much beyond their own reproductive success and fledging of offspring. To halt senescence and extend life span beyond the current maximum of 122 years and the longest current average of 83.9 years (Japanese women), it will be necessary to halt multiple insidious, cumulative, synchronous, and coordinated processes, the effects of which begin to show at about the fourth decade of human life and increase in prevalence and severity thereafter. Once these are halted, eliminated, or repressed, new later-acting processes may be exposed that curtail life span at some age just beyond currently observed limits (an idea revisited later). Rare accelerated-senescencing genotypes offer some indications of how chronological synchronicity might act. In both progeria and Werner's syndrome, individuals show many traits commonly seen with old age (i.e., loss of hair, poor skin elasticity, stunting, fragility, cataracts, and connective tissue hardening). These apparent single gene mutations lead to a wide variety of alterations although they affect only a single protein; so too may it be with alleles that bring on undesirable alterations with increasing age. Each will be likely to have multiple pleiotropic and epistatic effects that produce a cascade of alterations leading to senescence.

Chapter synopsis

Any, several, or all of the biological mechanisms reviewed here may influence age-related change in any molecule, cell, organ, or system. They all share the appearance of being inherent aspects of cellular metabolism and differentiation. All are correlated with age. Ongoing internal competition for scarce resources to either enhance reproductive function or increase somatic stability results in multisystem decline and failure for members of a species that lives sufficiently long to senesce. Each system represents an evolutionary tradeoff between function and stability. Major investments in the processes of growth, development, and reproduction (function) produce limited investments in maintenance and repair (stability) of a soma (a reproductive organ) ultimately susceptible to external insults that carries and reproduces the potentially immortal germ line. This tradeoff is likely to characterize all sexually reproducing species. At some point, investment of scarce resources in reproduction of the germ line (function) outweighs the benefits of somatic survival (stability) for a germ line capable of reproducing itself but a limited number of times in any particular environment (Kirkwood 1995). These mechanistic processes progress throughout adulthood, correlate closely with chronological age in populations, are observed universally among those who survive sufficiently long, result in progressive loss of function, are intrinsic to cellular patterns of reproduction, function, and respiration, and have deleterious results with respect to survival, making them all likely candidates for senescent processes. In this view, senescence is inherent to the molecules and processes that produce living organisms. Over time, these processes lead to the loss of molecular, organelle, cellular, and organ capacity.

Reduced infant and childhood mortality have ensured that most individuals in cosmopolitan societies survive to their sixtieth birthday: "twice the age required by our life history strategy for reproduction" (Olshansky and Carnes 1994, p. 21). This process has allowed underlying susceptibilities to CDCs that previously were not subject to selective forces to be exposed to, and perhaps be influenced by, natural selection. The evolutionary and biological theories reviewed in this chapter stem from a single premise – that reproduction and fledging of offspring (relative fitness) is the driving force behind evolution. Once this primary function is initiated, maintenance and survival of the soma become secondary to the more important function of providing for survival of the germ line (Kirkwood 1995). Among humans, cultural developments have produced individuals who survive beyond the ages of reproduction and rearing of offspring, i.e., to grandparenthood. These grandparents may enhance the survival and reproductive probabilities of not only their own offspring but also their grandchildren (i.e., their own inclusive fitness). Given these opportunities for late-life investment in their own total fitness, natural selection has likely acted to enhance late-life survival by eliminating early occurring senescent processes or by pushing them toward later decades of human life history.

For most species, a large reservoir of heritable genetic variation likely exists for late-life survival and physiological function. Extended somatic survival in humans and other species is associated with better immunological, repair, and maintenance systems. Enhanced activities of these processes seem to characterize not only humans, but also other relatively long-lived species (e.g., elephants and bats). This is particularly true among those animals in which the probability of reproduction and survival increases rather than decreases with age (e.g., sharks and tortoises). Any evolutionary model of human senescence that suggests a lack of selection for late-life survival over the past several 100 000 years is in error. Humans have greatly extended late-life survival and have had such at least since *H. sapiens* first roamed the planet. Among humans, multiple factors and mechanisms enhance biological vitality and reduce somatic frailty with respect to senescence processes. Cellular, metabolic, and physiological differences between those who survive to late life (i.e., ages of 75 and older) compared with those of the same cohort who succumbed before age 65, but after about the age of 40 or 50, are likely to reflect allelic variability affecting life span (Barzilai and Shuldiner 2001). Comparative studies of such short- and long-lived individuals will provide insights into the biology of human senescence and differential survival.

3   Human variation: growth, development, life history, and senescence

Many emphasize the uniqueness of humans among animals, while others see them as simply another mammal in the order primates. To understand current patterns of life and death, one must first examine human evolutionary history, the type of creatures from which we evolved, how their adaptations shaped their ancestors' physiology and function, and how these led to our current long lives and extended infant, child, adolescent, reproductive, and late-life periods compared with other similarly sized primates (Figure 3.1). Over the past several million years, hominids (hominins; members of the family *Hominidae* including bipedal primate mammals, and modern humans and their immediate ancestors and relatives) have comprised a primate group that was relatively terrestrial and mobile and likely exploited marginal environments for subsistence, while, over time, adopting a material–cultural life style (Dunbar 1988; Hunt 1994). Most primates are omnivorous, what set later hominids apart was the inclusion of gathering/scavenging/hunting in their behavioral repertoire. Given a reliance on marginal environments with scattered resources, early hominids may have experienced strong selective pressures for enhanced memory (remembering landmarks and resources), group cohesion (recognizing relatives and kin), and communication skills. Extension of life stages and life span improves such traits by allowing enhanced opportunities for learning through experience and for retaining group knowledge. These various trends coincided and co-evolved during hominid and, ultimately, human evolution. Specific patterns of inter-relationships among these trends remain unclear. However, bipedalism and free upper limbs were established well before any major increase in cranial capacity or the complex neural structures necessary for language acquisition developed.

Bipedalism is the hallmark of hominids. Studies among extant chimpanzees and fossil hominoids suggest that bipedalism may have evolved among our remote ancestors as a food-gathering "...adaptation that allowed for efficient harvesting of fruits among open-forest or woodland trees" (Hunt 1994, p. 183). Bipedalism likely evolved first as a conservative adaptation, allowing an

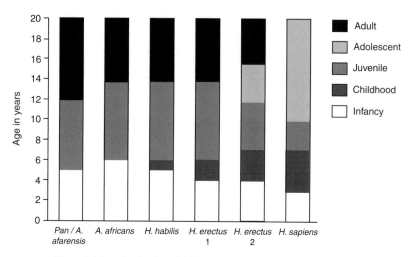

Figure 3.1 Length of various life history stages in human and other primates (after Schultz, redrawn from Smith 1991).

arboreal primate to continue to exploit the small fruits of trees once open forests developed (Hunt 1994). Selective pressures to increase cranial capacity or alter life history parameters would not have accompanied bipedality while it remained only an alternative postural feeding adaptation. However, once hominids came to depend on bipedalism as their primary locomotor pattern, selective pressures on pelvic orientation and size of the birth canal may have contributed to birthing less developed and more altricial infants. This slower development may have led to less developed (e.g., softer, thinner) longer growing cranial bones, potentially providing additional opportunities for neurological development and expansion both pre- and post-natally. Less developed neonates require greater parental care and supervision compared with their arboreal relatives and ancestors. Terrestrial bipedalism and foraging along with a greater need for constant parental care place additional selective pressures on abilities to communicate and recall previous events. This could have initiated positive feedback among increased complexity of neurological structures and function, and slowed growth and development.

In the ecological circumstances delineated, positive selection pressures for extending life span and improving neurological design and complexity could become self-reinforcing. If kindreds with more surviving elders are even slightly more likely to survive the vagaries of a constantly changing environment, over time such groups will leave more progeny carrying alleles associated with long-term survival. If longer surviving individuals also have greater opportunities to produce offspring or contribute to the well-being of their children's children, this further enhances their fitness and their kindred's inclusive fitness (IF). Either pattern increases selection pressures against early senescence and shorter life spans. Given ongoing, but unrelated, selection against those with poorer communication abilities and/or poorer long-term memory, selection pressures for increased cognitive ability and longevity could become self-reinforcing. Evolutionarily, other aspects of human physiology may have become secondary to trends toward increasing cognitive capacity and longevity. Alterations in patterns of female pelvic development associated with habitual bipedalism may have laid the groundwork for alterations in patterns of fetal growth and development that today separate humans from apes.

Over the evolutionary history of *Homo*, there has been ample opportunity for such selection to occur. Weiss (1984) suggested that as many as 3 million new mutants per locus over the entire human genome may have arisen since evolutionary separation of the hominoid (members of the superfamily *Hominidae*, including all tail-less primates – pongoids, hylobatids, hominids) and hominid lineages. He further cautions that humans are not particularly long lived for a mammal of our brain and body size and that there is no need to invoke special selection to explain our longevity, even though we live longer than our hominoid relatives. Weiss' (1984) extrapolations tend to support Cutler's (1980) contention that evolutionary changes in only a few loci may account for extended life span in humans and that regulation of life span may be a relatively simple process. Recently, Judge and Carey (2000) re-examined data on life span, and body and brain mass. They arrived at a conclusion similar to that of Weiss (1984): that the basic primate pattern of longevity predicts a human life span of 72–91 years. These estimates exceed the range of prior estimates based on such relationships by about 20 years, and are about 30 years short of the maximum life span yet observed for humans (122 years). Still, compared with similarly sized mammals, humans and apes live about four times as long as expected (Austad 1997), and primates are among the most long lived of mammals at any size.

Arguably, there are no uniquely human qualities. Instead, humans are unique in that they possess an array of interrelated adaptations that improve their adaptability; these include bipedality, large brains, dependence on vision, verbal communication, culture, manual dexterity, altricial-dependent infants, and long lives. Among primates, humans are by far the most odd in their patterns of growth, development, locomotion, and adaptations to the environment. No other primate remains unable to cling to its mother without aid long after birth, builds shelters, or consistently shares food. No other primate's head is so large compared with its body size at birth, or has offspring as furless and helpless as human infants. It is this unique amalgamation of biophysical and biocultural adaptations that patterns early life and structures the growth, development, and behavior of all human infants, children, and reproductive, and consequently post-reproductive, adults, and late-life survival. Still, any possibly unique human attributes are built on 60 million years (MY) of primate and 300 MY of mammalian evolution. The most recent estimates of DNA divergence show only a 1.24% between human and chimpanzee, compared with 1.62% for human–gorilla, 1.63% for chimpanzee–gorilla, and 3.08% for human–orangutan (Chen and Li 2001). Based on these data, and a 12–16 MY before present divergence for the orangutan clade, Chen and Li (2001) estimate 4.6–6.2 MY for the human/chimpanzee divergence. Thus, the veneer of "humanity" or "human-ness" only thinly overlays longer-term phylogenetic trends. Humankind's free use of their upper limbs to manipulate and transform the environment is what truly set them apart from other biological organisms.

As the sole extant, fully bipedal primate, our patterns of growth, development, and birth processes differentiate us at the skeletal, morphological, and culture-bearing levels from all other primates. Growth and developmental processes and culture also appear to be closely associated with our long-term vitality, extended life span, and late-life disease processes. Alteration of our species' life history, stretching human maturation and developmental processes to two decades over evolutionary time, has had significant influences not only on our growth and development but also on our patterns of adulthood and senescence. Much of this alteration of life history is likely to have occurred secondary to our species' elaboration of culture and creation of biocultural feedback loops. In rodent models, enhancement of life span is associated with an expansion of early life stages, such that the cessation of growth and reproductive maturation occurs at later stages in calorie-restricted animals than in their *ad libitum* fed litter mates. These processes in rodents may aid our understanding of how extension of early life stages became the norm among humankind relative to our earliest ancestors (australopithecines) and our closest modern relatives (chimpanzee, gorilla). This chapter will review concepts of homeostasis and stressors before examining how alterations in patterns of growth and development during hominid/human evolution have shaped our life history and predisposed us to late-life survival.

Homeostasis

A fundamental need in all organisms is the maintenance of homeostasis (a fairly stable internal environment fluctuating around multiple set points) in the face of detrimental environmental perturbations, while also reproducing. Alterations in any aspect of an organism's life history (e.g., fecundity, fertility, maturation rates, maturation periods, late-life reproductive potential) resonate across the life span and affect homeostasis and the allocation of remaining resources. Generally, alterations associated with increased fitness are retained and spread throughout the population. Those that decrease fitness are eliminated proportional to their relative effects on fitness. However, when effects on fitness are neutral, these variants increase and decrease randomly.

As mammals, humans have multiple systems for maintaining homeostasis that respond to both internal and external stress. The immunological, hormonal, integumental, cellular, sensational, and neural systems are all active participants. The central nervous system (CNS) is composed of multiple interacting control systems: the voluntary system that controls muscular movements; the autonomic system which controls the body's organs and internal functions; and the neuroendocrine–immune system that controls physiological processes, relationships between/across various systems, and interactions with the environment and culture. The autonomic nervous system includes the parasympathetic and the sympathetic systems; the parasympathetic controls normal body activity, while the sympathetic responds to stress. The former uses acetylcholine as its messenger between terminal neuronal axons and target cells, while the latter uses adrenaline and noradrenaline. Thus, what one stimulates, the other inhibits, and vice versa. Acetylcholine is also the common neurotransmitter within ganglia of the CNS.

The hypothalamic–pituitary axis (as part of the neuroendocrine–immune system) integrates internal responses to external stimuli by regulating hormonal outputs. The hypothalamus has direct neural connections via sensory neurons, and blood flows through the portal system directly from it to the anterior pituitary at its base. The hypothalamus also releases stimulatory neurotransmitters – e.g., amino acids (glutamate, glycine), bioamines (dopamine, catecholamines, acetylcholine), and peptides (atrial naturitic peptide (ANP), endothelin, cholecystokinin) – directly into the posterior pituitary, which is composed mainly of nerves terminating from the hypothalamus (Baxter 1997). Through these connections, the hypothalamus controls such functions as organismal temperature, maturation, reproduction, hunger, satiation, and emotion. Other parts of the neuroendocrine system control various aspects of homeostasis such as serum glucose levels and blood pressure. The hormonal system is integrated with our sensory array, while providing second-to-second regulation of the physiological milieu, e.g., pH, blood pressure, oxygenation, temperature, sodium. The neuroendocrine system also communicates with the immunological system which defends against invading pathogens and parasites and provides for self-recognition and defense against rogue cells that may replicate out-of-control or at inappropriate times.

In addition to the neuroendocrine–immune system, multiple additional physiological systems protect us from environmental stress and aid in maintaining homeostasis. For example, skin (the integument) creates a barrier between us and the external environment. It prevents invasion by parasites and pathogens, and retains moisture and dissipates heat, while allowing sufficient penetration of ultraviolet (UV) radiation to activate hormones but protecting us from excessive UV radiation. Cellular factors protect us from rogue cells and initiate cellular death (apoptosis). These homeostatic/defense systems work synchronically, providing most individuals with the opportunity for growth, development, maturation, and reproduction in a hostile world. For most, they generally serve well through their reproductive years. However, all show irreversible alterations with increasing age. Whether such alterations reflect underlying senescent processes or are themselves senescent alterations is often unclear. Reserve capacity (RC) is highly variable across individuals, within individuals across different systems and organs, and over time within the same individual and system, and it is subject to biological, environmental, and lifestyle/cultural/behavioral modifications.

Reserve capacity

Almost all physiological systems undergo continuous functional change throughout life. Almost all *overshoot their physiologically necessary capacity* (NC) during late youth and early adulthood. Total achieved capacity (TAC) for most physiological systems usually maximizes at about the age of maximum reproductive potential (MRP). TAC is often more than 100% greater than the capacity needed to maintain current homeostasis (e.g., NC) at the age of MRP. The difference between current TAC and NC equals the RC. RC may be of particular importance when studying senescence and age-related change among humans. Humankind's maximum achievable capacity for organs and systems evolved in response to harsher environmental settings; these likely required a greater NC in order to maintain homeostasis and/or limited individuals' abilities to attain greater TAC. Whether TAC has increased or the NC for survival has decreased, the overall difference (TAC − NC) appears to have increased among humans, leading to greater RC available for continued somatic maintenance. It is the remaining RC at older ages that promotes longevity among humans, their domesticated animals, and laboratory models when extrinsic threats are controlled. When disease, predation, and/or injury (stressors) are common, RC must be compromised to deal with them and little RC remains at mid-adulthood (30–39 years). Adequate food, lack of parasites and infections, no predation, and lack of injury (reduction of stressors) reduce expenditures of energy and resources in somatic maintenance, allowing retention of RC. Such RC may be invested in additional reproductive effort, but, when

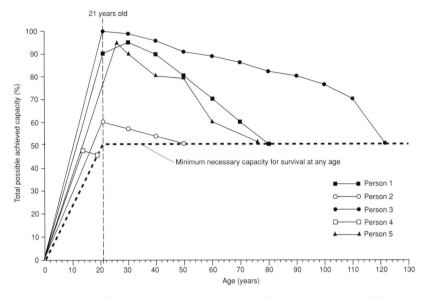

Figure 3.2 Hypothetical model for attainment and loss of reserve capacity (RC) with age.

reproduction is low, as in many of today's societies, remaining RC is available to maintain the soma at older ages (Fig. 3.2, Person 1). Conversely, some organisms may need a greater proportion of TAC simply to maintain homeostasis in their current environment and use their TAC more rapidly (Fig. 3.2, Person 2). Disease processes also lead to the loss of RC and increase the risk of additional morbidity and mortality. Some organs or systems within individuals may far exceed their needed capacity, thus overshooting to large RC (e.g., two, three, or even four times the NC); others may not have the ability to even attain necessary capacity, undershoot their need, attain only 50% of their current NC, and thereby fail to survive to reproduce at all (Figure 3.2, Person 4). How much capacity one must use for development *in utero* and during early life may also program physiological resource allocation so as to increase current capacity to achieve survival/reproduction and severely limit RC. In plentiful settings, TAC can far exceed the NC for reproduction and adult survival and contribute to longer life (Figure 3.2, Person 3). Loss of TAC and RC with age need not be linear over the life span; improved life style habits, altered environments, or medical intervention may slow loss or even improve function (Figure 3.2, Person 5).

To extend life span, an organism must be able to exceed the total functional capacity necessary to complete reproduction and investment in offspring.

Development of functional capabilities to maintain homeostasis is a target-seeking process, as is growth in general. If an individual is not operating above their current NC, they are at risk for loss of function. If current TAC is overextended in maintaining homeostasis and no RC remains, even a slight stress will push one below NC. RC is highly variable across individuals, within individuals across different systems and organs, and over time within the same individual and system, and is subject to biological, environmental, and life style/cultural/behavioral modifications – all conditions that a process of senescence should meet. Without a biological program allowing the storage of excess resources (those not currently needed to maintain optimal function), organisms would always be at their functional limit, and, thus, not able to even invest in offspring. RC is necessary for reproduction, and, after completing reproduction, for extending individual life sufficiently to care for offspring. This suggests that species who care for their young (e.g., elephants and humans) should tend to accumulate greater amounts of RC during early life than those with no or limited parental care following reproductive effort (e.g., salmon, some reptiles and fruit flies). RC may occur as additional functional units for organs composed of long-lived cells (e.g., myocytes, nephrons, spindle cells), additional proliferative capacity for rapidly dividing cells, increased body size to slow extrinsic damage, more efficient communication, neuroendocrine, or immune systems, greater storage of excess energy, or more mechanisms to delay intrinsic deterioration.

Most organ systems show large amounts of redundancy – e.g., kidneys, liver, lungs, immune, endocrine, brain, skin (Gavrilov and Gavrilova 1991). Humans, in fact, possess sufficient redundancy that most individuals may survive with but one kidney, half a liver, or a single lung. However, with age, redundant structures are lost from these organs, reducing both organ and overall RC and the ability to respond to environmental perturbations. Consequently, many elders need two to three times the amount of change in internal temperature or hypoxia to recognize physiologically fluctuations within their environment (see Beall 1994). Environmental stressors (anything that removes an organism from homeostatic equilibrium – e.g., heat, cold, hypoxia, noise, other organisms) that younger persons respond to almost instantaneously, causing but minor perturbations to their homeostasis, may, in the elderly, lead to long-term alterations in homeostasis and even death. Natural examples occur every summer and winter in the mid-latitudes, when more elderly are hospitalized with, or die from, heat stroke and influenza than any other age group except infants.

In general, elders tend to respond more slowly to a variety of stressors than do younger adults, probably because their stress-sensing systems have degenerated from their own optimums with age (Sapolsky 1990, 1992; Frolkis 1993; Beall 1994; Harper 1998; Beall and Steegman 2000). Patterns of physiological arousal and stress responses among elders also vary from those of young adults. Altered circadian rhythms, hypercortisolemia, slow responses to stressors, and slowed rebounds to previous homeostatic set-points probably reflect reduced abilities to sense departures or resetting at higher/lower levels due to the loss of functional capacity (Sapolsky 1990, 1992; Frolkis 1993; Harper 1998). For example, the hypothalamic–pituitary–adrenal axis (HPA) and the sympathetic–adrenal–medullary system appear to lose responsiveness and RC with age. In those with already compromised systems, RC may fall so low that eventually everyday stressors overwhelm control systems and prevent return to physiologically ideal levels. Consequently, as we age, new set-points at lower functional levels are likely to be established (Lakatta 1990) to prevent organismal death. Losses of RC are both cumulative and interactive. Altered responsiveness due to alterations in the sympathetic–adrenal–medullary system, leading to greater physiological arousal, may affect homeostasis in multiple other systems (Finch and Hayflick 1977). According to the current biological definition of senescence, loss of organismal RC across multiple interacting systems over the life span represents a senescent change. In general, biologically older individuals have less remaining innate capacity to monitor physiological stress, whether internally or environmentally induced, show slower response times to stress, respondless to deviations from homeostatic set-points once they are detected, and are more likely to succumb to low-grade stressors than biologically younger individuals. Furthermore, increased susceptibility to stressors secondary to long-term loss of TAC defines a senescent process that may be measured. Suggesting that maintenance of RC is important in retarding senescence is not new. Declining RC with age leaves organisms barely able to survive minor physiological perturbations from environmental stressors (see Fries 2001).

Stressors

Environmental, internal, and sociocultural factors that take internal systems away from their homeostatic range are termed "stressors". When encountered, such stressors elicit a physiological response – commonly, upregulation or downregulation of an internal system. Physiological mechanisms and systems that respond to stressors have intrigued generations of social scientists, psychologists, anthropologists, and human biologists (Selye 1956; McEwen 1998; McEwen and Seeman 1999; Stinson 2000). Humans respond via two major systems to destabilizing stressors: the sympathetic and parasympathetic nervous systems. The endocrine system is the main regulator of human stress response via the HPA of the sympathetic nervous system (Finch and Seeman 1999).

Once the brain recognizes a stressor, a series of events along the HPA axis leads to release of catecholamines (adrenaline and noradrenaline) and cortisol from the adrenal gland. This produces a stress response – vasoconstriction, increased heart rate, increased uptake of glucose, altered blood flow. The adaptive value of such responses – for example, "fight-or-flight" – for intermittent life-threatening stressors is obvious. In modern settings, encounters with other individuals and environmental stimuli produce constant physiological response and arousal, and thus "modern life" has been defined as a "stressor"(Selye 1956). Not all individuals experience the same stressors over their life spans, nor do all have physiologically identical responses to the same stressor. In particular, life style and psychosocial stressors are likely to have highly variable influences on individuals endowed with different and fluctuating homeostatic set-points, biological capacities, pre-natal/infant developmental experiences, sociocultural expectations and values, and life experiences.

Compared with life style and sociocultural stressors, the types of stressors the environment places on individuals and populations – heat, cold, hypoxia, solar radiation, diet, parasites, and infectious agents – are more predictable and measurable. Over the past century, human biologists have examined responses to environmental stress under the general rubric of "human adaptability" (Baker 1979, 1982, 1984, 1990). Individuals and populations adapt to their environments at multiple levels. Thus, for any particular stressor, adaptation is thought to follow a series of increasingly biological influences. Among humans, these levels of adaptation include individual behavior and culture, reflexes, accommodation, acclimatization (both short and long term), and genetic responses. Culture and culturally patterned individual behaviors and responses are humankind's first line of adaptability (e.g., fire, shelter, food processing, weapons, clothing choice, language) and are integral to their worldwide dispersal. Even before settled agriculture and metallurgy, traditional cultures allowed humans to reside in extreme environments (i.e., arctic temperatures, high altitude, desert dryness). Twenty-first century technology has hardly pushed these limits for self-sustaining communities. Although culture has been the mainstay of recent human adaptation, some high altitude populations probably show biological adaptations to hypoxia (Beall *et al.* 1994; Greksa *et al.* 1998).

Over evolutionary time, humans undoubtedly have depended on many levels of adaptation in addition to culture. Our multiple reflexes (innate responses to specific stressors) – e.g., pulling your hand away from the burn of a fire or closing your eyes in response to a draft of air – that protect us from harm are similar to those of many other mammals, as are, in general, our internal reflexes to low oxygen pressure, low pH, and high or low internal temperature. Reflexive systems provide for somatic maintenance with little investment in sensory monitoring and may be fine tuned to specific environmental stressors (e.g., hypoxia and heat). We also show a variety of short-term adaptations to noxious noises, odors, and similar mild environmental intrusions through a process of accommodation, such that the stressor is eliminated by moderation of our own awareness of the stressor's existence; that is, we physiologically ignore the stimulus by "getting used to it" – e.g., the ticking clock or humming light fixture or the smell of the tuna cannery, open cesspool, or oil refinery next door. Accommodation only works for non-life-threatening stressors. As stressors become more physiologically harmful, the body often must respond to more fundamental alterations in homeostatic ranges. The first of these responses is termed "short-term acclimatization" because the physiological alteration can occur within seconds up to a month or more, and is lost over a more-or-less similar length of time once the individual is no longer exposed to the stressor. "Long-term" acclimatizations are those responses to a stressor that occur over long periods of exposure measured in months and years, but that occur in one individual. The most extreme of these is "developmental acclimatization", wherein an individual is conceived, grows, and develops both *in utero* and after birth and attains adulthood while experiencing the stressor (Greksa *et al.* 1998).

How individuals respond to stressors involves multiple factors from sensory perceptions and hypothalamic signaling through physiological redundancy, RC, and developmental patterning. HPA reactivity and resiliency provide a good example of how various factors may affect one system as humans age. In humans, there appears to be little alteration in initial cortisol responses to stressors with age. However, recovery and resetting of the HPA is prolonged, allowing long-term exposure to cortisol. During pre-natal development, undernutrition is associated with increased levels of cortisol in the plasma, downregulation of receptors associated with the cessation of cortisol release, and reduced growth of a variety of organs including the liver and adrenal glands (see Barker 1998). Under basal conditions, few significant differences between men and women in plasma cortisol over the ages of 20–90 are observed (Finch and Seeman 1999). However, at younger ages, men show higher reactivity (measured as blood pressure and catecholamine increases) while after menopause women show higher morning cortisol, higher cortisol secretion rates, and higher 24-hour blood cortisol than men (Finch and Seeman 1999). Higher circulating cortisol may contribute to increased chronic degenerative conditions (CDCs) and morbidity with increasing age among post-menopausal women. With increasing age, loss of receptors due to intrinsic processes also occurs, leading to additional loss of control of circulating cortisol levels.

Among the elderly, for most systems, few differences in comparison with younger cohorts are observed under basal conditions of ambient temperature, stress, and well-being. Rather, on average, elders tend to show sluggish responses to stressors, suggesting that their systems remain operative and can respond to stress, but that they take longer to initiate. In addition, they may never return to basal levels, resetting either an elevated or depressed point. Thus, over time, not only is RC lost (used up), but more of the remaining capacity is lost with each additional stress than in earlier years. Consequently, values for multiple risk factors may move toward clinical cutpoints for disease states with increasing age.

Over the long term, our somas are constantly exposed to a variety of low-intensity stressors that are perpetually causing minor damages and dysfunction. Continued damage from temperature fluctuations, physical trauma (UV exposure, wind, cuts, bruises, strains), and toxic agents (pesticides, exhaust fumes, polychlorinated bipheny's (PCBs), lead, mercury, smoke, solvents) leads to a cumulative loss of cells and function. The totality of these stresses can neither be repaired nor circumvented forever, and ultimately must result in ongoing senescence and aging (Masoro 1996). Constant stress from such multiple insults, and subsequent metabolic responses, may lead to a slow general loss of the ability to compensate and to return to ideal homeostatic set-points, such that, over time, multiple systems (e.g., vascular, hormonal, neurological, metabolic, and immunological) show poorer internal control (McEwen and Steeler 1993; Seeman *et al.* 1997; McEwen 1999). These alterations are thought to produce a load on the soma that increases the probability of greater dysfunction, morbidity, and mortality, defining criteria for a senescent process.

A recent trend in stress research is to combine various risk factors (e.g., blood pressure, salivary cortisol, glycated hemoglobin, body habitus, adrenaline, cholesterol, serum lipids, genetic predispositions) into a measure termed "allostatic load". Allostatic load is then defined as a measure of the outcome of the variety of environmental, biocultural, and psychological stressors impinging on the individual over their life span (Seeman *et al.* 1997; McEwen 1998, 1999). Allostatic load is hypothesized to measure the cumulative wear-and-tear on the soma from maintaining homeostasis in the face of life-long stressors. Allostasis refers to the process whereby all the body's various systems (e.g., neurological, endocrine, immune, metabolic, respiratory, circulatory, cardiovascular, and lymphoid) interact to maintain physiological balance (McEwen 1998). Like the various syndromes (e.g., Syndrome X, New World Syndrome) discussed later, it is not clear that lumping configurations of risk factors into single measures of risk advances understanding of underlying genotype-specific causes. Rather, as with blood pressure and hypertension, a multitude of physiological checks and balances, along with genetically determined hormones, hormone receptors, and signaling peptides, are involved in producing each of the various factors making up this overall "allostatic load". Over time, most physiological measures show greater disregulation, more sluggish responses to environmental perturbations, less RC, and slower return to baseline levels. This has been termed wear-and-tear from the continual maintenance of homeostasis, in a disposable soma. Allostatic load may provide a first approximation for measuring this wear-and-tear model of stress. As such, it may also be a measure of the long-term loss of total functional capacity (FC) defined as a senescent process at the conclusion of the previous section.

Growth and development

It is during the processes of fetal and infant growth and development that organ capacity and function are established. These are the critical periods for the establishment of basic structures, capacities, and limitations because all later functions (e.g., survival, maturation, reproduction) depend on the quality of these basic foundations. So critical is this period of life that infant and child growth are often regarded as accurate measures of a population's overall health and well-being (Bogin 1988; Malina and Bouchard 1991; Bogin 1999), because infants are most sensitive to environmental hazards.

Pattern

Both intrauterine and post-natal growth are unique in humans (Bogin and Smith 1996; Leigh 1996; Bogin and Smith 1999). The fetal brain has two peaks of growth, one early in pregnancy when neurons are laid down and another nearer to parturition when supporting structures are developed; peak growth rates occur at about 18 weeks and 38 weeks, with the latter continuing to increase through the first 4 weeks post-partum before slowing; conversely, growth in length peaks at about the twentieth week and declines from then until birth (Mueller 1998). During the last trimester and first year of post-natal life, the human brain and cranium expand more rapidly than those of most other mammals. Having attained almost 50% of adult size by birth, the cranium expands to 80% by the age of 2, with almost 90% of all neural development complete by ages 5–6. Conversely, we are only about 20–25% of adult height and as little as 1–5% of adult weight at birth. The ability to crawl usually does not develop until several months post-natally, while the ability to walk commonly does not develop before 9–10 months and may not develop until almost 2 years in some, while the fully erect adult posture and final curvature of the spine is not completely attained until late adolescence.

The human growth curve shows several peaks and troughs over the ages between birth and the early twenties (Figure 3.3). This differs greatly from

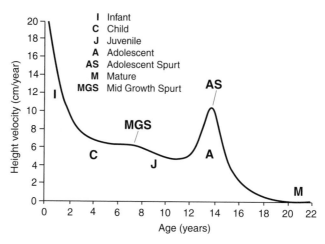

Figure 3.3 Human growth curve: velocity (redrawn from Ulijaszek *et al.* 1999, p. 91. Key: C, child; MGS, mid-childhood growth spurt; J, juvenile; AS, adolescent spurt; I, infant; M, mature; A, adolescent (redrawn with permission from Bogin 1999, Figure 4.4, p. 174).

the general ape pattern of relatively constant growth from birth through to attainment of adult stature, as is observed for most mammals (see Bogin 1998a, 1999). The rate (velocity) of growth in humans is most rapid during intrauterine development and the first months after birth, slowing from infancy through early childhood (ages 2–5), with a pre-purbertal peak (mid-childhood growth spurt; MGS) at about age 6–7, then slowing to a minimum (ages 8–10 in girls, 10–12 in boys) (Figure 3.3). This is then followed by rapid growth through adolescence (AS; 11–14 in girls, 13–16 in boys) and a slowing of growth into the late teens and early twenties as growth is completed and ultimately ceases. Adult stature is generally attained at ages 17.5–19.5 among boys and 15.5–18.5 among girls (Malina and Bouchard 1991, p. 54).

Among primates, both the mid-childhood and adolescent growth spurts appear to be unique to humans, as are the stages of life history during which they occur (Bogin 1998a). At about the time of the mid-childhood spurt, the brain attains its adult weight and the child enters the juvenile phase, which ends when the adolescent growth spurt occurs and puberty commences (Bogin 1998a). The rapid and sustained growth in skeletal size and mass during adolescence is a unique feature of human growth and development not shared with other apes (Bogin 1998a). Ulijaszek (1998a) suggests that humans are similar to other apes in the proportion of life span during which growth occurs (about 25% in humans, 29% in chimpanzee, 24% in gorilla, 29% in orangutan, and 20% in gibbon), and are quite different from rodents (6%), pigs (15%), or elephants (15%); humans, gorillas and chimpanzees also share similar gestation lengths

(9.5, 9.0, and 8.4 months, respectively). Time between birth and the attainment of MRP appears to be particularly long in humans. Humans attain 50% of their adult body mass about mid-way through the growth period (8.8 years) and after about 7.5% of their maximum known life span, whereas chimpanzees do so at about one-third of the way through their growth period (5.6 years) and after about 10% of their maximum known life span (Alexander 1998b). Thus, humans grow and mature more slowly than do other mammals and primates.

Humans also produce almost 2-fold larger offspring (3.2 kg) than other apes (1.5–1.8 kg) and wean them at half the age (24 versus 37–52 months), but human females are not sexually mature, on average, until 13 as compared with 7–10 years in other apes (Alexander 1998b). Humans are exceptionally late in developing all physical/motor features compared with all other mammalian infants, except marsupials with whom they may be usefully compared. Alexander (1998b) compared life history characteristics among eutherian mammals weighing at maturity 40–60 kg, a range that includes humans who fall at the upper end (56 kg). Human offspring are about as large as those of some ruminant offspring, who are capable of locomotion within hours of birth, generally are weaned between 3–6 months, and are sexually mature at 1–1.5 years. Among marsupials, infants are also born at an early stage of development (30 kg red kangaroo females produce 0.7 g offspring), are completely helpless, and spend the first months of neonatal life attached to a nipple (Alexander 1998b). Similarities to human offspring are obvious. The earliest primates may have shared with other arboreal tropical-forest dwellers (bats, tree shrews, *Arconta*) a precocial life history in a stable environment (Shea 1998). Humans show gestation lengths similar to other apes, but produce physically altricial infants similar to marsupials. Human infants are twice the size of other ape infants and more similar in size and proportion of mother's weight to those of physically precocial ruminants; human infants are as helpless at birth as are marsupial offspring. However, rather than first leaving the womb and crawling to and being constantly attached to their mother's nipple for 5–10 months, they are birthed with no self-mobility, but a well-developed ability to be heard, after which they are continually held to the breast by the free upper limbs of their mothers.

The total physical dependency of human newborns, their slow physical growth, both *in utero* and post-natally, lengthening of the human prereproductive phase, and the production of very large neonates represent a massive alteration in the life history of large-bodied primates. The causes of these multiphasic adaptations are unlikely to be simple, but rather are likely to involve locomotor, postural, sociocultural, and environmental alterations and adaptations. Shea (1998, p. 100) sees this as: "a matrix of retardation – birth rates, death rates and individual growth rates are only a fraction of the values found in other comparable sized mammals."

In most large-bodied primates, offspring quality outweighs quantity (a K-selection strategy) and each infant represents a significant investment of maternal resources, unlike species who reproduce frequent litters (an r-selection strategy). In primates, resources invested in somatic growth are "traded off against time, enhancing infant survival . . . and ultimately its reproductive success through extended . . . infancy and adolescence, relatively slow growth to puberty, long life spans, a relatively large body-size, and an expanded brain-size for body-weight" (Lee 1998, p. 102). Slowing of all early developmental life history phases would only be possible if there already was sufficient somatic reserve (RC) available to assure both opportunity for the completion of growth and enhanced reproductive success (RS) for organisms who devoted proportionally more fetal energy to neural development and less to physical development. Apparently, bipedal locomotion, an upright posture used to exploit scattered resources, and the availability of free upper limbs to transport and help suckle offspring, along with development of unique human patterns of group cohesion (language, along with food, childcare, and risk sharing among kin) and the use of tools and goods ("handaxes" and slings) to manipulate the environment, provided an ecological setting conducive to dependent offspring, long developmental periods, late reproduction, and kin-based care of offspring. As the dependency of infants increased, all other LH phases were reset in allometric step with these early-life alterations. Thus, slowing of growth and development ultimately set the evolutionary stage for the late-life survival we are able to experience today, and the unusually long post-reproductive survival of women.

Human life history (the set of co-adapted traits that determine mortality and reproductive patterns in members of a population; Sterns 1992; Sterns and Hoekstra 2000), and much of human culture, is based around the rearing of infants and children necessitated by our unique pattern of growth. Chimpanzee infants are more precociously "advanced" in their physical abilities (e.g., grasping, dexterity, mobility), with more developed skeletal and associated structures at birth, than are human infants. Throughout life, humans attain physical milestones at later ages than chimpanzees, as their reliance on adults remains high and their use of language and culture increases. Physical abilities among humans lag behind those of other large-bodied primates and mammals. Compared with our closest non-human primate relatives, physically humans are slower developing, altricial, and spend longer periods of their life history in infantile and child phases of life and proportionally less as reproductive adults (Figure 3.1). Still, although they have less precocial infants, humans do not breastfeed as long as other apes; rather, they show early weaning and supplementation, familial caretaking (siblings, relatives), and shorter birth intervals (Bogin 1998a).

Examples of pedomorphic features in adult humans include our relatively round and neonatal/infantile crania and facial features, undifferentiated and mobile shoulder joint, relative hairlessness (some more than others), and incisor-like canines. Examples of pedomorphism also include the retention of the ovum that formed each individual in the mother's womb for 15–40 years before fertilization, our slow fetal development and birth as a totally dependent individual (more like a premature infant or early-term fetus, more similar to a marsupial than a eutherian mammalian infant), our 20+ year period of growth and development, our retention of childhood features throughout life, and our slow decline in function with increasing age, with our mortality rate doubling only about every 8–10 years. Human offspring are by far the most physically altricial of all primates, and indeed all mammals, except the offspring of marsupials, whom in many ways human infants more resemble than they do the offspring of other primates or large-bodied mammals. In fact, hairless, altricial, dependent marsupial newborns may be valid models for understanding the immaturity and dependence of human infants, since they are totally dependent on maternal care for survival and are born in a state that would be considered premature for other mammals.

In contrast to their physical development, neurologically human infants are highly precocial. They show more rapid neocortical development, increased encephalization (ratio of brain to body weight), and more complex neural integration at birth and at all later ages than most other mammals including non-human primates. At birth, the human brain is about 25% of the adult size (approximately 357 g) and the head comprises about 25% of length (Davison 1987). During the final trimester of pregnancy, the rate of brain growth exceeds that of any other fetal structure. Prior to the third trimester, rates of growth in most other physical parameters exceed brain growth. Over the first 12–18 months of life, growth of the brain and cranium continue at the same high levels established during the third trimester, allometrically producing the round baby face and childhood cranial proportions. In other apes, the last trimester is not so exclusively devoted to neurological development and physical growth continues apace, thereby producing more physically precocious newborns. Since it is believed that humans developed from an ape-like ancestor that would have had such balanced fetal development (a plesiomorphic trait), human infants have been described as "secondarily altricial" (an autapomorphic trait); that is, humans are thought to have slowed their *in utero* physical development after their evolutionary split from other primates, while increasing their rate of neural development (both autapomorphic traits) (Smith 1986, 1991).

Among humans, gestation is approximately the same length as that of chimpanzees and gorillas (another plesiomorphic trait), while relative rates of growth have been reset, allowing humankind to develop a highly integrated neocortex. Among mammals in general, and primates specifically, the fetus determines the duration of pregnancy, activating an endocrine cascade leading to placental estrogen production (Nathanielsz 1996). Birth likely is initiated when fetal monitors signal that nutrients, oxygen, or some other critical factor is in scarce supply, indicating that the fetus has matured to the capacity of its uterine environment (Nathanielsz 1996). Humans, chimpanzees, and gorillas have about the same gestation lengths, suggesting either a lack of strong pressure to reset the length of gestation in hominids or an inability to alter this basic aspect of hominoid life history. Switching from skeletal to neurological investment, while increasing neither energetic needs nor length of gestation, apparently differentiates gorilla/chimpanzee from human fetal development. However, this also greatly alters rates and patterns of growth and development. Among mammals, primates generally show highly variable growth patterns. Gorillas show a sigmoidal pattern of weight attainment, while chimpanzees, humans, and male baboons show a complex (sinusoidal) series of accelerations and declerations with generally slow growth rates and extended average life spans (Shea 1998). Shea (1998) suggests this may, in part, be linked to the high energy demands of encephalization and/or to the need to minimize juvenile mortality in slowly reproducing species. Energy demands also seem to determine fetal growth and development and the timing of gestation (Nathanielsz 1996; Barker, 1998), providing a physiological link between slowed fetal and post-natal growth and patterns of distribution of limited resources to tissues during various phases of early life history. Ulijaszek (1998b) suggests that multiple molecules such as "genes with growth-promoting protein products" may respond to nutritional factors and mediate the actions of growth hormone or growth hormone receptor (GHR) or transcription of related DNA. A variety of other proteins (insulin-like growth factors I and II, insulin-like growth factor binding protein, cortisol) may also in part be up- or downregulated by nutrient levels, while the HPA axis further regulates growth and the distribution of nutrients and energy.

Mechanisms

Growth is a complex process with multiple regulators at the molecular, cellular, organismal, environmental, and, in humans, cultural levels (see Tanner 1981; Bogin 1988; Malina and Bouchard 1991; Ulijaszek *et al.* 1998a, 1998b; Bogin 1999 for general reviews). Growth and the scheduling of transitions across life history stages from early development through reproductive stages primarily respond to hormonal and neural regulation (Finch and Rose 1995). During fetal growth, the majority of growth regulators are synthesized locally, rather than being released into the systemic circulation by the endocrine system; these act as autocrine and paracrine hormones in close proximity to their point of synthesis (Hernández 1998). Such regulators act as embryonic inducers of morphogenesis, initiating cell differentiation programs that provide cells with new competences for responding to endocrine hormones and moving through subsequent life history stages (Finch and Rose 1995). Hormones are the messengers in a complex communication system integrating somatic responses with the environment and promoting growth and development. They, along with hormone-binding proteins, receptors, transcription factors, regulatory genes and secondary messengers such as cytokines, determine patterns, rates, and outcomes of growth (Finch and Rose 1995). Such factors include the somatomedins (insulin-like growth factors (IGF)-I and -II) which act as mitogenic factors and promote the differentiation of some cell types. IGF-II is particularly active during organogenesis, whereas IGF-I becomes important thereafter; plasma concentrations of IGF-I are also correlated with birth weight (Hernández 1998). The effects of IGF-I are regulated by nutrient transfer to the developing tissues – low nutrient availability reduces the production of IGF-I by the liver and of its various binding proteins (a minimum of six), one of which (IGFBP-3) correlates closely with fetal weight, while two others (IGFBP-1 and -2) show elevated levels during intrauterine growth retardation (Hernández 1998). The insulin/IGF-I receptor protein also represents the human homolog of the *daf-2* locus in *Caenorhabditis elegans*, which includes alleles associated with both longevity and reproductive delays (Tissenbaum and Ruvkun 1998).

IGF-I appears to be necessary for the optimal use of nutrients by and proper (equilibrated) growth of the fetus and placenta; further, IGF-I and its binding proteins require both serum glucose and insulin for their proper activity (Hernández 1998). Without proper levels and activities of IGF-I and its binding proteins, the nutrient distribution between fetal and placental tissue may become abnormal, as, for example, the differential allocation of resources to placental over fetal tissues seen in children of nutritionally stressed women (Barker 1998). Many proteins contribute to somatic growth and may respond to nutritional factors (Ulijaszek 1998b). For example, homeobox genes induce tissue- and organ-specific cell division and adhesion and stimulate cellular differentiation to specific tissue and organ types. Environmental influences on IGF-I, its receptors, and homeobox gene expression likely are examples of multiple opportunities that developing phenotypes have for responding to prevailing environments by resetting molecular, cellular, and physiological processes. Many such alterations may influence processes of senescence and, ultimately, organismic life span pleiotropicly.

Insulin also influences fetal growth. Insulin is synthesized by the fetal pancreas and thereafter regulates nutrient transport across cell membranes, while controlling the synthesis of peptide growth factors and their binding proteins (Hernández 1998). Growth hormone (GH), the most important hormone in post-natal growth, has little influence on fetal growth, since the fetus has few GH receptors; other hormones affect only specific structures, e.g., thyroid hormones (nervous system), androgens, and estrogens (skeletal and sexual systems) (Hernández 1998). Numerous additional "growth factors" – a variety of polypeptide-signaling molecules that bind to specific cell membrane receptors and promote mitogenic activity – exist in humans (Hernández 1998). Many of these have been conserved over evolutionary time and are shared across multiple phyla (see Zhao *et al.* 1998 for a review). The effects and concentrations of many of these molecules may also be altered by nutrient availability and endocrine factors.

In general, during post-natal life, GH (somatotropin) is the most important modulator of human growth and development, and is essential for normal growth (Malina and Bouchard 1991). GH is secreted by the pituitary in response to stimulation by growth hormone-releasing hormone (GHRH) from the hypothalamus. GH release is inhibited by somatostatin from the hypothalamus. GH decreases rates of carbohydrate uptake and utilization by tissues, enhances the mobilization of lipids from adipose tissues, and causes the production of somatomedins (growth promoters) by the liver that in turn stimulate the synthesis, mitosis, and build-up of tissues (Malina and Bouchard 1991). At puberty, GH interacts with testosterones and estrogens to initiate the adolescent growth spurt, prime development of secondary sex characteristics, and complete the final shaping of the mature adult. Twenty-four-hour concentrations of GH peak near the age that peak height velocity (i.e., highest rate of growth in stature) is attained, as do the peak circulating concentrations of somatomedian-C, which directly reflect GH levels (Malina and Bouchard 1991). Somatomedian-C stimulates the proliferation of cartilage cells at the growth plates of long bones, and thus linear growth in stature (Malina and Bouchard 1991). Receptors for GH are located on the membranes of almost all cells; binding of GH to cells initiates a series of steps that lead to increased gene transcription and protein synthesis (Hindmarsh 1998). GH apparently stimulates the release of IGF-I, which initiates local growth through paracrine action (Hindmarsh 1998).

All of the various endocrine organs contribute to growth and development: thyroid hormones (thyroxine, Tri-iodo-thyronine) increase oxygen consumption by skeletal and heart muscle, liver, and kidneys, potentiate the effects of GH, and accelerate biological processes (including growth and maturation); parathyroid hormone along with thyrocalcitonin from the thyroid maintains plasma calcium by regulating exchange between bone and circulation; the pancreas secretes insulin and glucagon which regulate carbohydrate metabolism (insulin also promotes protein synthesis and mitosis) steroid hormones from the adrenal cortex, such as aldosterone, maintain fluid volume, sodium, and potassium balance, while cortisol, a stress response hormone, enhances metabolism; and androgens and estrogens influence multiple aspects of physiological function and have anabolic effects on growth and secondary sex characteristics (see Malina and Bouchard 1991 for greater details). Among humans, and most sexually reproducing species, attainment of MRP follows closely after the end of growth. This occurs before the mitotic activity of ongoing growth begins to decline as stimulation from GH and related compounds is lost and cumulative and progressive declines in cell numbers and RC commence.

In humans, multiple genetic, cultural, and environmental factors influence growth, development, and, ultimately, life span. However, the timing and pace of growth and development are generally under neuroendocrine control. A species' life history pattern may depend on only: "...two main parameters of scheduling: (1) acquisition during development of cellular machinery that enables response to signals for the next life stage(s) *(competence)*, and (2) arrival of the stimuli" (Finch and Rose 1995, p. 3). Transitions across developmental stages are triggered by the neuroendocrine system – that is, hormones under the control of neural or similar pathways set the pace of life history (Finch and Rose 1995). Further, early rates of growth and development set the pace for the attainment of all later life history stages – that is, long periods of pre-reproductive maturation may necessitate equally long periods of reproductive maturity and/or parental investment. When parental investment is necessary to ensure sufficient reproductive success, even longer periods of maturity may be required. One way to ensure longer somatic survival may be extending the period of growth and development, thereby providing offspring with an RC beyond that needed for reproduction alone. Both hormone concentrations and tissue responses may be easily modulated and influence relative rates and patterns of development. A large variety of hormones modulate growth and maturation in humans (Malina and Bouchard 1991, list 22, Table 20.1, p. 344–5). Each is associated with a variety of receptors and activates various intercellular messengers. Given these multiple controls and activity levels, numerous avenues for alterations in the pace of growth and development following exposure to new stimuli are, and have been, available to evolving hominoids. Many such influences could have combined to produce slowed maturation, increased investment in neocortical structures, long-term dependence of offspring, and extended life spans among humans and other large-bodied hominids and hominoids.

Fetal origins of late-life disease

During the final decade of the twentieth-century, multiple links between pre-natal and infant development and late-life disease were observed (Barker *et al.* 1989a, b, 1990; Fall *et al.* 1992; Hales and Barker 1992; Barker *et al.* 1993; Law *et al.* 1993; Weder and Schork 1994; Dennison *et al.* 1997; Barker 1998). The most comprehensive of these studies are from the southern England district of Hertsfordshire (Barker 1998). In this region, intrauterine growth and development, and fetal undernutrition, as measured by birth weight and relative body proportions, are significantly associated with several late-life health outcomes. Obesity, hypertension, cardiovascular disease (CVD), non-insulin-dependent diabetes mellitus (NIDDM), Syndrome X, and hypercortisolemia apparently all share these same pre-natal predictors. Although numerous physiological and environmental factors (obstetric complications, infectious and parasitic diseases, starvation, famine), and sociocultural factors influence intrauterine and neonatal/post-natal growth and development, maternal malnutrition and its effects on the fetus apparently predominate with respect to influences on late-life disease (Barker 1998). It has been suggested that: "...the domain of (the) pre-natal environment could extend to the grand-maternal uterus because the prefertilization ovum that each of us arose from was born as a cell when our mothers were fetuses themselves...we were already typically more than 20 years old as a prezygotic cell before incorporating our paternal genes...The prezygotic domain of life history is completely unexplored for its influence on the individual" (Finch and Seeman 1999, p. 82). This domain no longer remains completely unexplored.

Barker and colleagues (see Barker 1998 for a comprehensive review; Dennison *et al.* 1997) in reviewing a variety of evidence suggest that numerous aspects of adult physiology (glycemia, cholesterolemia, blood pressure, cortisol responses, energy metabolism) are set, to a greater or lesser degree, in response to environmental factors encountered during gestation/infancy. A poor maternal nutritional environment triggers an increased investment of fetal resources into placental tissues, increasing nutrient uptake at the expense of developing fetal tissues. Reduced investment in somatic structures produces low birth weight infants, permanently reduces cell numbers in affected organs, alters the internal structures of affected organs, and resets various parameters of the hormonal system (Dennison *et al.* 1997). For example, among a sample of children born between 1935–43 at Sharon Green Hospital in Lancashire, U.K., those with the largest placentas and lowest birth weights showed the highest blood pressure (BP) as adults, while the length of gestation was unrelated to BP, suggesting no association with prematurity (Figure 3.4; Barker 1998). Before birth, low birth weight infants have already developed predispositions for the efficient use

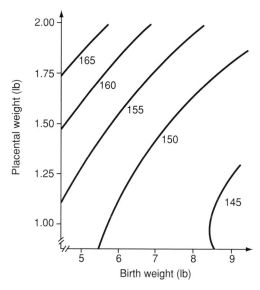

Figure 3.4 Adult systolic blood pressure (mm Hg) in relation to placental and birth weights (reprinted with permission, Barker *et al.* 1990, p. 260)

of energy, but have not needed to develop control systems for reducing these efficient response. When an improved nutritional environment is encountered post-natally, rapid weight gain and chronic degenerative diseases (CDDs) are thought to arise as sequelea of early programming for efficient use of the low nutrient environment encountered *in utero* (Barker 1998). Thus, programming of elevated adult BP may be more related to maternal diet during pregnancy than to any currently measured risk factors. In a study of children born during 1948–54 at the Aberdeen Maternity Hospital, mothers whose diets contained less than 50 g of protein, with high carbohydrates per day, showed elevated BP, as did those with over 50 g of protein with low carbohydrate intake, whereas those with high protein and carbohydrate intakes had the lowest BP (Campbell *et al.* 1996). It has even been postulated that the recent 20-year fall in cardio-vascular mortality in the U.S.A., Canada, Australia, and New Zealand (by as much as 25% for coronary heart disease (CHD) mortality from 1985 to 1990 among 24–75-year-old men and women may be linked to the improved fetal and infant growth and lowered infant mortality that occurred in these nations around 60 years ago (Barker 1998).

Poor maternal environments are an all too common occurrence in marginal environments. With chronic food shortages, extreme poverty, and nutritionally poor diets, restricted nutrition during natal and post-natal development pro-duces infants short for gestational age and/or who show low body weight for their length. In some cases, entire populations share in this phenomenon known as "stunting", where all individuals are short for stature, but are otherwise proportional in body size (Stini 1972; Scrimshaw and Young 1989; Frisancho *et al.* 1993). Given what is known of early life influences on longevity and late-life disease, it is likely that the same developmental factors associated with stunting influence life span. If stunting were beneficial, one might expect stunted individuals to show improved survival, as is seen with calorie-restricted rodents who are small compared with controls and show longer average and maximum life spans. If stunting is detrimental and reflects an accommodation to undernutrition and marginal well-being (Scrimshaw and Young 1989), and thereby increases frailty, stunted individuals would be expected to show poorer survival. Given the environments where stunting is most often reported – Amazonian native and hybrid populations in both rural and urban settings (Dufour 1992; Silva *et al.* 1995; Santos and Coimbra 1996; Fitton 1999; Silva 2001) and in many pockets of poverty worldwide (Mascie-Taylor 1991) – it is likely that multiple aspects of under- and malnutrition lead to stunting and shortened life span. Among Amazonian populations with more extensive contact with outsiders, those eating a variety of non-traditional foods generally are the tallest and heaviest, while those in more traditional environments are usually more "stunted". However, even among native South American groups with good nutritional status, adults remain short compared with the World Health Organization (WHO) standards, suggesting possible genetic components to their "stunted" stature (Dufour 1992; Johnston *et al.* 1999). Given changing nutritional environments, the fetal programming model suggests the hypothesis that an epidemic of NIDDM, hypertension, obesity, hypercholesteremia, hypercortisolemia, and CVD will follow dietary alterations currently associated with development in places such as the Amazon.

Many aspects of the endocrinological and biochemical control of fetal growth and development are well established and a variety of important regulators of growth and development (initiators, receptors, signal proteins) have been identified. During development, the effects of IGF-1, the main promoter of growth *in utero*, are regulated by IGFBP-3 in non-growth-retarded infants. However, with intrauterine growth retardation, the effects of IGF-I are regulated by IGFBP-1 and -2 receptors. IGFBP-3 activity is correlated strongly with fetal weight (Hernández 1998), a reduction in IGFBP-3 is used to benchmark intrauterine growth retardation. It is plausible that once IGFBP-1 and -2 structure the fetal system, these effects remain life long and, for some, may even be set in the ova of any female so affected. Thus, a lack of IGFBP-3 activity *in utero*, thereby reducing body size at gestation, may be followed by an increased activity of IGF-I throughout life. When exposed to a nutritionally adequate environment (an overabundance is not necessary given the mechanisms involved), this increased activity may influence adult morphology and disease risk. Given that IGF-1 is fundamental to the proper growth of the fetus and placenta and chronic undernutrition may permanently reduce IGF-1 concentrations (Dennison *et al.* 1997; Barker 1998; Hernández 1998), its associated binding proteins and their activity may be fundamental in structuring late-life morbidity from some chronic degenerative diseases following intrauterine growth retardation. For example, among both adults and children with either IDDM or NIDDM, low levels of IGF-1 are observed compared with non-diseased controls, even when glycemia is well controlled (Boullion *et al.* 1995). Both insulin and GH levels may also be reset by intrauterine growth retardation. Growth retardation is predictive of impaired β-cell development, and GH is essential for β-cell growth, while slow growth of β-cells leads to low insulin production (Dennison *et al.* 1997). Another model is that variable expressions of hormones and neural mechanisms during fetal and embryonic nutritional stress/growth retardation may become permanent, thereby programming the genotype toward an environmentally specific phenotype. Finch and Rose (1995) suggest: " . . . that the same genome may be programmed to yield alternative outcomes in development" due to the influence of neuroendocrine mechanisms (p. 8). Since many hormones and their receptors are active throughout our life history, pre-natal influences (e.g., undernutrition, malnutrition) may establish patterns of hormonal regulation or responsiveness *in utero* that are unalterable by post-natal experiences – a finding that appears to be true for at least hyperinsulinemia and insulin resistance (Law 1996; Dennison *et al.* 1997).

Other environmental insults during uterine development may produce outcomes similar to pre-natal undernutrition. Pre-natal exposure to tobacco is linked to low birth weight, criminal/anti-social behavior, and the development of externalizing behaviors throughout middle childhood, adolescence, and adulthood (Fergusson *et al.* 1998; Brennan *et al.* 1999; Ferguson 1999). Toxic aspects of smoke may produce stresses similar to those of poor nutrition *in utero*, thereby reducing investment in body size and key neurological structures. In addition to low birth weight, maternal smoking is associated with spontaneous abortion, poor perinatal status, and lower intelligence (Fergusson 1999). Intrauterine exposure to tobacco smoke may provide another model for understanding the fetal origins of late-life disease and the influences of early environment on subsequent life history and longevity.

Life history

As discussed previously, human patterns of growth and development are unique both pre- and post-natally. Variations from the non-human ape pattern most resemble those of chimpanzees and male baboons, who also show multiple accelerations and decelerations, but not in a "human" pattern. At birth, human infants more resemble the mid-term fetuses of non-human apes, or the help-less young of marsupials, than full-term primates. This pre-natal/natal pattern of slow maturation, delayed attainment of physical milestones, and rapid neurological development resonates across our species' life history, serving as a pacemaker of sorts for all subsequent phases. What we are as elders is highly dependent on what we were as adults, juveniles, children, infants, and fetuses, along with the various environmental conditions we experienced during these earlier phases of development. This life history perspective is not particularly new in gerontology, but it has received considerable recent attention (Sterns 1992; Sterns and Hoekestra 2000). Unfortunately, except for humans, their do-mestic animals, and laboratory animals, life history is difficult to reconstruct for most species. Today, many aspects of life history are fairly well documented for most humans (e.g., birth date, weight, and length, marital, occupational, and socioeconomic status, and ethnicity of parents), although the accuracy of these data varies widely across place, time, and individuals. However, other conditions surrounding birth – for example, parents' nutritional status, stress, and health, or weight gain of the mother – are seldom reported. Additional data on patterns of growth and development, childhood infections and injuries, age at menarche, or number of sexual partners are highly subject to recall bias and current sociopolitical/cultural circumstances. Other variables such as ed-ucation, occupation, adult height and weight, weight gain during adulthood, marital status, and number of offspring are more reliable and less subject to recall bias. Still, these data are not generally available for the large samples needed to examine life history events and life span.

For decades, gerontologists have insisted that longevity and health among elders are tied to factors occurring early in life that, in some fashion, preset the organism for a short or long life. Only in recent decades have integrated data on the same individuals from birth to death become available to test such hypothe-ses. The majority of such data that have been published come from western Europe and Scandinavia, particularly England and Finland. They show that events during fetal life do pattern, to some degree, adult physiological responses to both environmental stressors and excesses including a variety of metabolic conditions such as NIDDM, obesity, hypercholesterolemia, hypercortisolemia, and CVDs (Barker 1998). Similar trends may help to explain the consistent association of migration and *in situ* acculturation with hyperglycemia, obesity, and elevated BP observed among populations worldwide during the twentieth century (Baker 1984; Friedlaender 1987).

For women in most cultures, the number of children ever born or num-ber of pregnancies, major life history events, seem to exert strong influences on length of life. In many cultures without cosmopolitan medicine and modern methods of birth control, women are pregnant or lactating over most of their adult lives from menarche to menopause (Lancaster and King 1985). When this pattern is coupled with a marginal nutritional environment, a phenomenon of "pregnancy wasting" (when women never recover energetically from their previous pregnancy and lactation before becoming pregnant with another child) is often observed. In such settings, both maternal and infant mortality, along with spontaneous abortions, are high and female life expectancy is below that of men. The reverse of this has been observed in at least one study of U.S. women. In the Louisville Longitudinal Longevity Study (LLLS) of elderly women aged 65 and over, the majority of long-lived women never had any offspring and a significant proportion were never married (Crews *et al.* 1985). Although this was not a case-control study of longevity among childless women and women with children, it does support the hypothesis that having offspring may alter a woman's probability of long-term survival – perhaps the more children the greater the reduction in life span, or, conversely, not having offspring contributes to greater longevity. This is particularly interesting since, in the LLLS sample, the women were all born between 1880 and 1920, decades during which cosmopolitan medicine and antibiotics were in their infancy, birth control methods were less developed, and the nutritional adequacy of diets may have differed from today. Confounding the relationship between childlessness and longevity in the LLLS were two additional life history factors: these long-lived women were also well educated, almost all had completed high school, and many, college, and a significant portion (about one-third) had never married.

In general, more educated members of most populations show a higher life expectancy than those with less education (Pamuk *et al.* 1998). In addition, un-married women tend to live longer than married women in most cosmopolitan settings. Although educational attainment appears to be an important aspect of life history influencing longevity in cosmopolitan settings, it remains un-clear exactly how education influences life span. Education is likely to be a proxy variable for other, as yet unmeasured, phenomena. Since formal educa-tion (as measured by years of school completed) is a relatively recent cultural innovation, it is not likely to have been subject to selective forces. However, development of mental and physiological structures predisposing to individual abilities useful for attaining more years of education may have been selected as aspects of our species' life history for a variety of reasons over evolution-ary time (Dunbar 1988; Durham 1991). The abilities to use language, plan and organize our activities, understand and manipulate objects and culture, avoid social parasitism/exploitation, and to influence our own health and well-being are likely to have been long-term adaptive strategies among *H. sapiens*. Educational attainment may provide an important proxy for such abilities. In addition, better-educated individuals generally have greater access to health information, cosmopolitan medicine, and pharmacological products. Such access to scarce resources characterizes the elites of most present-day societies and likely did so for the elite members of earlier human societies. In cosmopolitan settings, the better educated are also more likely to engage in lower risk (military officers, doctors, lawyers, etc.) rather than high-risk occupations (infantry, dock worker, factory laborer), thereby reducing hazardous exposures, accidents, and risk behaviors. In earlier phases of human development, such innate abilities likely had different manifestations, perhaps providing better hunting and tracking skills, greater recall of where gatherable foods might be obtained, improved accuracy in tool manufacture, or better leadership and communication skills. The important point is that the intellectual structures and abilities underlying skills useful in our evolutionary past remain useful today. Attainment of these competencies is an integral aspect of human life history.

Cultural competencies

Humankind's ability to elaborate culture, develop technology, manipulate the material world, and improve their survival and reproduction has altered human life history since the first hominids adapted to bipedality. Prior to the evolution of bipedality, human upper limbs were not continually free to manipulate the environment, transport food and offspring, or grasp objects. Early in hominid evolution, infants were likely to be much more physically precocial, able to cling, climb, and brachiate at early ages (plesiomorphic) as do modern baby apes – e.g., chimpanzees, gorillas, orangutans, and gibbons. These traits develop at much later ages among modern humans (apomorphic). Although possibly not as precocial as other hominoid babies or their modern descendants, at the dawn of bipedality human infants were likely to have been born without today's large crania (brain expansion had not yet started) and females were likely not hampered by today's human pelvic constriction and large-brained offspring. As hominid evolution progressed, the gradual change to bipedalism placed new stresses on the hominid birth process and female pelvic structure. Subsequently, the later trend toward the encephalization of infants added to these stresses. One option would have been to lengthen gestation (as did elephants, to 18–22 months, with a 50–60-year life span). But human gestation length remains comparable to that of other apes (8–9 months). Another alternative is the production of smaller or less developed neonates. Following this latter strategy, humans now produce physically altricial infants who require years of parental investment.

Production of altricial offspring requires the re-scheduling of all subsequent processes of growth and development. Today, the whole process takes over 20 years of post-natal life. This is well after attainment of sexual maturity, and sets humans apart from other apes. It also leads directly to the re-scheduling/extension of all later phases of somatic life history. Slow maturation of offspring necessitates long periods for parental investment in offspring; this requires somatic maintenance of adults over an extended period of life. As noted earlier, such extended somatic maintenance may be easily achieved by redundancy of primary structures, leading to large RC, integrated defense systems to halt extrinsic attacks, and enhanced intracellular housekeeping capabilities to prevent damage (wear-and-tear) and eliminate wastes and byproducts of metabolism. The simplest way to ensure RS is to build sufficient capacity into needed structures and mechanisms that they overshoot that needed for reproductive success (Figure 3.2). It is this excess of RC that is available to extend life span when culture ameliorates environmental stress. Different individuals are endowed with variable degrees of capacity in various structures – some have low RC across multiple systems and die young, others have high RC across multiple systems and become centenarians.

Pedomorphism

Human pedomorphism, the retention of infantile-like features into adulthood, exemplifies interacting cultural, genetic, and environmental influences on human life history. Pedomorphism in humankind far exceeds that of any other primate species, a point often illustrated by a comparison of human and common chimpanzee profiles as infants and adults. A human adult looks more like its infant form than an adult chimpanzee looks like its infant form. A comparison of both with their close relative, the bonobo, suggests that bonobos may show an intermediate degree of pedomorphism. An evolutionary trend toward pedomorphism and resultant altricial infants may have occurred at the base of hominid evolution, although, based on fossil dentitions, the fully human pattern of tooth eruption and, by inference, the juvenile maturation did not occur until much later (Smith 1986, 1991; Anemone *et al.* 1996). This perspective suggests that selective pressures affecting growth and development that produced pedomorphic characteristics have resonated across the entire human life span and affect multiple aspects of life history.

During human evolution, behavioral, cultural, and biological responses to changing environments converged to produce the bipedal, verbose, culture-bearing, large-bodied, and pedomorphic primate we know today as human. Although numerous aspects of human anatomy, from our round faces and

mobile upper limbs, digits, and shoulders to our hairlessness and body proportions, all show pedomorphism, it is our extended period of dependency that
presents the most obvious example. Our overall growth rate is not remarkable
among primates; however, the normal length of the period, over 20 years in
some populations, is quite remarkable. The common chimpanzee is usually
sexually mature, fully grown, and reproducing at less than half of that age. An
understanding of how human pedomorphism evolved is crucial to understanding human life history. Examining why early life periods were extended and
multiple infantile features are retained into adulthood will aid in understanding
not only senescence but the more important question of why we live as long as
we do. One answer to the first question has always been that the "need" to learn
language and culture pushed humans toward an extended period of dependency.
But evolution does not respond to "needs" and it is difficult to imagine how an
increased reliance on language and culture necessarily leads to less physically
developed and more dependent neonates.

A more likely suggestion is that cultural, biological, and environmental pressures converged at crucial points in hominid evolution to slow certain aspects of
development and thereby allow extension of critical periods for development of
additional features and structures (competencies) that enhanced attainment of
MRP. What sets hominids apart from all other primate species is their habitual
bipedalism and their degree of encephalization. In the beginning, bipedalism
most likely evolved as a feeding adaptation to reach fruit-bearing branches
and/or to transport food to a safer place to eat (once having obtained it). When
bipedalism became the primary form of hominid locomotion, there was no need
for extended infancy and childhood to learn language and culture. Hominids
were not yet hampered by large-brained infants nor was there a need to develop
an expanded neocortex, along with its more complex neural integration. However, along with bipedalism, a second and equally important trend did emerge.
The upper limbs, previously subservient to the body's weight, were emanci-
pated from their role as supports and, at least partly, from their role as feeding
implements. Forelimb freedom set the stage for constant manipulation of the
environment, the development of material culture, and eventual expansion of
the neocortex. As the hands developed new functions in food acquisition and
the hind limbs became the sole support of hominids, it is unclear exactly when
brain expansion began in the hominid lineage. But it is clear from the fossil
record that by 2 MY ago larger-brained hominids already had begun to appear
(e.g., *H. rudolfensis*). This expansion of neocortical areas has continued up
through modern humans.

Expansion of neocortical areas allowed the evolving hominid line to develop
the higher order neurological-based integrative functions that today characterize
human culture and behavior. In modern humans, post-cranial development is

most rapid during early gestation and proceeds to attainment of almost full birth length during the first two trimesters, actually slowing after about week 20 and declining from then until birth. At first, brain development lags behind that of the body, but then proceeds very rapidly during the middle of the second and the third trimester. After birth, the same rapid rate of brain growth continues over the first 12 months of post-natal life, during which the brain attains roughly 70% of its largest adult size. This pattern does not characterize any other modern ape. Whatever "caused" this human pattern of gestational development to emerge also led to the pedomorphism seen in today's humans, the resetting of all later stages of human life history, and our current life history and manifestations of senescence.

Whatever "caused" the pongid–hominid split and led to bipedalism and free upper limbs set the evolutionary stage for the later elaboration of human culture, language, altricial newborns, and infant dependency. Early incipiently bipedal proto-hominids were likely to have been similar to pongids, producing precocial infants capable of clinging to their parent (a plesiomorphic trait) as she swung/walked/waddled/strode from tree to tree searching for, and carrying, fruits and nuts to eat (Smith 1986, 1991). However, as dependence on bipedalism increased, the upper limbs were more often available for new tasks. Mothers likely more often carried infants to where they wanted them, allowing less physically developed (more altricial) neonates and infants to survive without needing to cling to the parent. Alone, such carrying behavior is not likely to have produced rapid changes in limb development, since both clinging (precocial) and non-clinging infants (altricial) would probably survive equally well. However, with continued development of an upright posture and free upper limbs, several trends converged – elaboration of the neocortex increased verbal communication; and elaboration of material culture reduced gestational investment in motor development, and extended post-natal growth and development. All of these could be satisfied through carrying behavior. Thus, free upper limbs in part set the stage for the development of altricial infants and the slow development we observe in modern humans. As the upper limbs became more capable of elaborating material culture, the neocortex expanded to include new neurological complexes allowing development of culture and cultural elaborations such as woven reeds and grasses, animal hides, or leaves of trees along with edibles and tools. This suite of interacting characteristics produced additional pressures on fetal development and, ultimately, the neural structures underlying language and complex forethought. Increased bipedalism altered the human pelvic dimensions and orientation, while elaboration of the neocortex led to an increased investment in neural development and a larger proportion of fetal energy directed toward the neocortex, but less for physical development. These biocultural interactions reset all phases of human life history, from the pace of growth and development to senescence. Because they may carry their offspring, bipedal parents are highly mobile, as are marsupials with similarly altricial newborns. As with marsupials, human infants have little need for precocial development of limbs and muscles. This provides opportunities to invest scarce fetal resources into neocortical tissues and neuronal integration. Trends toward slowed development and pedomorphism may have occurred fairly early in hominid evolution (2 MY ago). Examination of dental development patterns of modern apes and humans along with those of fossil hominids (australopithecines, erectines, and archaic *sapiens*) indicate that neither the human pattern of tooth eruption nor child/juvenile dependency were fully developed before the *H. erectus* stage of hominid evolution, or perhaps even later (Smith 1986, 1991; Anemone *et al.* 1996).

Eventually, these various trends (i.e., free upper limbs, bipedality, altricial infants, verbal interactions, brain expansion) converged to produce a physically modern human with the physiological structures to allow human cultural competencies. This led to a specific culturally created microenvironment to which our ancestors adapted and which we continue to create today. Culture allowed all humans in effect to experience a similar microenvironment (i.e., sufficient food and group food-sharing; comfortable skin temperatures due to body coverings, fire, and lodgings; as infants, shelter, food, transport, parental/relative/grandparental care, and relative safety/security, and, as adults, companionship, reproductive success, and old age). Cultural adaptations produced a relatively homogenous ecological/environmental niche in which humans could reproduce, develop, grow, mature, and repeat this cycle despite the specific physical/biotic environment. Any homogenous environment sustained over a long period of time leads to reduced genetic variation. Adaptation to a specific set of environmental stressors over time may lead to a type of evolutionary strategy commonly referred to as K-selection. K-selected species (elephants, gorillas, humans, whales) usually experience a relatively constant environment, tend to be large, have few births over their life span, experience single births, and live longer than r-selected species. The latter commonly reside in rapidly changing and/or transient environments, reproduce early in life, produce litters, and generally have shorter lives (mice, mushrooms, insects, and bacteria). In general, all existing species of primates are more K-selected than r-selected, with one or two offspring per litter, some parental care, and relatively long lives.

Resetting growth

Resetting of post-cranial fetal development to a slower pace produced newborns resembling the fetal stages of other apes, or perhaps the live born pups of some marsupials. This alone necessitates extension of post-natal development and growth. A variety of processes could have been reset during fetal development to allow more cranial and neurological development. For example, overall length of gestation could have been extended (as with elephants). However, gestation length varies only slightly across a wide variety of existent primate species: wild baboons, 175–180 days (Berkovitch and Harding 1993); humans, 260–280 days; gorillas, 260–280 days; and chimpanzees, 240–260 days. This suggests that large-bodied primates may already have been at about the upper limit for length of gestation even before the evolutionary radiation of hominids (a plesiomorphic trait). Increasing the pace of intrauterine growth would also allow for larger neonates. However, bipedal locomotion and pelvic remodeling led to restrictions on the pelvic canal and neonatal size. Thus, the options available for resetting fetal development and allowing greater neurological development among evolving hominids were limited. The most viable, apparently, was to reduce investment in post-cranial somatic structures, thereby producing more altricial and dependent newborns. Humankind has extended the period of fetal-type development of post-cranial elements into the first year of post-natal life; their final stage of growth does not even commence until the mid-teen years (the adolescent growth spurt) and only then do humans finally catch up in growth to their ape relatives (Smith 1993).

Extending fetal-type development to the post-natal period produced infants who are completely dependent on others for transportation for their first 2 years of life, who are unable to obtain their own nourishment without aid for 5 or more years, and who do not complete growth for over two decades. Human infants have about 50% of their cranial size at birth, while chimpanzees and bonobos have about 75%. However, humans have much larger heads relative to their bodies and much less limb and grip development than their ape relatives. Since, at birth, human infants are about 25–50% less developed cranially and even less physically than other apes, it is no surprise that they require more time to complete all subsequent life stages. This trend toward encephalization of neonates and infants set humankind, along with some earlier hominids, apart from our nearest evolutionary relatives as early as 2 MY ago. Modern gorillas and humans provide an ideal comparison. Gorilla males outweigh the average human male by about 2.5 to 1, as do females; however, their maximum life span is only about one-half (60 years) that of humans (122 years), while the brain volume of an adult human is about twice that of a gorilla. Between humans and the common chimpanzee, average weight comparisons are more equal, although the largest chimpanzee weighs quite a bit less than the largest human (200 versus > 600 pounds). However, small adult chimpanzees and small humans weigh about the same (80 versus 100 pounds), although chimpanzees only live about 40% (50 years) of the human life span, and possess only about one-third of our cranial volume.

Another indication of slowed development in humans compared with other apes is our unique pattern of dental development (Smith 1986, 1991; Anemone *et al.* 1996). Chimpanzees have advanced development of their molars compared with their anterior dentition and calcification of adjacent crowns occurs contemporaneously; in humans, anterior dentition develops before posterior and molar development is staggered (the 6-, 12-, and 18-year molars) (Anemone *et al.* 1996). Comparing eruption patterns of chimpanzees, fossil hominids, and humans provides an estimate of when hominids first showed patterns approaching those of modern humans (Smith 1986). Based on available fossils, all australopithecines show more ape-like patterns, while erectines show an incipient human pattern (Smith 1986; Anemone *et al.* 1996). Alterations in dental eruption are likely to indicate significant changes in environmental pressures and species' adaptations. A lack of need for large cheek teeth to process coarse foods in early life would allow for slower dental development and suggest increased parental investment in provisioning or processing foods. Slowed dental development may also follow slowed overall (altricial) development or provisioning and processing could lead to greater altriciality. Tooth development and its delay in humans relative to apes and early hominid fossils is congruent with humans' overall altricial physical development at birth.

Delayed development provides a childhood stage to human life history. This appears to be unique to humans. All primates show infant stages (ages 0–1 year in humans), but, in other primates, this tends to develop directly into juvenile (the period of sexual maturation in apes; in humans, children are aged 1–4 years, juveniles are aged 5–12, and adolescents are 13–18) and reproductive adulthood (in humans, ages 19 until menopause for women and about age 60 for most men). After reproductive adulthood, other apes succumb to senescent and disease processes, while humans enjoy a period of late-life adult survival, followed by senescent degeneration. Childhood (ages 1–4 in humans) and adolescence (ages 13–18) are unique life history stages found only among humans (Bogin 1988; Bogin and Smith 1996, 2000; Bogin 1999) Childhood, occurring between infancy and juvenilehood, is a time of learning and attainment of competencies that allows children to mature physically, mentally, and emotionally before experiencing puberty, sexual maturation during adolescence, and parenthood.

It remains unclear exactly when during hominid evolution this specific pattern of growth and development came to characterize humankind's ancestors (Smith B.H. 1986, 1991, 1993; Bogin and Smith 1996; Mann *et al.* 1996; Thompson and Nelson 2000; Kondo *et al.* 2000). Evidence indicates that early hominids (australopithecines and habilines) and modern apes share patterns of growth and development, particularly tooth development and eruption patterns (Smith 1986, 1992; Anemone *et al.* 1996). Interpretations of *H. erectus* data suggest they also show advanced post-cranial development relative to dental development, i.e., the ape-like pattern (Smith B.H. 1993). Together, these data suggest that the human growth pattern did not become similar to that of today's modern humans until the advent of early modern *H. sapiens*, although not all branches of the *sapiens* lineage may show this pattern (Thompson and Nelson 2000).

Neanderthal dental and femoral development patterns suggest they were more advanced in dental maturity or slower in post-cranial development than either early modern or present-day humans (Thompson and Nelson 2000). Based on dentition alone, it has been argued (Smith 1991) that dental development in Neanderthals was the same as modern humans. However, analyses of femoral growth and dental maturation in the same individuals show that achievement of proportional femoral length by dental age in Neanderthals is significantly below, and that for gorillas is well above, the curve representing modern humans (Thompson and Nelson 2000). Gorilla and *H. erectus* share a pattern of rapid proportional femoral growth at early chronological ages, while modern humans show suppression of growth during childhood followed by an adolescent growth spurt (Thompson and Nelson 2000). This pattern has been described as "catching up" (Smith B.H. 1993). The Neanderthal pattern of growth and development does not fit with either of these two models (Thompson and Nelson 2000). The Neanderthal pattern of relative dental and femoral development may be as unique as is that of modern humans.

Combined, these findings suggest that sub-specific variability in proportional growth and development may occur rapidly on an evolutionary time scale. Neanderthals only came into existence between 250 000 and 150 000 years before present (YBP) and are not thought to be observed after about 30 000 YBP. These results further suggest that the selective pressures on *H. sapiens neanderthalensis* and *H. sapiens sapiens* (as reflected in their relative patterns of growth and development) were different and may themselves have diverged as the two groups did. More rapid attainment of adult post-cranial size relative to dental maturation may have contributed to *H. sapiens sapiens* survival into the modern era, while Neanderthals did not. Thompson and Nelson (2000, p. 490) suggest that Neanderthals were characterized by "either slower or delayed linear growth, or advanced dental development relative to modern humans". Alternatively, they may have experienced both delayed growth and rapid dental maturity. This suggests that early development of their dentition to process foods and slower growth may have resulted to accommodate Neanderthal's poorer nutritional inputs and coarser diet. It also suggests that parental investment among Neanderthals may not have been as fully developed as in *H. sapiens sapiens*. Given less parental investment, rapid dental development, and delayed growth, Neanderthals may not have had the opportunity to develop the large physiological organ reserves during growth that characterize modern humans and promote late-life survival. Slowing growth while advancing maturation of dentition may have been a reasonable evolutionary strategy for a species exploiting marginal environments; however, restricted RC may also have limited opportunities for post-reproductive survival. Modern humans and their lineal ancestors show more rapid linear and slower dental development, suggesting they may have maintained growth on a less coarse diet. A longer growth period provides the opportunity to establish a larger RC and, ultimately, greater survival. Both adult and immature Neanderthals also consistently show proportional differences (shorter distal segments) in their long bone lengths compared with modern humans (Kondo *et al.* 2000). This further suggests these two sub-species were not only distinct, but also differentially adapting to environments with divergent constraints on their patterns of growth and development.

Given that patterns of dental maturation, rate of attainment of adult stature, and limb proportions differ across taxa of contemporaneous hominids, it seems likely that other aspects of life history (e.g., total development, infant growth rates, age at weaning, length of infant dependence and childhood, age at puberty, presence or lack of adolescent growth spurt, attainment of MRP, age at first birth, and age at attainment of adult stature) also differed. Such life history traits still vary somewhat across modern human populations and are very different from non-human primates. Among baboons, age at onset of puberty, duration of post-partum cycling prior to conception, length of gestation, and age at weaning all appear to respond to resource availability (Berkovitch and Harding 1993). Variations in life history patterns observed among Neanderthals, erectines, and modern humans provide support for the suggestion that "... life history evolution among taxonomic groups may sometimes be chaotic, which would frustrate strong inferences by the comparative method in the study of life histories between taxonomic groups" (Finch and Rose 1995, p. 1). This may be particularly true in hominid groups like the australopithecines that encompass a wide variety of local types over their 4–5 MY evolutionary history or the two identified sub-species of *H. sapiens* (*sapiens* and *neanderthalensis*) that evolved over the past 250 000 or so years and likely co-existed in some areas for numerous generations. Wide variation in rates and timing of growth across species and populations and the variety of patterns within even single species suggests that there may be neither a "single human growth curve" (Bogin 1988) nor a single "life history for the *Hominidae*" (Smith and Thompkins 1995); rather, sub-species and even different populations within a species may show variable patterns. One key to understanding variable development and reproductive schedules among apes and humans is to determine how the timing and quality/quantity of hormonal cascades differ across species.

Among humans, growth and development are so gradual that offspring are capable of developmentally adapting to the environment throughout fetalhood, infancy, childhood, and adolesence. Intrauterine development responds to the availability of calories, protein, fats, vitamins, and minerals along with ambient temperatures and availability of O_2. During their 20+ year development, humans continually respond to environmental factors, temperature, availability of oxygen, solar radiation, and nutritional inputs. Limb proportions, shape, body length and size, chest width and depth, skin pigmentation, amounts of fat and lean tissues, and patterns of hormone release are all plastic. Extensions of growth and development over decades allow humans a wide latitude in developmental response to environmental stressors – much more so than other extant non-human apes. Humankind's ability to alter internal and external phenotype in response illustrates our adaptability. Variable average ages at menarche and menopause among women illustrate the temporal variation (adaptability) of timing of human life history events. These range from about 12 and 50 years in some U.S. sub-groups to 18 or more and as low as 45 or less years among some high altitude women. Another is the age of completion of linear growth (stature) which also shows wide temporal variation. This ranges from the early twenties to the early thirties among men in U.S. cities compared with those raised at high altitudes such as Nepal and Peru. Wide temporal ranges for life history events illustrate the success of humankind's biocultural adaptations at providing for immature offspring over their long-term dependence and semi-dependence.

Reproduction and life span

Sex differences in life span

Sex differentials in human longevity favoring women in more cosmopolitan settings are well known (Hazzard 1986; Holden 1987; Smith D. 1993; Crose 1997). Often neglected are extensive data showing that, for most species, females do not necessarily outlive males, nor have women outlived men in most historical populations, nor do they in many current human populations (Gavrilov and Gavrilova 1986; Macintyre *et al.* 1996). Currently in the U.S.A., U.K., and many other societies, women enjoy about a 5- to 7-year higher expectation of life at birth (about 79 years) than do men (about 72 years) (Rakowski and Pearlman 1995). In a number of populations, men and women exhibit about the same life expectancy (e.g., Sweden, India, Kuwait), while, in some, men still survive, on average, somewhat longer than do women (e.g., Somalia, Ecuador). The human female longevity advantage is of relatively recent genesis and characterizes only the most recent cosmopolitan societies, with well-developed maternal and child care. Where women tend to outlive men, boys and men tend to have higher death rates than women and girls throughout the life span. Death rates tend to vary most between the sexes at and immediately following birth (the perinatal period), during adolescence (when accidents and trauma kill more boys/men), and during the reproductive years (when girls/women succumb to maternal-related causes), and following the vigor of the reproductive years at mature adult ages (when men die more rapidly from CDDs). After about age 70, the female mortality advantage, when there is one, with respect to CDDs, decreases. At around age 85, mortality rates converge and virtually identical rates characterize the sexes as they succumb to life's rigors in a manner reminiscent of water glasses in a restaurant (Figure 1.1B). In the U.S.A., women who survive to age 65 may expect to live an additional 19.4 years, and men only 15.0, but women who live to age 85 may expect 9.1 more years and men 7.6 (Rakowski and Pearlman 1995, Table 1).

Multiple hypotheses have been advanced to explain why, in modern settings, with well-developed public health care systems, women tend to outlive men (Smith D. 1993). Genetic variability in sex chromosome complement (XX/Xy) and loci located on the X- and Y-chromosomes have dominated explanatory models of sex differences in longevity (Smith D. 1993; Crose 1997). These are based on observed sex differentials in conditions such as hemophilia, color blindness, and xeroderma pigmentosum due to hemozygosity of X loci in men. Such X-linked genetic traits appear in mammalian males at a rate equal to the allele frequency for that trait in females. In females, this is true only for dominant traits, while recessive traits occur at a rate equal to the square of the allele frequency. For any recessive conditions, if the allele frequency is 1% in females, 1% of males are affected but only 0.01% of females are. However, instead of an observable chronic condition such as hemophilia, some X-linked alleles may confer only slight sub-clinical decrements in overall physiology. Over the life span, these may lead to greater physiological stress, wear-and-tear, or disregulation in those with hemizygosity (men) than in those with homozygosity for non-detrimental alleles or heterozygosity (women). Given the number of known X-linked conditions (over 150 loci associated with named clinical conditions have been identified on the 1997 map of the X-chromosome), a large reservoir of frailty-enhancing alleles may be available on the X-chromosome. That there are frailty-enhancing alleles observed at so many X-linked loci also suggests the converse – that there are alternative alleles at these loci predisposing to greater vitality, along with a majority that likely have moderate to neutral effects on vitality/frailty. The X-chromosome provides a basis for estimating the minimum number of loci with detrimental effects on physiology to be expected on the other 22 human chromosomes. Since the X is an average-sized chromosome, thousands of loci with alleles contributing to frailty are likely to be distributed across the whole genome.

Many additional factors are likely to have contributed to sex differentials in survival. For example, contrary to earlier models, the X- and Y-chromosomes are not completely incompatible; rather, they show significant pseudohomology. There is sufficient sequence similarity for them to bind as tetrads during cell division for gamete production, for them to form bridges, and for them to recombine. Crossing-over may produce Y-chromosomes carrying loci or pseudo-loci that produce defective or no protein, contributing to male frailty. At least two Y-chromosomes variants – greater heterochromatin in the long arm and a deletion at the distal end of the long arm – reportedly correlate with greater longevity in men from a Russian village and an Amish kindred, respectively (Kusnetsova 1987; Smith D. 1993). In the Amish, men with the deletion Y show greater longevity (82.3 years) than either their spouses (78.5 years), female kin (77.4 years), or other Amish men in the same local area (Smith D. 1993). Both the deletion and the added heterochromatin occur in the distal portion of the long arm of the Y-chromosome. This is also the general area where loci responsible for masculinization of the fetus are located. The distal deletion responsible for greater longevity among the Amish kindred must be distal to these masculinization loci or masculinization would not occur. These Y-chromosome findings suggest that some factor or factors that may be involved in the normal process of masculinization (predisposing men without such Y-chromosome mutants to earlier mortality).

Another factor proposed to affect male survival is that all human embryos are predisposed toward the female phenotype; that is, if none of the loci associated with maleness is ever activated, the somatic anlagen unfolds as female. Thus, masculinization is imposed on a basically female embryo. This process may alter survival potentials through any number of pathways. For maleness to be achieved completely, loci both on the Y-chromosome and elsewhere must be activated in a specific sequence to retain male and eliminate female structures in the bisexually potent embryo. Numerous defects in this process occur (e.g., testicular feminization, pseudohermaphroditism), wide variability in the final phenotype is tolerated (the range of normal is broad), and multiple sub-clinical errors may be easily incorporated into the soma during this process. Embryonic masculinization may also explain in part why men are less buffered from the environment both *in utero* and during life than are women (see Stinson 1985). Male embryos and fetuses invest more effort (e.g., nutrients, proteins, energy, metabolic activity) into producing their somas and differentiating them to males than do females. This may, in part, explain why males lag behind females during intrauterine development and post-partum maturation. Males also must differentiate within a narrow window of opportunity (the first 14 days of life), producing proteins and hormones not only to masculinize, but also to halt the natural progression of their embryo to a female phenotype. These activities occur so early in development that they are energetically expensive and may go awry to such an extent that the primary sex ratio must be very high to ensure sufficient male births. The primary ratio is often suggested to be around 2:1 male conceptuses to female, but may actually be much higher, although the sex ratio is only 1.05:1.00 at birth. Thus, maleness itself is a hazard for survival to birth, whereas femaleness is less so. Further, during embryonic and fetal development females may be better able to buffer environmental stressors than males, who are investing energies in becoming males.

Another possible contributor to poorer long-term male survival may be that cells with less heterochromatin (45X₋) replicate more rapidly *in vitro* than do cells with a normal karyotype (46XX), while those with more heterochromatin (47XXy) replicate slower. Differential cell replication rates may account for the shorter stature and developmental dysregulation characteristic of those with 45X₋ chromosome complements and, similarly, the physical features of trisomy 21 may be in part due to slowed mitosis – in the former case, by producing more and smaller cells more rapidly, and, in the latter, fewer and larger cells more slowly than normal during development (Hauspie and Susanne 1998). Supportive evidence for this effect possibly occurring in normal men is the finding that those with 47Xyy constitutions average about 13 cm. above the stature of their 46Xy fathers, which also is about the average difference between men and women (Hauspie and Susanne 1998). It is possible that men whose smaller y chromosome carries much less heterochromatin than the women's X may achieve more rapid cellular mitosis (even with two Y-chromosomes) and produce somatic cells more rapidly than do women. To the degree that cellular replication influences life span, this also may limit men's ability to respond to stresses or changing environmental conditions and leave men more susceptible to degenerative processes with increasing age.

Testosterone may also influence male life span and provide an example for antagonistic pleiotropy (Crews and Gerber 1994; Finch and Rose 1995). As the major androgen responsible for masculinization and development of sexual characteristics at puberty, testosterone is necessary for the attainment of reproductive capability in men and ultimately their reproductive success. As necessary as testosterone is for reproduction, men with higher levels of testosterone may be at greater risk for neoplastic diseases of their reproductive organs, CHD, and stroke. Testosterone is associated with everything from aggressiveness and lack of impulse control to anti-social behavior and violent causes of death (suicide, homicide, accidents) related to greater risk-taking behavior. Testosterone is also believed to act in concert with the neuroendocrine–immune system in responding to stress, by potentiating the HPA axis to increase the release of stress hormones (Schulz-Aellen 1997). It is this close association with the stress response that may explain testosterone's influence on CHD and stroke, along with its associations with aggressive and impulsive behavior. Conversely, high estrogen levels are thought to protect women from heart disease, stroke, and cancer of their reproductive organs and breasts until menopause, after which the loss of estrogen and relative increase in testosterone exposure is thought to increase risks for these same conditions. Estrogen has ameliorative effects on BP, lipid profiles, bone loss, and general body habitus. However, at menopause, these effects are lost and women become more masculine in their risk factor profile, an example of an age-specific alteration in physiology inherent to women's ovaries. Any or all of these physiological mechanisms may act to reduce longevity in men compared with women when adequate nutrition and maternal and health care are available.

Among women, changing secretory patterns and altered activities of ovarian steroids modulate at least some aspects of senescence and life span. For example, following natural menopause, women with early menopause, age 44 and younger, show higher mortality rates from all causes of death over their remaining life span than do those with late menopause, ages 50–59 (Snowdon *et al.* 1989). The ovarian steroids, estradiol and progesterone, show complex patterns of activity during the life span. Peaks of secretion are observed during uterine development and at puberty, and during completion of sexual maturity, but during infancy and childhood they drop to barely detectable levels. Following sexual maturation, estrogens remain high through the third and fourth decades of life, but decline again as fewer primary oocytes remain and menopause occurs throughout the fifth decade. Following menopause, both estrogen and progesterone drop to barely detectable levels and remain there. Over a woman's life span, many of their cells and tissues ebb and flow in concert with these ovarian hormones. For example, vaginal, uterine, and breast tissues exhibit hyperplasia during puberty and hypoplasia following menopause. Concurrently with this hypoplasia, women become more susceptible to circulatory diseases and neoplasms, both of which are suspected to result, in part, from basic senescent changes. Thus, internal physiological alterations arising following menopause appear to be senescent processes (Snowdon *et al.* 1989; Austad 1994). Understanding women's reproductive senescence and the subsequent physiological alterations that increase mortality risk should be a priority of those interested in senescence. Few other aspects of human physiology fit as well the criteria and definition of senescence as does cessation of reproductive function.

Reproductive senescence

Menopause (cessation of monthly menstrual cycling and the absence of menses) marks completion of the reproductive span in women. All women who survive sufficiently long experience menopause. The nature of human menopause has been of long-term interest to both evolutionary and gerontological theorists (Lancaster and King 1985; Snowdon *et al.* 1989; Pavelka and Fedigan 1991;

Austad 1994; Wood *et al.* 1994; Peccei 1995a, b; Leidy 1999; Leidy-Sievert 2001; Peccei 2001b). One view is that menopause evolved through natural selection to curtail reproduction at some point in the life span of women. Another is that menopause is the outcome of relaxed selection pressures for continued reproduction beyond the same point. The first view implies that total loss of reproductive potential provides an evolutionary advantage. The second suggests that women's ovaries already remain active well beyond a reproductive span that likely averaged only 15 or so years, often ended before age 30, and seldom continued beyond age 40 or so years during most of hominid evolution. It is not likely that prior to the very recent epoch, the last 10 000 years or so, that many women even survived to become menopausal. Thus, it is unlikely that prior to recent millennia any positive selection for menopause could have occurred during hominid evolution, and it is unlikely that any ever has. Among modern-day subsistence agriculturalists and nomadic "hunter–gatherer" (or more precisely "gathering–scavenging–hunting") groups, a substantial proportion of women survives to menopausal ages. However, today's marginal populations do not, in any sense, reflect conditions that are likely to have prevailed during most of hominid or even human evolution. The gradual cessation of ovarian cycling after about age 40 and progressive decline through the fifth decade of life seen in women appear to be characteristic of most extant long-lived mammals (elephants, apes, whales) (Snowdon *et al.* 1989; Austad 1994, 1997; Packer *et al.* 1998; Leidy 1999). Menopause appears to be an inherent aspect of the architecture of the ovaries of mammalian species that complete oogenesis before birth (Pavelka and Fedigan 1991). Slowed cycling and menopause appear to be pan-mammalian characteristics resulting from allometric relationships and vary in concert with other aspects of life history (Wood *et al.* 1994; Leidy 1999). In particular, the number of primary oogonia and their intrinsic frailty lead to a specific rate of loss (atresia) with time and likely in part determine the age of menopause in concert with alterations in release of hypothalamic factors. In general, human menopause appears to reflect the ultimate outcome of basic aspects of mammalian ovaries, none of which continues to produce mature ova past about 50 years of age (Packer *et al.* 1998; Leidy 1999). In this view, menopause in women is simply an artifact of biocultural processes that have extended human life beyond the biological reserve of mammalian female reproductive organs (Packer *et al.* 1998).

The timing of menopause is determined by the number of primary oocytes laid down during fetalhood, their rate of loss over the woman's life history (Leidy 1999), and hypothalamic signaling. Primordial germ cells do not share the same life history as somatic cells (Kirkwood 1977, 1995; Leidy 1999); they develop exclusively from the yolk sac of the embryo, and, by the fifth to sixth week post-conception, they migrate to the ovary, become oogonia (transitional diploid cells capable of repeated mitotic divisions), pass through the first prophase of mitosis to produce oocytes, and are then held in stasis for the next 15–50 years, after which they either experience atresia or are released through ovulation and complete meiosis to produce fertilizable ova (Leidy 1999). As primary oocytes within the ovary mature, they release rising amounts of estrogens, thereby maintaining menstrual cycling via the hypothalamic–pituitary–ovarian axis; once there are insufficient follicles to produce the amount of estradiol needed to maintain signaling to the pituitary or hypothalamic messages cease, cycling ceases and menopause ensues (Leidy 1999).

At menopause, ovaries no longer retain sufficient primary oocytes (primary ovarian follicles capable of producing a mature ovum) to produce a physiological level of estrogen sufficient to complete follicular maturation. This age-related follicular atresia (degeneration of oocytes by shrinkage of either the oocyte or follicle, atresic follicles may continue to produce progesterone through lutinization (Leidy 1999)) of oocytes is well documented, with numbers declining from about 7 000 000 in the 5-month-old female fetus to 2 000 000 at birth, to 400 000 at puberty, to 165 000 at ages 25–31 years, and finally to 11 000 at ages 39–45 years (estimates through puberty reviewed by Leidy 1999; those after puberty from Finch and Hayflick 1977, pp. 318–56, who also reported 3 500 000 for 4-month-old fetuses and 733 000 at birth, about one-half and one-third, respectively, of Leidy's more current estimates). The number of oogonia laid down in the fetal ovaries is species specific and likely to be under genetic control, as is atresia (Leidy 1999). Thus, close to the point of their MRP (estimated at about 20–25 years), women have at least 40 times more primary oocytes in their ovaries (but only 2.3% of their original fetal maximum) than they do as they approach menopause at about age 45, when they have only 0.2% of their fetal maximum remaining.

In general, the majority of data suggest that menopause is an epiphenomenon of evolutionary pressures in ecological settings where early female reproduction (ages 12–13 years) and death during the third (twenties) and fourth (thirties) decades of life were the norm. This is in stark contrast to the ages of 15–18 for first reproduction and frequent survival to menopausal age so often cited among today's traditional living groups by those arguing for an adaptive value to menopause (see Snowdon *et al.* 1989; Austad 1994, 1997; Alvarez 2000). If an epiphenomonon of other evolutionary pressures, it is expected that the cumulative incidence curve for menopause would have to follow (after accounting for RC) a curve similar to that of cumulative mortality. In generations/eras leading up to and following the epidemiological/demographic transition, this relationship is likely to have shifted. However, it may still be observed to some degree with mortality data from less developed populations today or in earlier decades. In modern clinical settings, the earliest natural occurrence of menopause is

(A)   **Percentage of post-menopausal women in the U.S.A.**

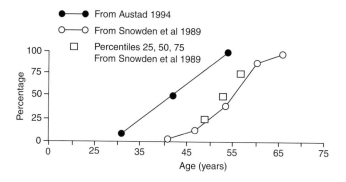

(B)   **Percentage of women deceased**

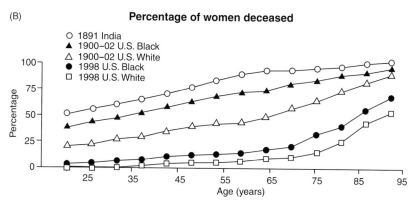

Figure 3.5 Menopause (A) and mortality (B).

in a woman's twenties with this frequency increasing exponentially thereafter (Figure 3.5A). Although termed premature ovarian failure (POF) before age 40, there is a continuum from POF to the arbitrary clinical cut-off point for early menopause at age 40+ (Leidy 1999). Among earlier hominids, age at death and menopause likely co-varied such that the proportion decreased/increased more rapidly than did the proportion post-menopausal, as it does among modern apes. As human patterns of mortality changed toward longer-term survival in more recent eras, the age distribution of the menopause has not altered greatly (Figure 3.5B). Human ovaries, as did those of most mammals, are likely to have evolved in settings where late-life survival, let alone reproduction, was unlikely, came with substantial risk, and provided little relative selective advantage in competition with conspecifics, most of whom generally reproduced early in life, died before completing their fourth decade of life, and seldom survived to show reproductive decline.

It has been suggested that only women among the primates experience true menopause as part of their life history. Although other apes also show quite long life spans – common chimpanzee and bonobos ($\bar{x}$ ~35, maximum ~ 50), gorilla ($\bar{x}$ ~ 35, max ~ 50), orangutan ($\bar{x}$ ~ 30, max ~ 45) (see Harvey and Bennett 1983; Harvey *et al.* 1987) – none average beyond the 50-year barrier for true menopause. Furthermore, all types of apes show severely reduced fertility at older ages, reminiscent of the reduced fertility of women in their late thirties and forties prior to menopause (Caro *et al.* 1995). Other aspects of life history also differ between apes and humans. For instance, age at first birth for common chimpanzees at Bossou, Guinea is 12–14 years and the birth rate peaks at ages 20–23 years (0.333/year), but falls in later years "until menopause at just over 40 years" (Sugiyama 1994). Although the age at first birth of these apes is close to that for some gathering/hunting groups, the age of 40 for "menopause" is a decade earlier than usually reported for women (Sugiyama 1994). The inter-birth interval among these chimpanzees was 4.6 years overall, but only 1.5 years when infant deaths within 3 years were excluded; about 72.7% of infants survived to age 4 and about 71.4% of those to ages 4–7 years (Sugiyama 1994). Inter-birth intervals for gathering/hunting women are more of the order of 2 years with 80% survival to age 4. Maximum life spans among the various ape species appear to cluster at about the same age at which, on average, women experience menopause. The human reproductive span and that of our nearest ape relatives are about the same length; what differs is our extended late-life survival (Hawkes *et al.* 1998).

An alternative suggestion for human menopause is that, as with chimpanzee life history, at some point in hominid evolution ovarian function and life span were more tightly age linked, and that, over time, as infants became more dependent, women who experienced relative POF had greater lifetime RS (Peccei 1995a, b; Leidy 1999). Finding that POF acts like a Mendelian trait provides some support for this model; however, evolutionarily, menopause also seems to have tracked both the evolution of dependent offspring and our increased life span (Peccei 1995a, b; Leidy 1999). That a Mendelian trait such as POF continues to segregate in human populations at polymorphic frequency suggests less than total selection against the early cessation of reproduction, and perhaps little differential fitness associated with the early cessation of female reproduction. Age at menopause (including at POF) shows wide variation (mid- 20s to 60 years), as is common for a quantitative trait with high heritability and little effect on differential fitness, similar to senescence and life span themselves. The two primary determinants of menopause, number of primary oocytes laid down in the fetus and the rate of atresia, also show significant degrees of genetic determination (Leidy 1999). They are both highly variable, and likely have little long-term effect on inclusive fitness. Furthermore, if menopause were locked in "a dynamic process" with increasing life span (Peccei 1995a, b), when and why did it stop? If it stopped because the limits of ovarian function were reached, this is just the fragility model and finite resources of oocytes remain the cause of menopause. The weight of available data on health declines, clinical findings, and mortality shows that menopause is a detrimental life stage associated with increasing loss of function and homeostasis compared with the previous high estrogen stage of female life history (Snowden *et al.* 1989; Austad 1991).

Menopause appears to be an indicator of ovarian aging that is closely associated with senescent processes affecting other systems and tissues that lead to an increase in the probability of death with increasing age (Snowden *et al.* 1989). Contrary to the suggestion that "postcycling survival depends upon slowing of aging in all somatic systems except reproductive" (Alverez 2000, p. 436; see also Pavelka and Fedigan 1991, 1998), Snowden *et al.* (1989) suggest that declining reproductive function is a senescent process closely associated with general somatic senescence, so much so that early menopause (before age 40) carries a 1.95 odds ratio for death within 6 years among Seventh-day Adventists compared with those experiencing menopause at the common age of 50 years. Thus, rather than being disconnected from somatic processes of senescence, ovarian senescence leading to menopause and a low estrogen state is itself a biomarker of senescence. The later natural menopause occurs, the longer a woman is likely to live.

Today, the median age at menopause is about 50 years (Leidy 1999). Menopause may still occur at earlier ages in more stressful environments such as at high altitude, in conditions of nutritional deprivation, and in areas where women continue to be either pregnant or lactating for most of their reproductive period. The latter condition likely characterized most women who ever survived to experience menopause (Lancaster and King 1985). Thus, at best, menopause is an allometric outcome of the intrinsic processes of oögenesis in a long-lived species that became long lived because of biocultural adaptations having no or little effect on lifetime fitness and maintenance of reproduction. As such, the ongoing tendency to link menopause to post-reproductive survival in evolutionary models seems inappropriate. They are likely to be unrelated biologically since culture produced post-reproductive survival in females of a species whose ovaries were already characterized by a finite functional period. Further, menopause appears to be no more than a marker of cessation of menstrual cycling, statistically unrelated to reproductive span, completed family size, or fitness (Wood *et al.* 1994; Leidy 1999). Most women actually cease reproduction well before they complete menopause, with many women in natural fertility settings entering menopause following the birth and lactation of their final child without ever entering a last menstrual cycle. Rather than post-reproductive survival, which implies a dependence between the cessation of reproduction and late-life survival, the less gender-specific term "late-life survival", applicable to both men and women, is more accurate and useful in studies of human life history.

Late-life survival

An extended period of post-menopausal (late-life) survival characterizes the life history of women in most societies (an apomorphic trait). If curtailment of reproduction during later life were associated with increased inclusive fitness (IF)[1], natural selection could produce a post-reproductive period (Hamilton 1966). However, it seems unlikely that curtailment of reproduction could ever lead to greater total fitness. It is more likely that, through reproductive benefits to an organism with an already finite reproductive period, late-life survival might increase IF. Unlike women, men may sire offspring in their sixth and later decades of life, although this ability is greatly reduced. Still, in many modern cosmopolitan societies, with adequate nutrition and health care, men tend to die at younger ages than women. It is puzzling that post-reproductive women tend to outlive men, who remain at least somewhat capable of producing offspring, even when both sexes have adequate nutrition and health care. One theory is that women generally invest more in each offspring than do men, and women who produced in late life may have failed to fledge (to rear an offspring until it is prepared to live without parental investment) their late-life offspring. The death of the Gombe chimpanzee Flint upon the death of his mother Flo, although he was well over the "normal age" for fledging, provides a clear example of how this may occur (Goodall 1986). Thus, the theory goes, late-life reproducing females may have had more offspring, but fewer of these offspring survived to themselves reproduce, thus curtailing any trend toward late-life reproduction. A more likely model, based on the general female mammalian tendency to show reduced fertility in the fourth and fifth decades of life, is that humans have always experienced an upper age limit to female reproduction, as do all mammals (the plesiomorphic condition). Thus, human menopause may not be the derived trait, but rather a continuation of what appears to be the rule for large-bodied (>50 kg) mammals.

---

[1] Inclusive fitness (IF) was defined by Hamilton 1966 as: "an individual's own reproductive success stripped of all those components due to the influence of its relatives, plus that proportion of relative's reproductive success that was due to the influence of the individual in question once it has been devalued by the coefficient of relationship between them."

Undoubtedly, once they occurred, post-reproductive females in conjunction with multiple offspring per reproductive female would enhance opportunities for grandmothers and extended investment of women in their kindred's inclusive fitness – the Grandmother Hypothesis (GMH) (Hill and Hurtado 1991; Hawkes *et al.* 1997, 1998; Alvaerez 2000; Peccei 2001a, b). Grandmothers know not only who their children are, but also who their maternal grandchildren and great-grandchildren are. Furthermore, at least to some degree, these elderly females are more closely related genetically to their matrilineally derived kin, who all carry the matriarch's X-chromosome, which boys sired by her sons do not (Katz and Armstrong 1994). This extra relationship due to the X-chromosome may provide a fairly strong genetic basis for both cross-cousin mating and matrilineal investment by females (Katz and Armstrong 1994). Women providing ongoing investment specifically to their own descendants may increase their reproductive value and produce positive selection pressures on any alleles that enhance post-reproductive/late-life survival (see Swedlund *et al.* 1976; Mayer 1991; Wood *et al.* 1994; O'Connor *et al.* 1998). Hawkes *et al.* (1998) suggest that: "Mother-child sharing was instrumental in developing the "long postmenopausal" life spans of humans and the cessation of female reproduction." However, cessation of female reproduction is common to mammals and needs no explanation, and food sharing cannot be a complete explanation for late-life survival in humans. First, provisioning is but one aspect of the total investment of mothers in offspring. Second, neither extended post-reproductive survival nor menopause has yet developed in other mother–child food sharing species (i.e., wild dogs, hyenas, numerous species of birds, or killer whales). Although the type and extent of mother–child food sharing is much greater in humankind, this behavior alone does not lead to enhanced survival of mammalian females. Furthermore, patterns of grandparental investment today do not reflect those occurring when late-life survival evolved.

To the degree that social organization allows these older generations to contribute to the well-being and reproductive success of carriers of their genes, positive kin selection for long life occurs. Today, larger proportions of grandparents, great-grandparents, and even great-great-grandparents have direct influences on the survival and reproductive success of their descendants than ever before. This is reminiscent of the GMH – that menopause evolved to increase opportunities for women to invest in their later-born offspring, grandchildren, and other matrilineal descendants. However, this would make grandmothers both the selective force and the trait produced by selection, i.e., grandmothers influenced their IF thereby leading to themselves. More likely, cultural evolution led to expanded life spans, which eventually surpassed the ages that the ovaries of women had ever possessed the capacity to continue producing fertilizable ova.

When few lived beyond four decades, insufficient numbers of grandparents would have been available to push selective pressures against short survival through nepotism and IF or loss of reproductive capacity. More likely, cultural developments provided the first opportunities for the late-life survival of women and men. Once cultural development provided settings wherein grandparents contributed to the RS of their offspring, relatives, and the local deme, the stage was set for selection against any alleles that failed to promote late-life and post-reproductive survival. In the GMH (Peccei 2001a; Leidy 1999; Leidy-Sievert 2001), grandmothers are both the cause for and effect of late-life survival. Culture was likely the prime mover, starting and maintaining pressure toward survival beyond that needed for RS. However, selective pressures likely were, and still are, against short life spans, rather than for longer-term survival. Only when culture developed to the point of providing opportunities for late-life survival would older individuals be available to invest in their kin. By enhancing their own IF, elders would place selective pressures on shorter life spans, and kindreds and demes segregating alleles associated with enhanced late-life survival could be more successful than those without.

To test predictions based on the GMH, Alvarez (2000) reviewed life history traits for 16 primates. Life history adjustments on a basic primate pattern could provide for slowed somatic senescence, while not favoring longer fertility among women (Alvarez 2000). If fertility declines with age in other apes occur at about the same age as in women, the question to be answered is not why reproduction declines to zero (a plesiomorphic trait), but why continued late-life somatic survival occurs (an apomorphic trait) (Hill and Hurtado 1991; Hawkes *et al.* 1998; Alvarez 2000). Among women, age at first birth is adjusted to the entire life span, birth rates are higher than expected based on age at maturity, and infants are weaned earlier and, perhaps, at lower weights than expected from the general primate model (Alvarez 2000). Although Alvarez's (2000) models support the GMH, they also assume there exist "postcycling females . . . senior women (who) benefit pregnant and lactating kin and their dependent offspring . . . (suggesting the) hypothesis that the reproductive benefit older women achieve through younger kin could have favored the accumulation of any small changes countering somatic senescence late in the life span" (Alvarez 2000, p. 444). However, this GM model (as are all GMH models) is premised on the existence of that which it seeks to explain, the existence of postcycling somatic survival among women. Similarly, their model " . . . assumes that the long juvenile period of humans is a consequence and not a cause of longevity. The reproductive benefits achieved by long-lived grandmothers through younger aged female kin favor delayed maturity." Two questions must be asked: where did the long-lived grandmothers come from to begin their investments that lead to delayed maturity of offspring and, if there already were postreproductive long-lived women who were successful, why would the juvenile period then need to be shortened? The GMH as stated is circular. Post-reproductive female survival must first exist for sufficient grandparents to be available to invest in their offspring's offspring to promote long-lived women who then invest in themselves. More parsimonious and less circular explanations for altricial newborns and dependent offspring are available. Resetting of growth rates to accommodate the pelvic alterations involved in bipedalism and to provide the necessary neocortical development and integration required of a species dependent on hand–eye coordination, group living, culture, and language, and the development of altricial offspring likely occurred long before late-life survival became a common phenomenon among humans.

A simpler beginning for selection pressures favoring late-life survival is suggested by monkeys and apes who may have greater reproductive rates when they aid fledging their offspring compared with conspecifics lacking helpers. Offspring of individuals receiving parental assistance may grow faster and be weaned at earlier ages, allowing females to reproduce more frequently (Lee 1998). Among early hominids and humans grandmothers were not likely to have been widely available to provide maternal assistance with offspring. Matrilineal kin, older and younger sisters, and aunts are not available in patrilineal societies. Only pre-reproductive sons and daughters might be. Others with similar degrees of genetic relatedness to offspring (paternal) as grandmothers probably were available. An older sister or aunt is more likely to be available to invest in a female's offspring than her own mother. Given that maternal relatives other than grandmothers were more likely to be available for helpers, the effect of grandmothers would be even more limited. Further, "helpers at the nest" do not necessarily improve infant survival or reproductive rates in non-human primates or other mammals (Caro *et al.* 1995; Crognier *et al.* 2001).

We may confidentially assert that the cultural forces/processes that shaped the human average and maximum life spans we observe today have never been experienced by any other hominoid species. Thus, it would seem unlikely that any ape species should share the survival advantage and a menopausal process similar to that of women. Finally, although the theory that the extra hands of grandmothers as "helpers at the nest" may improve the reproductive success of women and survival of their offspring has received some empirical support (see Mayer 1994; Hawkes *et al.* 1998), this may not be a general primate or mammalian characteristic. For example, among tamarin groups in Brazil, Dietz and Baker (1993) observed that the number of extra helpers was unrelated to female reproductive success. Similarly, among neither baboons nor lions does the presence of grandmothers correlate with reproductive success (Packer *et al.* 1998). However, Crognier *et al.* (2001) report data suggesting a possible association of "probable helpers" with higher fertility and child survival among Berber women of Morocco. Last, variable estrus patterns, seasonality/non-seasonality of copulation and reproduction, and the presence or lack of menopause all characterize various primate groups. For example, among savanna baboons (*Papio cynocephalus* subsp.) of Africa, non-seasonality of reproduction is a general characteristic, apparently because of extreme dietary diversity in these groups (Berkovitej and Harding 1993). These authors suggest that non-seasonality of reproduction should be more likely among species with wide dietary diversity rather than those with a restricted diet, characteristics of hominid diets. Among Golden Lion tamarins (*Leontopithecus rosalia*), avoidance of reproduction and weaning in the dry season apparently has structured post-partum estrus to allow two litters over a 7-month wet season (Dietz and Baker 1993). Obviously, environmental pressures have shaped multiple aspects of female primate life history and reproduction, but true menopause appears only to be observed in the most long-lived of primates.

Reproductive senescence in men is not as precipitous as the pattern observed in women. Some men apparently have sired children past their eighth decade of life, but, for most, reproductive success likely declines after the age of 50. Although spermatogenesis apparently continues into the eighth and later decades, sperm per unit volume of semen drops from 100–200 million per ml to 30–60 million or less per ml, while motile and well-formed sperm decline in frequency (Finch and Hayflick 1977). Among men aged 50–59, only 50% of their seminiferous tubules have spermatids, compared with 90% among 20–39-year-old men; after age 70, sperm production declines by about 10% per decade (Finch and Hayflick 1977). Testosterone production peaks during the third decade of life, in conjunction with attainment of MRP, and then declines gradually over the remainder of the life span. Some men maintain high levels throughout life. Such longitudinal patterns of change indicate that declining reproductive function is concomitant to growing older in men; however, andropause (a cessation of androgen production) may be a misnomer for this gradual decline. Whether or not declining testosterone is due to senescent changes in the testis has not been well investigated, but, given available information, this seems likely. The same processes that lead to reproductive senescence in women also may operate to some degree in men. For most of hominid and human evolution, the reproductive potential for men likely decreased slowly over the first decade or so following attainment of MRP, but then dropped steeply during the fourth decade of life as extrinsic risks increased the probability of death and few individuals survived into their fifth decade of life. This suggests that over most of hominid evolution, men have likely made their major genetic contributions to future generations during their third and fourth decades of life. However, they likely also had some degree of reproductive success at later ages, such that the ability to produce viable sperm (a small cost) is retained through later years. Compared with the cost of oogenesis, ovulation, and pregnancy, male investment in spermatogenesis is minuscule and of little long-term consequence compared with the costs of male–male competition for access to females. Little selective pressure to halt spermatogenesis should ever occur, since even one occasional late-life offspring would make the investment worthwhile for all men, given the wide variation in male RS.

By their nature, females invest more in producing and rearing offspring than males, while males invest more in maximizing access to mates (Dunbar 1988), including competition with other males. Most mammalian males have a re-stricted window of reproductive opportunity during their prime reproductive years due to male–male competition, generally higher morbidity and mortal-ity, and female choice, whereas most mammalian females reproduce early and throughout their life span. This is particularly true in group-living species such as lions, gorillas, baboons, chimpanzees, whales, and humans. Therefore, se-lection pressures on females are likely to allow them to maintain their somas longer than males who not only show higher mortality rates (while obtaining mates) but who also usually do not care for offspring, given that parental in-vestment (the cost of producing, caring for, and fledging offspring) for males often ends with insemination. In primate groups, some males adopt a strategy of "opportunistic mating" (e.g., baboons, macaques) (Dunbar 1988), while in others (e.g., gibbons, marmosets) males invest heavily in offspring care and have RS equal to females. It is possible that males who rely on opportunistic matings experience less adult mortality due to reduced male–male competition, while those who invest considerable care in offspring may experience selection pressures similar to females for somatic maintenance. Consequently, males us-ing these reproductive strategies may experience longer reproductive periods and life spans than their conspecifics.

Another general finding is that rates of senescence are higher in organisms with higher fertility rates and lower in those with lower fertility rates (Hamilton 1966; Rose 1991), illustrating that life history characteristics such as reproduc-tion, longevity, and patterns of senescence are intimately related. Experimental data from roundworms, fruitflies, rodents, and mosquitoes bred for extended life spans document strong associations between reproductive output and so-matic longevity (Rose 1991). In general, extensions of average and maximum life spans in animal models are associated with reduced and/or delayed fertility, while the best way to develop long-lived animal strains is to select only those in-dividuals capable of reproducing at ages above the average age of reproduction in the baseline population (see Rose 1991). Life span extension in rodents by dietary restriction is also associated with delayed reproductive maturation and late fertility, suggesting that caloric restriction may enhance longevity through its influence on growth, development, and maturation (reviewed in Chapter 5). Such results suggest evolutionary tradeoffs between early rapid reproduction and continued survival, as predicted by models such as a disposable soma and antagonistic pleiotropy. Given different reproductive strategies among men (higher possible total fertility, low investment in any one offspring) and women (low lifetime fertility, high investment in each offspring), men may be more at risk and have higher rates of senescence secondary to their more opportunistic reproductive strategies and higher potential fertility. However, any change in life history characteristics that extended life span after the age of MRP and menopause for women could have a greater effect on the RS of men given their greater variation in RS. If male survival had positive consequences on late-life reproduction and men's total fitness, this could place strong selective pressure against brevity of life in men and push late-life survival. Possible contributions of late-life RS among men have been overlooked for their influences on fitness in almost all models of human life span.

Biocultural factors

Culture is the hallmark and mainstay of human adaptability. The homogenizing effect of culture on both humans and their environments has produced at least three major trends that partly underlie patterns of senescence, life span, and longevity in modern humans. First, over the past several million years, natural selection on humans has occurred within culturally created ecological niches that provided a relatively constant microenvironment. Because of this cultural microclimate, biological alterations may have been slowed and it is likely that substantial genetic variation for functions and proteins currently non-essential is retained in our gene pool. Numerous traits or factors easily altered by cultural behaviors may now be non-essential but retain underlying genetic variability. Variations in genetic propensities for body morphology and the close association of body shape and size with climate may reflect some such factors. Quantitative phenotypic characteristics with high heritability (defined in the additive sense: variation due to genes/total variation) – for example, height, weight, skin pigmentation, IQ as measured by standard tests, cholesterolemia, BP, glycemia – may not have been strongly influenced by natural selection toward an ideal value because they generally have been non-essential for RS or are compatible with RS across a broad range of variation. This is likely to be the reverse of selective pressures maintaining – for example, blood pH, body temperature, or minimum heart rate – which must be fairly constant across all individuals and are likely to be subject to strong stabilizing selection.

If during most of hominid evolution, few individuals survived much beyond the average age at death, longevity would not influence greatly the RS and IF of individuals. Given that longevity and RS were not associated over most of hominid evolution, there should exist a relative abundance of additive genetic variance (high heritability) for both longevity and nongevity ("nongevity", the opposite of longevity, dying at an earlier age than the original (wild) type, less than the average life span). Conversely, if longevity greatly improves the RS or IF of individuals, most people should live about the same length of time and heritability of longevity should be relatively low. In the first case, longevity should be higher in families wherein grandparents and earlier generations were more long lived and lower in those with short-lived ancestors (high heritability) and the overall correlation of relatives longevity lower than in the latter case. If the latter, correlation of longevity of grandparents and parents with their descendants should be very high due to low heritability (little remaining additive genetic variance) and little variance in longevity should be observed. Data analyses have led to both outcomes being reported for different samples. Longevity of parents and offspring and of grandparents and grandchildren have been believed to be only poorly to moderately, but still significantly, correlated (0.10–0.40) (Finch and Tanzi 1997). However, recent data suggest a larger maternal genetic effect on life span (heritabilities: mother/daughter 0.66, mother/son 0.84) compared with paternal (father/daughter 0.44, father/son 0.48) (Korpelainen 2000). Korpelainen's (1998) heritability estimates outrange previous estimates because they are based solely on offspring who survived to age 40. Both processes may be at work and in some evolutionarily stable strategy (ESS) among humans. Heritability of longevity is low to moderate, suggesting that longevity is related to total fitness (RS + IF) in humans and that there are additive factors segregating within kindreds and genes that promote longevity.

Secondly, humans, although living in a "culturally created" ecological niche, have remained biological generalists. For humans, culture, biology, and environment have interacted so as to allow human organisms to experience the same basic microenvironment while residing in a variety of macroenvironments. Humankind's propensity to convert a bewildering variety of plants, animals, bacteria, and fungi to energy and offspring is a hallmark of our unique type of generalist strategy. This bioculturally determined propensity to process and eat anything may have influenced this hominid's need for an expanded neocortex, neural network, and language skills. Our neural processes require high nutritional, gestational, and developmental inputs. As neurological structures became important for survival as a mobile generalist, birthing of physically poorly developed fetuses requiring longer periods of growth, development, and maturation outside the womb became the norm. This lengthened infancy and extended the juvenile growth period through the first, and eventually the second decade, of life likely precipitated the lengthening of all later life stages and the maintenance of RC and the soma into late life (not the reverse as suggested by Alvarez 2000). This outcome has only now at the beginning of the twenty-first century been realized for any large proportion of humankind.

A third biocultural trend, obvious among modern human populations, is that today natural selection apparently acts more by reducing reproductive success after MRP than by influencing survival to reproductive age in many populations. That is, there has been a trend toward improved survival of infants and children, at least among the elite in most societies in recent millennia, and for these same individuals to show better survival overall. To the degree that length of survival increases RS, positive selection for extended life span occurs and the process becomes self-reinforcing as more individuals with a genetic propensity for longevity leave more offspring and increase their total fitness.

In support of these ideas are findings that, for every organism examined, "non-essential" genetic variation related to individual and general population longevity has been reported. In the laboratory, almost any organism can be placed in an environment that significantly increases population average and maximum life span, reduces early life fecundity, and/or extends fecundity to older ages. Over several generations, these organisms can be selected so as to have an average life span that approaches, and may exceed, the maximum life span of the wild type. Maximum life spans may increase to twice those be-fore such "artificial" selection (selection of types based on human intervention to promote some phenotypic aspects rather than reproductive success/fitness). However, this "life span advantage" is not seen in the wild, presumably be-cause such genotypes are outcompeted by less long-living genotypes carrying other alleles that provide a reproductive advantage over these "longevity alle-les" in the natural environment. Heritable variation for longer life is available to wild-type *Drosophila*, nematodes, rodents, and primates. In natural settings, where these organisms have adapted over evolutionary time, these "longevity genes" represent "non-essential" genetic variation – "non-essential" because, in the wild, by the time this variation is expressed, survival and reproductive competition at younger ages have already determined which organisms pass the most genes on to the next generation. In these species, alleles promoting extended life are mainly under the influence of random genetic processes. Their frequencies change due to linkage with other alleles influencing early life sur-vival/reproductive success, or linkage to those that show neutrality with respect to natural selection because selection for extended life span currently is not a viable (i.e., stable) evolutionary strategy for the population.

Among humans, and perhaps other more K-selected species, less "non-essential" variation for longer life may remain because some such variation may be "essential" for the extended life spans already enjoyed by these species.

That is, what may once have been "non-essential" life span-enhancing variants for our hominoid and early hominid ancestors may, over the past 2 million years or so, have been sculpted by natural selection to now be essential for the 100-plus year life spans of modern humans. In this case, culture may have been to humans what the laboratory environment has been to fruitflies and rodents – a protected environment in which those genetically endowed with a longer life have had the opportunity to successfully reproduce and outcompete those destined to have shorter lives. Among humans, alleles contributing to the extension of life may already to some extent have reaped the benefits of selection against alleles contributing to early mortality through cultural interventions to prevent infant, child, maternal, and adult mortality. Culturally molded environments capable of sustaining large numbers of individuals and cultural factors that have allowed grandparents to survive to invest in their descendants may also have increased the representation of life span-enhancing variants. Such life span-enhancing allelic variants may already be the norm (wild type) at multiple human loci.

As humankind developed culture, their biology remained that of a generalist: a bipedal omnivore capable of eating roots, fruits, grains, leaves, nuts, insects, carrion, and just about any other animal. With the continued development of culture from the earliest hominids through modern *H. sapiens* came a rapid increase in encephalization and development of the neocortex. This was in part accommodated by an increased head size of neonates and reduced development of precocial traits, rather than by altering the length of gestation. Thus, over the last 2 million years or so, human intrauterine development has been rescaled such that cranial capacity increases more rapidly, limb dexterity and grip less rapidly, and the degree of progress toward sexual maturity at birth is lower than that of either of our two closest relatives (chimpanzees and bonobos). It is clear that this resetting of growth and development during fetal life, infancy, and childhood has profoundly influenced our species life history and created a pedomorphic species; how this resetting affected longevity and life span processes is less appreciated or understood.

In more recent centuries, culture has in part ameliorated the actions of natural selection. Today, rather than survival to reproductive maturity being the major factor influencing RS, for many populations it is post-maturity reproduction that most influences differential RS in more cosmopolitan settings, since the vast majority of individuals born now survive to the age of reproduction. Although most individuals in more cosmopolitan settings survive to reproductive adulthood, not all reproduce nor are all equally capable of reproducing. In fact, as many as 10–15% of adults may be infertile, while another 10–15% never have any offspring for various reasons. In fact, some estimates suggest that as few as 50% of individuals in any one generation may contribute genetic material to the next generation. It is also possible that, particularly for men (and males of other species also), length of life may influence opportunities to reproduce such that those who survive longer have more opportunities. If longer-lived men enjoy greater RS, any genetic propensity toward longer life would be positively affected by natural selection, while any alleles reducing life span would be selected against. It is also likely that longer-surviving parents and grandparents may have greater opportunities to "invest" in the care, well-being, and/or survival of their progeny, their offspring's progeny, and their kindred than do shorter-term survivors. Positive selection for longevity will occur to the extent that such longevity is genetically inherited and such investment increases RS and the inclusive fitness of individuals carrying such "longevity" alleles. To the extent that any alleles conferring enhanced longevity on its carrier allowed investment in its "identical by descent" counterparts in related individuals, the kindred would increase its representation of "longevity" alleles. This process would increase the local population frequency of these alleles and ultimately lead to their wider distribution throughout the species (see Lasker and Crews 1996 for a model).

Chapter synopsis

Longevity in modern humans is largely secondary to evolutionary trends that produced brainier, but less completely developed, more dependent, and slower developing offspring than did our ancestors or do our modern relatives. Culture is the hallmark and mainstay of human adaptability. Still, factors affecting rates, timing, and allocation of scarce resources during intrauterine development are found to influence longevity and health among today's humans, strongly supporting suggestions that human patterns of slow fetal and infant development have shaped our species longevity by requiring the extension of both early and later-life stages. This extension of life stages is similar to what is seen among calorie-restricted rodents and non-human primates, *Drosophila* selected for longevity, and species maintained in hypothermic settings. The main difference is that humankind achieved these endpoints through their unique biocultural adaptations to a wide variety of environments. These produced unique evolutionary pressures on fetal, infant, and child growth and development, delayed intrauterine development, and decreased maturity of neonates, for whom the first year of life closely resembles late fetal development. This coupled with slow growth allows time for expansion of the neocortex with some structures of the cranial bones not closing until the twenties. Eventually, reproductive success came to depend on the elaboration of culture, communication, and language.

The homogenizing effect of culture on both humans and their environments has produced at least three major trends that partly underlie patterns of senescence, life span, and longevity in modern humans. First, over the past several million years, natural selection on humans has occurred within culturally created ecological niches that provided a relatively constant microenvironment. Because of this cultural microenvironment, biological alterations may have been slowed and it is likely that substantial genetic variation for functions and proteins currently non-essential is retained in our gene pool. Numerous traits easily altered by cultural behaviors may now be non-essential but retain underlying genetic variability. Variations in genetic propensities for body morphology and the close association of body shape and size with climate may reflect genetic factors. Quantitative phenotypic characteristics with high heritability (defined in the additive sense: variation due to genes/total variation) – for example, height, weight, skin pigmentation, IQ as measured by standard tests, cholesterolemia, BP, glycemia – generally are compatible with RS across a broad range and are not as strongly influenced by natural selection toward an ideal value. This is likely the reverse of, for example, selective pressures maintaining blood pH, body temperature, or minimum heart rate, which must be fairly constant across all individuals and which are likely to be subject to strong stabilizing selection.

Today, rather than an expectation of life of about 20 years, characteristic of most of hominid/human evolution, we can expect to live around 75 years and hope for 100. Rather than having a high mortality among infants and children aged 5 years and less, today almost 99% of children born in cosmopolitan settings survive to their first, and 95% to their fifth, birthday. Over recent centuries, increased human life expectancy resulted almost solely from biocultural and behavioral developments (e.g., nutrition, public health, hygiene, antibiotics), rather than from new biological or genetic adaptations. The most rapid increases in life expectancy followed the industrial revolution and were secondary to the development of experimental science, public hygiene, and health care. These processes led to life expectancies of 45 years in the nineteenth century, 55 years in the early twentieth century, and over 80 years in selected segments of contemporary societies in the twenty-first century.

Among humans and most other mammals, attainment of MRP follows closely the end of growth. This is a period of maximum RC and physical fitness, but it is also characterized by a general decline in mitotic activity, loss of stimulus from growth hormone and related compounds, and increased loss of cells. Over time, such alterations/losses along with internally and externally generated stress lead to the loss of cells and RC of organs and physiological systems. These alterations represent senescent changes in the degree to which they lead to or allow chronic and progressive loss of function or disease. Increased life expectancy has exposed multiple innate susceptibilities to a variety of chronic debilitating diseases among long-lived humans. This argues strongly that biological processes have not yet caught up with our biocultural alterations that have extended our life. Many of these innate susceptibilities to disease appear to be a legacy of our evolutionary trajectory, reflecting adaptations to earlier selective pressures. Some such susceptibilities may be traceable to our phylogenetic background as mobile, omnivorous, arboreal primates (plesiomorphies). Others reflect our 5–6 My of evolution as bipedal hominids (sympamorphies), still others reflect more recently developed dependencies on culturally elaborated social systems for survival and reproduction (apomorphies). Although many physiological functions decline with age, they do not do so in a constant or consistent manner (Gavrilov and Gavrilova 1991; Finch 1994). Rather, some physiological functions may improve with age (Gavrilov and Gavrilova 1991; Mayer 1994), while failing functional reserves may be improved at older ages secondary to alterations in activity and dietary patterns. Mortality at 85 years and above is due to multiple insults (Susser *et al.* 1985; Crews 1990b); and, in support of the RC model developed here, loss of organ reserve is not clearly responsible for any of these insults. After 85, survivors are all "high-flyers", like Person 3 in Figure 3.2. their organ reserves have always been above those of most others. Thus, their group average at and after age 85 years may still remain above that of the average of all individuals at younger ages (age 65). Among the oldest-old (ages 85 plus), only the best endowed remain as heterogeneity in survival becomes ever more dependent on biological variation. Those with susceptibilities to CDDs and infections/parasitic disease have already succumbed. For these nonagenarians, the entire system is declining in concert and slowly from its peak function. Mortality takes on a pattern similar to water glasses in a restaurant at about this age. Most mortality before about age 85 may be due to inherent aspects of human physiology and specific genetic propensities, while thereafter it is the multitude of senescent processes and physiological resources that determine mortality and survival.

In today's human societies, many survive sufficiently long to be grandparents. As with most species, reproduction and late-life survival are closely linked but co-varying life history parameters. Unlike any other species, humankind culturally manufactures its environment, providing opportunities for individuals to express their innate vitality as late-life survival. It is likely that some of the previously random variation for life span-enhancing alleles is, today, necessary to provide our current extended life spans and may be more or less fixed in human populations. Unfortunately, most remaining genetic variation for differential late-life survival is probably non-essential; that is, these alleles probably act at ages beyond which natural selection has a strong effect on RS. Most remaining (non-essential) genetic propensities for longevity and non-gevity, although originally arising in particular kindreds, are today likely randomly distributed throughout the species. Culture has provided new opportunities for long life and late-life survival for humans, the only known cultural animal. However, we do not all share equally in these genetic predispositions nor are all humans born into environmental and cultural settings which enhance one's probability for long life.

4   Human variation: chronic diseases, risk factors, and senescence

Humans interface with their physical environment through a complex series of biocultural mechanisms. This range of interaction allows a wide variation in phenotypes. Humans use culture so extensively that inclinations for complex symbolic communication and patterns of social interaction may be genetically programmed as are the physical and neurological structures on which these activities depend (Durham 1991). Inter-individual communication allowing experiences to be retained across generations provides opportunities to inhabit a variety of ecological settings. Therein, local populations may develop both biological and social/cultural/behavioral variability while responding to environmental and sociocultural circumstances. Such local adaptive responses allow populations to differ significantly in phenotype from others residing elsewhere (Crews and Bindon 1991). All populations tend to maintain relatively constant living environments that provide human needs for food, shelter, reproduction, and infant/child care, and promote survival of their members. This similarity across multiple external environments allowed humans to maintain a 99.5% genetic similarity across their range. Although cultural systems alter local populations by restricting mate choice and concepts of ideal mates, thereby producing socioculturally determined selection, wide-ranging human populations remain a single interbreeding species, unlike many wide-ranging non-human groups (Lasker and Crews 1996). Humans all share the same physiology and basic life history, and remain susceptible to the same internal processes of senescence. With larger numbers of individuals surviving to their later decades of life, this similarity is revealed as a senescing soma that loses function across multiple integrated systems associated with chronic degenerative processes. Local selective pressures (e.g., malaria, cold, hypoxia, solar radiation, plant toxins, viruses, bacteria) produce variations in behavioral, sociocultural, and biological responses across populations, but the underlying degenerative processes remain the same.

Chronic degenerative conditions

Background

For those aged 65 and older, heart disease (40%), cancer (21%), and stroke (8%) account for 69% of all mortality (Rakowski and Pearlman 1995). Chronic morbid conditions – arthritis, hypertension, hyperglycemia, hyperlipidemia, and sensory impairments, along with tissue and bone loss – also are common. At the same time, organic brain disorders may affect 20% of those over 65 and 50% of those over age 85 (Rakowski and Pearlman 1995). These chronic degenerative conditions (CDCs) result from interactions among multiple genes, environments, and cultures. Differences in diet and nutrition across individuals and populations illustrate such interactions. What one eats depends on both cultural choices and environmental limitations. For many in cosmopolitan populations, culture determines diet. For those living in marginal settings with limited resources, diets are restricted to locally available items. The nutritional and energy content of even the same food varies widely from region to region. Because nutrient content and availability vary, one may ingest any range from adequate calories with inadequate nutrients, through to adequate nutrients and low calories, to too little, or too much, of both. Once ingested, multiple digestive enzymes, cell receptors, intracellular cytokines, transcription factors, and transporter proteins with inherent and environmentally determined variability allow individuals to extract differing amounts of calories and nutrients from the same diet. Substances in some foods inhibit or enhance the absorption of nutrients from others. Inducible enzymes further alter nutrient extraction, and metabolic processes are upregulated or downregulated by gene, environment, and culture interactions. In some, the result may be hyperglycemia, while, in others, the result may be hyperlipidemia or hypertension.

This mix of predisposing factors and individual variation in metabolic responses makes CDCs difficult to understand and control. Continuing with dietary influences, overnutrition, particularly fat intake, may lead to hyperlipidemia. This may be ameliorated or exacerbated by apolipoprotein phenotype and/or obesity. Insulin and leptin levels, along with culturally determined food choices and ideal body types, also moderate obesity. Hyperlipidemia and obesity act synergistically to produce hyperglycemia, atherosclerosis, and high blood pressure. These in turn are modulated by multiple alleles. Obesity, hyperlipidemia, hyperglycemia, and elevated blood pressure are independent risk factors for heart disease and stroke. A similar model applies to cancer, where ingestion of carcinogens and anti-carcinogens in foods is related to cultural activities and environmental sources. Higher proportions of saturated fats in the diet are associated with a variety of cancers. Similarly, a variety of food preparation methods (e.g., curing/salting with nitrates, smoking, barbecuing) create carcinogens in and on foods, while other dietary components (e.g., fiber, vitamins, antioxidants and their precursors) reduce the risk of cancer. Such effects are ameliorated or enhanced depending on whether individuals carry resistance or susceptibility alleles.

Human biological variation related to CDCs has evolved over 5–6 million years, a span of time that has included a variety of prevailing global, continental, and local environmental pressures and the spread of humans across the globe. Obviously, human external physical features and allele frequencies observed across geographical regions have responded to local and regional environments. Prevailing cultural competencies, beliefs, and restrictions have also structured human variation and, to a great degree, limited biological responses to multiple environmental pressures. Human variability, established over millions of years of evolutionary time, may be little related to the current ecological circumstances of contemporary populations. Over time, hominids/humans elaborated culture from the earliest "hand-ax" to the latest robotics, in addition to physical and genetic variation, in efforts to maintain their somas and reproduce their germ lines. During this time, culture has become as much of an "environmental pressure" as were those pressures that culture was developed to eliminate. Early in human evolution, "culturally elaborated" risks likely included exposures to parasites in food and water, restrictions on mate selection, religious ideologies, social stratification, warfare and other types of organized violence, limited diets, and, eventually, agriculture and village life leading to our current industrial cosmopolitan societies. Examples of modern "culturally elaborated" stressors abound – e.g., antibiotic-resistant types of tuberculosis, *Staphylococcus*, and *Streptococcus*, mutated anthrax and plague bacilli, pollution, global climate change, habitat destruction, and overpopulation, over-crowding, and malnutrition. Despite multiple cultural elaborations (apomorphic traits), humankind's biology remains that of a moderate-sized, polygamous, omnivore (plesiomorphic traits). Today, bipedal humans are capable of creating a wide variety of material culture and technologies with their free upper limbs and expanded neocortex (apomorphic traits).

Description

The majority of late-life diseases fall under the rubric of CDCs. These include all processes leading to progressive deterioration as measured on a continuous scale in clinically, metabolically, and physiologically important traits (Crews and James 1991; Crews and Gerber 1994; Gerber and Crews 1999). Any condition or syndrome that afflicts an individual over an extended period of time while simultaneously decreasing physiological function is a CDC (Crews and Gerber 1994). Neither age at onset nor severity of the condition precludes this definition. CDCs are generally diagnosed when some "risk factor" (e.g., serum glucose or cholesterol, bone density, blood pressure, T cells, hematocrit) crosses some clinically predetermined boundary. Boundaries for CDCs are often based on epidemiological and clinical data showing differential morbidity and/or mortality among those with elevated or depressed values for some key risk factor.

Cut-points for CDCs change as diagnostic expertise, population trends, or methods of measurement and clinical evaluation techniques improve (Crews and Gerber 1994; Gerber and Crews 1999). Serum cholesterol, glucose, and blood pressure levels promoted for U.S. residents provide examples of such changing criteria. In the 1960s and 1970s, 240 mg/dl was the average total serum cholesterol (total-c) level for adults in the U.S.A. Consequently, a level of ⩽240 mg/dl was seen as desirable and the goal that those at higher levels should strive to attain. As population average serum cholesterol declined, additional research showed high coronary heart disease (CHD) mortality even at serum cholesterol levels below 240 mg/dl. Consequently, the desirable level was set at <200 mg/dl (American Heart Association 1988). Today, with simple techniques available to determine lipoprotein subtypes along with total-c, all three cholesterol fractions (total-c, low-density lipoprotein (LDL) and the high-density lipoprotein (HDL)) are now monitored. The new desirable levels are less than 180 mg/dl, less than 120 mg/dl, and 35 mg/dl or greater, respectively (National Cholesterol Education Program Expert Panel 1993; Gerber and Crews 1999). Boundaries for diagnosing clinical hypertension have also declined, from 160 mmHg for systolic and 95 mmHg for diastolic hypertension (Rowland and Roberts 1982) to 140 mmHg and 90 mmHg (Joint National Committee V 1993), over about the same period. Similar changes have occurred in diagnosing diabetes as research has provided a better understanding of the detrimental effects of hyperglycemia. Rather than needing two measures above 200 mg/dl following a glucose challenge of 75 g (one at 1 -hour, the other at 2-hours' post-load), a single fasting measure above 140 mg/dl is sufficient (WHO 2002). Elevated serum cholesterol, glucose, and blood pressure are established risk factors for cardiovascular diseases (CVDs) (including all cerebrovascular diseases, hypertension, heart disease, and circulatory conditions). CVDs are insidious in onset, and often are not recognized until an ischemic attack. Defining disease as cut-point on a continuous risk factor distribution suggests that extremes of normal physiological function represent disease processes. Defined thusly, the distinction between normal and disease is not as clear cut as it is for influenza or tuberculosis.

Many other CDCs, including most neoplasms, slow virus infections, human immunodeficiency viruses, and autoimmune diseases, are only diagnosable once they have progressed to clinically recognizable pathology (Crews and James 1991; Crews and Gerber 1994). These conditions are not dependent on progressive or continuous risk factors in the same sense as are CVDs. Rather, they have a specific initiator at some definite point in the past, perhaps due to a somatic mutation event, exposure to a vector, or an inherited DNA variation, and may progress in subclinical fashion over much of a person's life span. Such CDCs generally result from complex interactions between genotypes, environments, and culture that begin as early as the conceptions of parents and grandparents, and then progress and change over the entire individual life span (see Barker 1998; Finch and Seeman 1999). Among the most important long-term predictors of CDCs may be how the developing fetus responds to its intrauterine environment. Somatic responses to the environment during intrauterine and early post-natal growth may physiologically program humans for increased risk of adult-onset neoplasms, CVDs and cerebral diseases, various neurological syndromes (e.g., Alzheimer's and Huntington's diseases), hypertension, hyperglycemia, hyperlipidemia, and diabetes mellitus type II. Osteoporosis, rheumatoid arthritis, multiple sclerosis, many of the autoimmune disorders, and a variety of neurological problems (e.g., amyotrophic lateral sclerosis) also may show association with early life and post-conception events. Such phenotypic plasticity in responding to prevailing conditions was likely to have been a major advantage to hominids experiencing relatively constant environments, preparing newborns for survival in a harsher or milder setting. In rapidly changing environments, fetal and infant developmental plasticity may be short circuited and lead to detrimental late-life outcomes.

Evolutionary perspectives

Based on available skeletal data, CDCs probably were rare during most of human evolution. This was largely because most people died at relatively young ages (under 40) (Loth and Yaşçan 1994). Since the Paleolithic, particularly in recent millennia and centuries, new life styles and longer life expectancies have placed more people at risk for living sufficiently long to develop CDCs (Cohen 1989; Crews and Gerber 1994; Gerber and Crews 1999). Obviously, human culture has had greater influences on the expectation of life during this period than have genetic alterations. Genetic factors were likely to have been more influential during earlier epochs of human evolution. Human physiology remains attuned to Paleolithic and Neolithic life styles, diets, and hazards. Modern life styles with refined, high-calorie foods, sedentary activities and entertainment, thousands of persons per km^2, and novel infectious organisms create new evolutionary pressures that play out against this Neolithic heritage (Coon 1965; Eaton and Konner 1985; Eaton *et al.* 1988; Armelagos 1991; Crews 1993a,b; Crews and Gerber 1994; Marks 1994; Gerber and Crews 1999; Molnar 2000).

Most of our current late-life CDCs are expressed during phases of our life history that seldom were attained during earlier epochs of hominid/human evolution. This led Weiss (1989a) to question if known CDCs influenced evolution of the human life span. Currently, predispositions to many CDCs do not manifest completely until well past life expectancy estimates for earlier phases of human evolution. Over most of hominid evolution, they would have exerted little, if any, influence on either individual or inclusive fitness (IF). Therefore, current CDCs have had little influence on the evolution of human senescence, and no selection to postpone their deleterious effects would have occurred during hominid/human evolution. However, once propensities to CDCs reduced the IF of those carrying them, selection against such alleles would be quite effective. CDCs that most curtail life span today commonly affect individuals in their sixth and later decades of life, well after the bulk of reproductive effort is complete for most. Few hominids or humans ever lived so long before recent millennia. Until recent generations, alleles predisposing to CDCs had little or no impact on fertility or IF. CDC-predisposing alleles were generally passed on to succeeding generations. Given 3 million new mutants per locus over the entire human genome during hominid evolution (Weiss 1989a), alleles with late-acting detrimental effects have had ample opportunity to accumulate. The allele for Huntington's disease (an autosomal, dominantly inherited, progressive neurological condition caused by expansion of a CAG trinucleotide repeat on chromosome 4 to more than 35 copies) provides an example. This disorder commonly strikes individuals in their mid- to late-forties, but in succeeding generations tends to occur at earlier ages as repeat length increases. When only a few individuals per generation lived to 40, this condition was passed on generation after generation with the disease seldom observed, even as the allele attained fairly high frequencies in various local demes.

Risk factors

When some aspect of human physiology, morphology, genetics, culture, or environment co-varies with risk for a particular disease or syndrome, it is termed a risk factor. As with phenotypes such as pigmentation and body proportions, many physiological "risk factors" (e.g., body habitus, alleles, blood pressure, glycemia, diet, pathogens, cholesterolemia, heart rate) vary widely between individuals and within populations. Risk factors may also vary within any individual from second to second, hour to hour, or over longer time periods. As an example, average blood pressure varies across population, ethnicity, sex, age, time of day, and place of measurement. Additionally, an individual's blood pressure responds to environmental and psychosocial inputs in a variety of fashions (rising at work for some, falling for others, or falling at home for some, rising for others) and may vary widely from minute to minute (James and Bindon 1991). Many aspects of physiology are highly variable and environmentally labile, providing individuals with flexibility in response to changing environmental circumstances. Other physiological measures (e.g., pH, core temperature, oxygenation of the blood) also vary, although less widely and even small proportional alterations in these may lead to loss of function and death. Alterations in pH or core temperature in response to environmental perturbations indicate a systemic loss of homeostasis. These are not risk factors; rather, they measure the soma's inability to maintain necessary capacity and its decline toward total system failure. Risk factors increase susceptibility to disease. Disease leads to loss of function and ability to maintain physiological set-points for such measures as pH, core temperature, and oxygenation of blood.

"Risk factors" include an endless variety of genetic, environmental, behavioral, and cultural factors. Exposures to silica, asbestos, cigarette smoking, unprotected sex, ingestion of soy sauce, alcohol, saturated fats, and salted and smoked foods may all increase individual susceptibilities to CDCs. Risk factors vary widely and may be measured as continuous (age, blood pressure) or discontinuous (allele, genotype, sex). Both retrospective and prospective studies are used to determine risk factors. Those with specific conditions or who die early are considered cases, while those without the condition or who survive are controls (Sing *et al.* 1985). Physiological, genetic, or morphological differences between the cases and controls are viewed as risk factors for the specific condition under study.

Multiple aspects of environmental, genetic, physiological, and sociobehavioral variation are risk factors or modulate various risk factors for CDCs. These act in concert, at odds, synergistically, and multifactorially to produce variation within and between individuals and populations in all somatic systems. Clinically, it is often only after overt disease occurs that malfunction is sufficiently severe to recognize a CDC. Hyperlipidemia, hypertension, hyperinsulinemia, obesity, nitrates, tobacco smoke, chemical exposures, fat-rich/fiber-poor diets, alcohol consumption, lack of physical activity, and overnutrition are frequently numbered among the major risk factors for CDCs. In representative samples, they are associated with greater risks for CHD, CVD, cancers, and/or diabetes. Such alleged risk factors are expressed most fully in today's more "cosmopolitan" societies, among whom environmental circumstances are more determined by social and cultural priorities than actual physical environment. Life-long development of risk factors and subsequent CDCs in part reflects the multiple processes underlying senescence. These include all the senescent processes already discussed. As with senescence itself, study of risk factors and their progression over the life span cross-cuts all areas of gerontology, geriatrics, human biology, and anthropology. Anthropologists who take cross-cultural and cross-population approaches to risk factor progression have shown that multiple aspects of the social milieu (e.g., cultural conformity, life style variation, social stress) alter physiological risk factors (Dressler *et al.* 1982; Bindon *et al.* 1992; Bindon *et al.* 1997; Dressler *et al.* 1998).

Firmly lodged in cosmopolitan culture is the notion that risk factor modification is beneficial – beneficence that will be measurable as a reduced risk of CDCs and a longer life span (Paffenbarger *et al.* 1986; Muldoon *et al.* 1990). In the arena of public health, however, debate continues about the extent to which risk factor modifications improve either quality of life or life expectancy (Oliver 1992). Even in studies designed to reduce serum cholesterol (by diet and/or pharmacological means), reductions in mortality from coronary diseases were offset by deaths from other causes, while mortality from all causes remained constant or actually increased (Muldoon *et al.* 1990). Prospective studies suggest that some physiological tradeoff occurs whereby lives saved from coronary events, for example by lowering cholesterol, are still lost through increased mortality from cancer and exogenous causes of death (including suicide, accidents, and homicide) (Muldoon *et al.* 1990). This should be expected if mortality risk is actually proportional to systemic somatic breakdown, such as that which is captured by the concepts of frailty, allostatic load, or wear-and-tear, rather than single independent risk factors.

Additional research suggests that, regardless of populations and/or cultural setting, physically active individuals have lower risk factor levels (e.g., cholesterol, blood pressure, serum lipids and glucose, obesity) than their sedentary counterparts. Still, when compared within any one population, the most physically active show only 1–2 years' higher life expectancy than the most sedentary (Paffenbarger *et al.* 1986). Multiple prospective studies of risk factor reduction and case/control series show that "improved risk factors profiles have neither a synergistic nor an additive effect as predicted by the risk factor model, but instead follow an apparent law of diminishing returns" (Olshansky and Carnes 1994, p. 69; see also Oliver 1992). Reductions in specific risk factors do not occur in isolation. They are only a single physiological alteration in a heterogeneous background of genetically, socioculturally, and environmentally influenced susceptibilities to, and protections from, CDCs. Genetic propensities to disease may be unaffected by or respond easily to changes in life style. Some may respond well to drug regimens, while others may not. Specific behaviors/lifestyles may enhance some genetic predispositions, while reducing others. Cigarette smoking greatly increases risks for lung cancer. However, only those unlucky enough to both carry biological predispositions to lung cancer and smoke may suffer such consequences. This may also apply to high blood pressure (hypertension (HT)) and salt ingestion. It appears that only those with predispositions to salt-sensitive HT may respond to high-sodium diets with elevated blood pressure. Conversely, those who maintain low-calorie, high-nutrient diets may retain physiological function (reserve capacity (RC)) and maintain low risk factor levels (e.g., Seventh-Day Adventists), leading to fewer CDCs and longer life.

In addition, many risk factors along with their associated morbid conditions often appear together. For example, hyperinsulinemia, hypertension, hyperlipidemia, and obesity frequently occur together in the same individual along with diabetes, atherosclerosis, and CHD. This constellation of risk factors has been labeled "Syndrome X" (Weiss *et al.* 1984; Raeven 1988; DeFronzo and Ferrannini 1991; Ferrannini *et al.* 1991). A slightly different constellation, including gout, gallstones and hypertriglyceridemia, among Native Americans is known as "New World Syndrome". Some have suggested that a single or, at most, several risk alleles greatly increase risk/susceptibility to such metabolic and physiological problems and "cause" these syndromes. More likely, such "syndromes" represent a multitude of genetic subtypes with multiple loci and alleles with direct, secondary, and multilevel effects. Labeling a constellation of variable clinical measures as a syndrome fails to identify any underlying genotype or specific phenotype. This leads to the lumping together of different predisposing environments and genotypes into a single phenotypic endpoint. Clinically defined HT is an illustrative example of such lumping. To determine "normal" versus "hypertensive", a continuous variable, blood pressure, is divided at a cut-off point. High blood pressure may result from a multitude of physiological alterations, genetic predispositions, and dietary factors. A number of candidate intermediate phenotypes for elevated blood pressure have been identified (Kurtz and Spence 1993; Lifton 1996; Crews and Williams, 1999). Hypertensive subtypes include low renin salt-sensitive, salt-sensitive without low renin, high angiotensinogen/low renin/high ACE, insulin-resistant, high plasma renin activity, and low birth weight/high cortisol groups (see Luft and Sharma 1993; Soubrier and Bonnardeaux 1994; Lifton 1996; Barker 1998; Crews and Williams 1999). Each subtype likely represents very different combinations of gene–gene, gene–environment, and gene–culture–environment interactions, all of which lead to the same endpoint: elevated blood pressure.

Specification of intermediate phenotypes also will provide more homogeneous subgroups for the epidemiological study of most CDCs and senescence (see Kurtz and Spence 1993; Lifton 1996; Crews and Williams 1999). These will allow more precise specification of alleles and loci predisposing to subtypes of CHD, CVD, cancer, hyperglycemia, or dementias. Recognized "syndromes" and diseases are encoded by multiple pleiotropic, thrifty, epistatic, thrifty/pleiotropic, direct and indirect, and late-acting predispositions which enhance and limit risk factor progression and modulate senescence. At conception, individuals all receive variable genetic potentials and predispositions that are played out in specific environments and particular cultures. Some traits promoting somatic maintenance and survival may be irrelevant because current conditions favor reproductive effort over somatic maintenance. Others may represent adaptations to earlier stressors no longer encountered, but still carry benefits for or have been co-opted to benefit somatic survival or maintenance. During fetal and infant development many systems are flexible, responding to the prevailing intrauterine environment. In some individuals, systems may be exhausted during early life and the organism will fail to survive to birth or reproductive age. In others, fetal programming may produce better adaptations to the prevailing environment, allowing the organism to survive and reproduce better, thereby preserving any genetic tendencies toward environmental responsiveness. The following sections briefly review some of the molecular, physiological, environmental, and sociocultural/behavioral risks that predispose humans to the CDCs that pace human senescence. Many of these add to the burden of wear-and-tear that leads to the loss of RC. Most often, CDCs are approached as problems in human physiology because clinical and medical applications and the need to cure patients drive research. This has produced a basic understanding of how the soma's mechanical and metabolic systems show progressive dysfunction with increasing survival beyond the minimal necessary life span.

Physiological risk factors

Many risk factors for CDCs are measured on continuous scales and compared between cases and controls or across samples from different populations. Biological anthropologists have particularly concentrated on variability in body morphology and physique, measures of internal homeostasis, blood pressure, cortisol, sodium, catecholamines, calcium, lipids, glucose, and somatic responses to external stressors, hypoxia, heat, cold, and solar radiation.

Body habitus

Among the most obvious aspects of human variation are body shape and size. From the Bantu and Nilotic peoples of Africa to the Eskimos of Alaska and the Ituri and Turkana of Africa, to the morbidly obese in cosmopolitan settings (>600 pounds), human morphological variation is amazing. Given the basic similarities involved in human growth and development regardless of population, the range of variation in body sizes is astonishing. Compared with other species of large-bodied primates and mammals, variation in human pigmentation is much less remarkable than the variability in size, shape, and proportions. Overall associations of body proportions (limb lengths to body size) and shape (short and round/tall and linear) with temperature extremes and the influence of shape and proportions on temperature homeostasis are well established (Katzmarzyck and Leonard 1998). Although female choice and sexual selection cannot be ruled out for any overt physical features, external aspects of human variation including body shape and pigmentation have responded to multiple environmental pressures.

Body types and proportions are generally thought to reflect selective pressures a group has encountered during its development as a unique population in a particular ecological setting. Among humans, certain body shapes are more prone to specific illnesses and CDCs. Aspects of body shape, size, or proportion consistently associated with disease endpoints are termed "body habitus". Body habitus responds to genes, to environmental, maternal, and intra-uterine factors, to pre-natal and post-natal nutrition, and to sociocultural practices and behaviors. Environmental pressures experienced *in utero* and during infancy pattern adult body habitus and predispositions to CDCs. From gestation through maturity and reproductive adulthood, inter-individual variation in body habitus is broad and increasing; during later decades of the life span, variation declines. At about age 18, body mass index (BMI) shows fairly high heritability (0.82), while, among those aged 81+ years, it falls by about 20–25% to a more moderate 0.63 (Carmichael and McGue 1995). In this large cross-sectional sample of twins, selective mortality over the life span appears to fall more heavily on those concordant for either high or low BMI. Adult body composition, morphology, body proportions, physique, and skeletal ossification also respond to personal behaviors and activities throughout the life span. Even at older ages, muscle mass and bone density may be increased with a regimen of physical activity and proportions of fat to lean mass may be altered with hormone supplements.

Body composition and proportions are continually changing over the human life span as various life history stages are completed and new ones initiated. For example, at birth, we are the brainiest we will ever be; at any other point in our life span, our brain will be a decidedly smaller proportion of body weight. Throughout infancy, the human encephalization ratio remains high, eclipsing that of all other mammals. At attainment of the maximum reproductive potential (MRP), about 20–25 years or so today, we are as tall and our muscle mass is about as large as either will ever be. Only continued high levels of physical activity will allow us to maintain this level of muscle mass past our prime reproductive years. Height itself is destined to begin shrinking in just a couple of decades as gravity and the intrinsic wear-and-tear from bipedality in conjunction with fragility of the human spine and skeletal supports reveal themselves as vertebral compression with microfractures and bone loss. As age increases, from about the mid-thirties to the late sixties, human body size tends toward greater robusticity, increasing proportions of body fat, and lower fat-free mass; those who survive into their seventies and beyond show lower percentages of body fat and less variation in aspects of body habitus (Garn 1994). Among "healthy, white women" aged 20–80 years, protein and mineral content decline linearly (although cross-sectional) with age, while fat content increases, such that 40–60-year-olds are significantly fatter than 20–39-year-olds, but those aged 60–80 years show similar profiles to 20–39-years-olds (Aloia *et al.* 1996). As muscle mass and mineral content of the soma decline with increasing age, muscles tend to weaken and bones tend to demineralize in many elders. Maintenance of high levels of physical activity and exercise seems to offset some portion of such losses, but not all (Garn 1994). Since losses of muscle and bone mass with age increase the risks for CDCs and death, they represent ideal candidates for senescent processes.

Among the most frequently and consistently measured somatic risk factors for CDC are height and weight. Body weight is consistently associated with morbidity from numerous CDCs and low weight appears to be an increased risk for neoplastic and infectious/parasitic diseases at any age. Too much weight for body size and there is an increased risk for CVDs, diabetes, and HT. Weight is a proxy variable, representing fatness or obesity when in excess and low energy or protein intakes/stores when low. Fatness, the amount of adipose tissue carried within/on the body, is poorly correlated with either BMI (weight (kg) / height (m^2), Quetelet's Index) or body weight. Subcutaneous fat is commonly measured with the use of spring-loaded calipers that measure the width of a double layer of skin and adipose tissue at specific sites (see Bogin 1999 for details). Ratios of skinfolds compare relative fatness across body segments, the subscapular/triceps ratio compares trunk fat to limb fat, while the abdominal/triceps ratio compares abdominal to limb fat. Persons with larger subcutaneous fat deposits in their trunk areas (e.g., suprailiac, mid-axillary, subscapular skinfolds) are at a higher risk for CHD, diabetes, CVD, fractures, and HT. The fat/lean ratio, based on computerized axial tomography scans of body sections, compares storage with metabolically active tissue. Higher ratios are associated with morbidity and mortality from multiple CDCs, and co-vary with other risk factors – age, sex, blood pressure, glycemia, cholesterol, and birth weight. Commonly described patterns of fat placement also include pear/feminine/peripheral (fat on hips and buttocks) and apple/masculine/centripetal (fat interabdominally and on the shoulders). In settings without caloric restraints, intra-abdominal fat increases throughout adulthood and correlates with an increased risk of CHD, CVD, hypertension, and diabetes (Yeung 1997). Both cross-sectional and cohort samples show that after about the sixth decade of life, fatness declines as survivors tend toward a leaner body habitus and coefficients of variation for BMI, weight, and subscapular skinfolds decline (Garn 1994; Aloia *et al.* 1996). Obesity is defined as being overweight for one's skeletal size. Skeletal size is commonly measured as stature, although additional measurements of frame size (e.g., biaromial and suprailiac widths) may provide more precise estimates. BMI is commonly thought to measure relative overweight and occasionally has been called, albeit incorrectly, an index of obesity. BMI values make no distinction between lean and fat tissue, and measure neither fatness nor obesity. Persons with very different physiques and fatness levels have identical BMIs. Still, BMI cut-points for obesity and desirable BMI levels have been proposed (Seltzer 1966; Andres 1980; Hodge *et al.* 1996). At one time, a BMI $\geqslant 30 \text{kg/m}^2$ was thought to represent obesity (Seltzer 1966). Later, the U.S. National Nutrition Board proposed lower cut-points; 27.8 for women and 27.3 for men. Currently in the U.S.A., BMI averages about 27 kg/m^2. Examination of mortality data suggests a U-shaped relationship between BMI and life span. Survival is best and life span longest over the range of about 20–26 kg/m^2 in most samples. Individuals with very large (BMI $\geqslant 30 \text{kg/m}^2$) or excessively small (BMI $\leqslant 18 \text{kg/m}^2$) BMI are the least likely to survive to late ages. In cosmopolitan settings, persons who survive to older ages (65+) tend to show more moderate BMI on average than those who do not survive late adulthood (45–64 years).

Although not an accurate estimate of fat mass, BMI has been used as such in population and clinical research (Crews and MacKeen 1982; McGarvey *et al.* 1993; Comuzzie *et al.* 1995; Hodge *et al.* 1996). Multiple genetic, environmental, and cultural factors determine fatness and obesity. Well-established risks include leptin, number of fat cells, hyperinsulinemia, hyperglycemia, hypertriglyceridemia, calorie and protein intakes, and genotypes. Variation in both insulin (chromosome 11) and *apo B* (chromosome 2) alleles are linked to differences in fat mass (Comuzzie *et al.* 1995). Women with a clinically determined myocardial infarction (MI) tend to have lower apolipoprotein B levels, include more angiotensinogen converting enzyme (ACE) D/D homozygotes, and to show a slimmer body habitus than those without a previous MI (Schuster *et al.* 1995). Alleles promoting low serum apo B apparently promote a leaner body habitus while also increasing risks for CVD in women homozygous for the ACE deletion. Those with high serum *apo B*-promoting alleles, but not ACE D/D homozygotes, may be more obese and at risk for diabetes and elevated blood pressure. Multiple gene–gene and gene–environment interactions influence development of an obese body habitus and increase the risk for MI and other vascular diseases. Regulation of energy metabolism and storage are complex physiological processes that respond rapidly to altered environmental and cultural circumstances. In general, free-ranging organisms are predisposed to ingest and conserve calories which are rare in the wild. When excess calories become available, mammals are physiologically predisposed to store them in fat cells. When calories are scarce, mammals become very efficient users of ingested calories (>85%). When calories are abundant and sufficient fat stores are laid down efficiency declines (<50%).

When a continual surfeit of calories is available, humans show a wide variation in amount and percentage of body fat, with the very lean below 10% and the morbidly obese above 80%. Humankind's omnivory, efficient metabolism, and propensities to store extra calories as fat were designed to deal with leaner, high-fiber, low-calorie diets, not today's overabundances. Humans have few metabolic options for handling excess energy once it is ingested. It can be stored, extractive efficiency may be downregulated, and metabolic activity may be upregulated. Nevertheless, whenever energy input exceeds demand, most somas will store the excess until storage tissues reach their maximum. However, some individuals continue to maintain low fat stores even when ingesting a surfeit of calories, suggesting innate propensities toward either low extraction or poor energy conservation. Individuals with such inabilities to extract, store, or conserve excess energy when it became available were likely also to have been less reproductively fit over much of human evolution. Today, such allelic variation may provide a late-life survival advantage in settings where obesity-related conditions are a major killer of those who survive to late adulthood (45–64 years) and old age (65+) and leaner individuals tend to survive more frequently into late life (75+). It is not clear whether fatness and obesity are senescent processes. Both accelerate a variety of senescent processes, e.g., generation of reactive oxygen species (ROS), mutations of mitochondrial DNA (mtDNA), and loss of bone. Both also are reversible and neither is necessarily progressive or chronic if reversed. However, if not reversed, fatness and obesity are chronic, progressive, and degenerative, leading to early mortality and shorter life span.

Skeletal risk factors

Bones alter in both size and shape during growth and development, and in density, mineral content, resilience, and strength throughout life (Garn 1994; Aloia *et al.* 1996). Adequate amounts of calcium, phosphorous, vitamin D, physical activity, and exposure to ultraviolet radiation (sunlight) are necessary for proper bone mineralization throughout growth, development, and adulthood. In addition, a wide array of physiologically active substances are needed for proper calcium metabolism and bone development (e.g., parathyroid hormone, calcitonin, growth hormone, insulin-like growth factors, insulin). These, along with their various receptors and intracellular messengers, produce a broad array of possible responses to environmental stimuli. With high variability in response and signaling networks, bone growth may respond to selection pressures just as does the neuroendocrine system, a major modulator of bone growth. Intrauterine undernutrition leading to low body weight predicts lower peak bone mineralization, greater fatness, and an increased risk for early mortality (Cooper *et al.* 1996).

At attainment of adult size and MRP, human bone is at its densest and strongest during the life span. At about age 40, bone density begins to decline as testosterone and estrogen, promoters of bone growth and stability, decline. Loss of bone and osteoporosis (bone resorption without replacement) lead to fragility of the skeleton and an increased susceptibility to fractures of the hips, wrists, and spinal column. Throughout life, new bone is constantly added, while older bone is removed from skeletal structures. During growth, this process leads to reshaping and remodeling as bones lengthen and strengthen, and accretion outpaces resorption. Peak bone mass is achieved at about the age of MRP when bones are maximally mineralized and strongest. As early as the fourth decade of life, resorption may exceed accretion in some, leading to increasingly demineralized, brittle, and osteoporotic bone (Drinkwater 1994; Ziegler and Scheidt-Nave 1995). Such alterations in bone microarchitecture increase the risk of fracture, which leads to loss of function, activity and mobility. This then alters other risk factors (e.g., fat mass, BMI, blood pressure) and increases risks for decubitus ulcers (bedsores), skin tears, rashes, septicemia, other infections, and ultimately mortality.

Osteoporosis is a skeletal disease of chronic low bone mineralization. Often found in the elderly, osteoporosis alters skeletal architecture and causes loss of resiliency to strain. This extreme type of bone loss is more frequent among women (Lynn 1994). At all ages, men have larger and denser bones than women. This is partly due to men's higher testosterone levels (a promoter of bone accretion) and loss of estrogens at menopause by women, producing even more rapid bone loss thereafter (Ziegler and Scheidt-Nave 1995). After the attainment of MRP, loss of bone mass is not linear nor are the rates of loss equal for men and women. Prior to menopause, women already lose bone mass slightly faster than men; following menopause, they lose bone mass 2–3 times faster (Stini 1990, 2002). Bone mass increases in response to diet and exercise during growth and development, reproductive adulthood (20–44 years), and late adulthood (45–64 years), and perhaps even later. However, among elders (ages 65+), neither the amount of nor the rates of bone loss appear to be greatly modifiable by life style or dietary changes. Neither aerobic or resistance exercise nor the ingestion of large doses of calcium, vitamins, or other dietary supplements appears to alter rates of bone loss or replacement. Estrogen replacement therapy reduces the rate of bone loss in post-menopausal women, as may growth hormone in either sex, but does not aid in recovery of lost bone. Because of their lower peak bone density and greater loss of hormonal regulators, post-menopausal women suffer higher rates of bone loss, fractures, osteoporosis, and related mortality than same-age men. After age 65, osteoporotic hip fractures in women are 3–4 times as frequent as in men and skeletal and osteoporotic changes are major contributors to functional limitations and disability in the elderly (Boult *et al.* 1994; Mulrow *et al.* 1994). Osteoporosis increases the risk for hip fractures, and, subsequently, muscular atrophy, bedsores, secondary infections, and mortality (Cummings and Nevitt 1989). Osteoarthritis may occur in concert with osteoporosis and lead to even greater loss of locomotor function and ability to care for one's self.

Individuals homozygous for mutant vitamin D receptor alleles may be at a particularly high risk for fractures. Among "Caucasian" women, the BB genotype is associated with lower adult bone density and osteoporosis may show 10 years earlier than in those with more common alleles (Morrison *et al.* 1994). Heritability of propensities to osteoporosis and degenerative bone disease suggest that multiple alleles which increase risks are segregating at a variety of human loci (Krall 1997; Livshits *et al.* 1999). Additional genetic propensities are suggested by ethnic/racial variability in the incidence and prevalence of osteoporosis. "Caucasian" and "Asian" women are at highest risk, while "African-American" and "Hispanics" have a lower risk (Ziegler and Scheidt-Nave 1995). Observed risk levels correlate with variable rates of skeletal and physical maturation and maximum bone diameters in African-American children and adults compared with European-American samples. Differences persist throughout life as samples classified as "white" show greater bone loss, less dense bone, more osteoporosis, and greater urinary excretion of calcium than those classified as "black" at later ages (Anderson and Pollitzer 1994). With increasing age, all adults, including healthy elders, have lower bone mineral and protein levels than at younger ages (Aloia *et al.* 1996). However, the range of bone loss is wide and some may maintain high density even into very late life (100+). Bone loss and osteoporosis are chronic, progressive, irreversible, and degenerative, and result in the loss of both function and life, fitting major criteria for a senescent process.

Blood pressure

Blood pressure (BP) is a cumulative measure of function and homeostasis across the entire cardiopulmonary system. BP is one of the best-documented risk factors, after age, sex, and weight, for premature mortality. BP varies widely across individuals, is influenced by biological sex, increases with age, and may vary by as much as 100 mmHg within one individual over the course of a single day (James and Baker 1990; Crews and Williams 1999). Genes contribute to adult variation in BP measurements (Crews and Williams 1999) and predispositions to high or low BP may be set *in utero* in response to the maternal environment (Barker *et al.* 1990; Barker 1998). Environmental and sociocultural factors influence BP (Dressler 1991; Bindon and Crews 1993; Dressler 1996; Bindon *et al.* 1997; Dressler *et al.* 1998), as do physiological factors, age, sex, weight, height, diet, culture, body habitus (Mancilha-Carvalho 1985; Silva *et al.* 1995; Fitton 1999; Silva 2001), and perhaps cultural history (Wilson and Grim 1991; Curtin 1992). With increased somatic survival, the human circulatory system loses functional integrity and BP increases due to both intrinsic and extrinsic factors. This functional loss is much less in populations living their traditional life ways than in modern cosmopolitan settings, in part, because so few individuals survive sufficiently long to show senescence and those who survive are exposed to less stress and wear-and-tear on their regulator systems over their life spans.

To achieve clinical consistency and in order to aid in determining clinical treatment regimens, BP elevated above a particular cut-point has been defined as HT (systolic BP $\geqslant$ 160 or 140 or diastolic BP $\geqslant$ 95 or 90). The clinical utility of dividing a continuously distributed physiological measure into a dichotomous indicator of a CDC (HT), as opposed to those without HT (normotensives, NT), is obvious. So defined, HT is relatively rare in most populations retaining a majority of their traditional life ways, but quite common among middle-aged and elder members of more cosmopolitan societies, although less common with increasing age. As discussed elsewhere, HT may not be an ideal outcome variable for epidemiological or genetic analysis (Crews and Williams 1999). For instance, BP is a continuously distributed and independent risk factor for CVDs, transforming it into a dichotomous variable eliminates information on variation. Labeling high BP as a disease implies that some cardiovascular conditions are secondary manifestations of HT, but others are not, and gets us no closer to the actual disease phenotype. High BP may result from a variety of etiological factors – diet, body habitus, life style, sex, genes, environment, cancer, or any combination thereof. Labeling these all as HT lumps together individuals with variable and perhaps multiple underlying predispositions into a single class, thereby reducing the power of genetic and epidemiological studies (Crews and Williams 1999). BP is regulated by complex systems of checks and balances that are likely to vary widely from person to person (James and Baker 1990; Crews and Williams 1999). Different subsystems and aspects of regulation decline at variable rates and may compensate one for another, but eventually, with continued somatic survival, components reduce function and eventually fail. Men and women diagnosed with HT often show clear physiological differences and variable etiologies. Men present with higher vascular resistance, lower cardiac index, and more frequent and severe left ventricular hypertrophy; women show lower indices of resistance and a higher cardiac index even when their BP is no higher than that of men (Alfie *et al.* 1995). Observed physiological variation may be related to estrogen's stimulatory effects on nitric oxide (NO) production, the serum levels of which correlate with BP (Alfie *et al.* 1995). NO is but one component of the complex hormonal/metabolic system regulating BP and multiple additional gene products and metabolic factors are likely to influence sex differences in BP (James and Baker 1990; Crews and Williams 1999).

Barker *et al.* (1990) estimate that the perinatal environment may add as much as 25 mmHg to adult systolic BP and diastolic BP, placing it among the most influential risk factors for elevated BP. "Essential hypertension" may also result from allometric relationships among general somatic growth (particularly height), growth of the kidney, kidney function, and renal homeostasis in environments with a surfeit of caloric intake during fetal and post-natal development (Weder and Schork 1994). The basic hypothesis is that renal defects producing HT are secondary to accelerated intrauterine growth and development in high-calorie modern settings (Weder and Schork, 1994, p. 151). Weder and Schork do not see the "thrifty genotype", as described by Neel (1962, 1982) for NIDDM, fitting data on HT. They suggest that alleles promoting high BP could be retained without there being specific alleles for HT or genetic disequilibrium in the distribution of alleles regulating BP. Loci contributing to or limiting kidney growth and function and acting to maintain renal homeostasis may increase blood flow as an indirect pleiotropic effect. In earlier ecological circumstances, no selective pressure would have been exerted to retain alleles limiting BP's allometric increase with body size during growth and development. The high-calorie, rapid-growth environment experienced by infants and children in today's societies was never experienced by any large segments of earlier human populations. In environments of such abundance, BP may increase in response to the need for increased renal flow unfettered by specific genetic limiters to either BP or stature. High BP in adults and CHD may then arise as allometric (pleiotropic?) responses to rapid and sustained linear growth.

This allometric process, along with underlying developmental influences, may help to explain why BP is related to adult stature in some populations (Samoan, New Guinea, and Native American) and ecological settings, but not others (European-American or African-American) (Mancilha-Carvalho and Crews 1990; Crews 1993a; Harper *et al.* 1996). Elsewhere, stature is inversely associated with coronary mortality (Marmot *et al.* 1984). Taller stature is positively associated with BP in more traditional-living samples undergoing rapid transitions to more modern ways of life and changing nutritional patterns. This seems to support the early-life programming model. Many adults now aged 50+ in such samples today would have encountered much different nutritional environments as fetuses, infants, and children. The possible influence of stature as an independent modulator of BP and other risk factors is seldom determined in population research. Stature is generally treated as a confounding variable and controlled statistically, ignored altogether, or only used in construction of indices such as BMI (which were developed specifically to be uncorrelated with stature). The influences of neonatal factors on stature and those of stature on BP, CVD, and diabetes suggest that stature has both direct and indirect effects on BP, and thus length of life (Crews and Losh 1994).

When first observed, "age-related" increases in BP were believed to be an inevitable consequence of getting old (see Oliver 1961). Once numerous examples of non-Western societies where BP did not increase with age were observed, it became clear that elevated BP was not a necessary concomitant of long life (Oliver 1961; Friedlaender 1987; Crews 1993a,b; Silva *et al.* 1995; Harper *et al.* 1996). Populations with low BP and/or no increase of BP with age are found throughout Africa, South America, Asia, and the Pacific Islands. Their one commonality is retention of traditional life styles, including diet and activity patterns, into the modern era (Oliver 1961; James and Baker 1990). Apparently, when acculturation to more cosmopolitan/western/modern ways of life occurs, many populations show a concomitant age-related elevation in BP (Cassel 1976; Mancilha-Carvalho 1985). Today, BP may vary so greatly across populations that the mean of one population (U.S. African-Americans) may be near the maximum observed in another (the Yanomami of Brazil).

The exact reasons for BP variation between individuals and across populations are difficult to determine. Risk factors for elevated BP include life style and behavioral differences that lead to sociocultural stress, effects likely to be mediated by the neuroendocrine system through the release of cortisol, adrenaline, and noradrenaline. It is anticipated that numerous alleles contributing to BP regulation unique to various subpopulations and kindreds within larger populations will be identified, as will multiple instances of gene–gene and gene–environment interaction (Crews and Williams 1999). BP is not a single variable; it is the complex outcome of multiple factors. BP shows fairly strong heritability (0.44 systolic, 0.34 diastolic), and shared environment effects (0.27), and responds rapidly to stressful sociocultural circumstances (James and Baker 1990; Hong *et al.* 1994; Crews and Williams 1999). As with many risk factors, heritability estimates for BP decline with increasing age (Hong *et al.* 1994), suggesting that more frail BP genotypes die earlier as life-long environmental and cultural interactions determine survival.

Arteriosclerosis and atherosclerosis are common among survivors to late adulthood and old age in cosmopolitan societies. Atherosclerosis (fatty deposits on the intima of arteries) may occur secondary to nutritional patterns common in cosmopolitan societies, particularly high levels of animal fats, cholesterol, and protein. Atherosclerosis appears to be a result of life style and environmental variation rather than being an aspect of biological senescence. However, arteriosclerosis (the hardening of the arteries with increasing age) may be an inevitable consequence of late life and represent an overt manifestation of senescence (Arking 1991). Arteriosclerosis occurs commonly and spontaneously in non-human primates (e.g., chimpanzees, macaques), suggesting it may be a common aspect of senescence in large-bodied primates (a plesiomorphic trait). Arteriosclerosis and atherosclerosis represent different etiological processes. The latter appears more environmentally determined, while the former appears to be an intrinsic aspect of arteries composed of smooth muscles with a damageable intima subject to severe stress and shear forces that produce wear-and-tear of non-replaceable tissues. Persons with elevated BP generally do not live as long as those with more moderate BP levels and late-life survivors tend toward more moderate BP levels than those observed among middle-aged samples. Arteriosclerosis along with damage to the arterial intima and heart tissues from high BP are chronic, irreversible, progressive, and degenerative processes. Although BP can now be controlled with multiple drugs, the underlying damage is not repairable and leads to an increased risk of death.

Lipids

Fatty acids/esters in the blood (e.g., total-c, LDLc, HDLc, triglycerides, very low-density lipoprotein (VLDLc)) modify risks for a multitude of CDCs and act synergistically with BP, age, sex, and body habitus to increase somatic frailty and reduce survival. Patterns of metabolic response to high-fat diets and lipidemia are dependent, as are most risk factors, on a combination of genetic (e.g., which *apoE* alleles one carries), developmental (e.g., fetal nutrition), environmental (e.g., diet), and cultural (e.g., workload) factors. Modification of high-risk lipid profiles (e.g., fasting total cholesterol $\geqslant 200$mg/dl, LDLc/HDLc $\geqslant 4.0$, triglycerides $\geqslant 300$, HDLc $\leqslant 35$) by alterations in diet and activity are easy to achieve for most individuals in samples from cosmopolitan settings. This suggests large environmental influences on serum lipid levels in most persons. However, there are large subsets of individuals in these samples, as many as 20–25% in clinical trials, with elevated lipids who do not respond to diet and activity intervention. For these individuals, only pharmacological intervention reduces lipidemia to an acceptable level. This suggests strong genetic influences on lipid levels in many individuals.

Family and twin data show higher heritabilities for lipid levels, up to 80% for total-c, 76% for LDLc, and 66% for HDLc, in cosmopolitan settings than those observed for BP or life span (Snieder *et al.* 1997). Similar to BP and body habitus, but contrary to life span itself, heritability estimates for serum lipid levels are lower among elders. This suggests larger influences of culture and environment on lipids with increasing age (Snieder *et al.* 1997), and selective mortality of those with the most extreme values. Along with other risk factors, the fetal/infant environment is likely to influence life-long responses to dietary fats and predispose to higher or lower serum lipids. During adulthood, serum total-c, LDLc, and apolipoprotein B (the apolipoprotein that aids in transport of LDLc) are highest in lower, compared with higher, birth weight/girth groups, while abdominal girth at birth is a stronger predictor of lipid profiles than is BMI (Barker 1998). Apolipoprotein E protein or allele types also influence lipid levels: those producing only apolipoprotein E4 show the highest total-c and LDLc levels, while producers of only apoE2 protein show the lowest levels. Greater total-c and LDLc increase risks for atherosclerosis and elevated BP, ultimately leading to CHD, stroke, MI, and shorter life. *ApoE*4* alleles are significantly less frequent among survivors to older age than among younger cohorts of the same populations. Hyperlipidemia is a chronic condition leading to progressive and degenerative deterioration of the vascular system, which promotes obesity and atherosclerosis, and increases the risk of death. Some part of the damage caused by hyperlipidemia appears to be reversible through exercise, dietary alterations, surgery, and pharmacological interventions. Not all persons experience hyperlipidemia as they age; however, those who do, tend not to survive as long as those who do not. Loss of lipid homeostasis with increasing age appears to be a detrimental senescent process revealed now that members of the human species survive sufficiently long in high-fat/-calorie settings to succumb to lipid-related chronic degenerative diseases.

Glycemia

Circulating glucose levels are influenced by multiple genetic and non-genetic factors. Glycemia (serum glucose level), specifically the non-enzymatic glycation (NEG) of macromolecules due to high glucose leading to the generation of advanced glycation endproducts (AGEs), is proposed as a candidate mechanism for senescence in long-lived cells and tissues (Cerami 1985, 1986; Cerami *et al.* 1987). Data showing that glycated hemoglobin (Hb_a) is associated with elevated risk ratios, similar to those for BMI and BP, for total mortality in both diabetic and non-diabetic adults support the suggestion that glycation may influence CDCs and senescence (Khaw *et al.* 2001). Lack of increased glycation of hemoglobin and albumin in non-cosmopolitan settings (McLorg 2000) suggests that neither NEG nor AGE are universals in human senescence. Still, AGEs are likely to contribute to the loss of function of individuals residing in cosmopolitan settings, accumulate in a chronic, progressive and irreversible fashion, and increase somatic susceptibility to mortality.

Like BP, glycemia responds to a constant barrage of dietary, environmental, genetic, and sociocultural modulators and may range widely over a 24-hour period in normal (lows of 50 mg/dl or less following a 10–12-hour fast, to highs of 199 mg/dl or more at 2 hours following a challenge) and hypo- or hyper-glycemic individuals (lows of 20 mg/dl, highs over 300 mg/dl). This continuous variation is often dichotomized as normal (normoglycemic, glycemia ⩽140 mg/dl fasting or 2 hours after a glucose challenge) and diabetic (hyperglycemic, glycemia above 140 mg/dl fasting or 2-hours post-load) (WHO 2002), and sometimes trichotomized (normal, impaired glucose tolerance, and diabetic). The hyperglycemic phenotype has several identified subtypes based on etiology and age of onset. The major division is between type I and type II diabetes (WHO 2002), previously identified as insulin-dependent (IDDM, juvenile) and NIDDM (adult-onset) (Crews and MacKeen 1982; Harris 1983; WHO 2002). Type II diabetes often responds to insulin therapy, while type I may have its onset other than during the juvenile stage of life (e.g., among Pima Indians and Naruans). Other subtypes of hyperglycemia include maturity-onset diabetes of youth (MODY), which segregates in a number of kindreds, gestational diabetes, which affects pregnant women in settings of high calories and high weight gain during pregnancy, and a variety of neoplastic and genetically related conditions. The incidence patterns of these subtypes indicate linkage of glucose metabolism to risk-enhancing alleles at multiple loci, along with gene–environment interactions that influence risk. In more cosmopolitan settings, type II affects a large proportion of elders (ages 65+). In the U.S.A., type II peaks in prevalence at about 40% in African-American women aged 60–65 years. Prevalence is only about 20% in those aged 80+ years, perhaps selective mortality. In more traditional settings, hyperglycemia does not appear to be a major problem (Friedlaender 1987; McLorg 2000). However, among transitional (those being affected by monetary-based and media-dominated cosmopolitan culture, i.e., the most rural Western Samoans), "modernized" (those participating fully in cosmopolitan, but retaining their ethnic identity and culture, i.e., American Samoans, Pima Indians), and cosmopolitan populations, hyperglycemia, obesity, hyperinsulinemia, and high BP are closely linked and identified as Syndrome X (DeFronzo and Ferrannini 1991).

Monozygotic twins show concordance rates above 45% for type II and 90% for type I diabetes, indicating genetic influences on both. Discordant monozygotic twins with type II diabetes have significantly lower birth weights than their co-twins, suggesting non-genetic *in utero* influences (Vaag *et al.* 1996; Hales 1997). Type II incidence clusters in families and kindreds. Familial clustering and twin concordance may result from post-natal or shared intrauterine environments rather genes alone (Law 1996). Poor maternal nutrition leading to poor fetal development appears to alter pancreatic control of insulin release, promote insulin resistance, and predispose to hyperglycemia in adults (Law 1996). Poor glucose homeostasis and insulin resistance in infants and children also result in type II diabetes in adults. In the general population, adults whose birth weight or weight at 1 year was low have impaired glucose tolerance (IGT) and type II more frequently than those of average weight or more (Law 1996; Barker 1998). Even during childhood, low-birth weight infants show higher plasma glucose levels (Law 1996). Adults whose mothers were exposed to famine during late pregnancy also show more impaired glucose tolerance and insulin secretion than those not exposed (Ravelli *et al.* 1998) and fetal and infant growth retardation are consistently related to insulin resistance and "Syndrome X" (Barker *et al.* 1993; Hales 1997).

Serum glucose correlates with multiple aspects of adult morphology and physiology. Obesity is the strongest predictor of hyperinsulinemia and hyperglycemia, and all are associated with higher BP. Dietary factors and physical activity patterns may influence glycemia independent of fatness and obesity. Diets balanced in calories, carbohydrates, fats, and protein are associated with moderate glycemia. Those with excess calories, fats, or carbohydrates promote higher glycemia, increased risks for early mortality, and reduced life span. Physiological alterations seen with hyperglycemia mimic those seen in senescing organisms, but occur at earlier ages. In humans and laboratory rodents, accumulations of AGEs, generation of ROS, wear-and-tear, loss of cells in non-dividing organs, cross-linking of molecules, and systemic dysregulation of homeostasis are accelerated in the presence of hyperglycemia. Humans have no *in vitro* mechanisms to eliminate AGEs once they are formed (Cerami *et al.* 1987). Thus, any degenerative alterations resulting from AGEs will be cumulative, irreversible, and progressive. Advanced glycation of macromolecules in somas that regularly experience high circulating glucose levels is likely to contribute to senescent damage in a variety of cell types and organs. A lack of increase in glycated hemoglobin and albumin with age in populations living more traditional life styles likely reflects differences in the gene–environment relationships associated with different life styles, along with calorifically and nutritionally different diets.

Molecular risk factors

Molecular risks for senescence encompass everything from DNA variants (e.g., alleles, restriction fragment length polymorphisms, quantitative trait loci, tandem repeats, microsatellites, mtDNA), to proteins produced by the soma (e.g., insulin, cell receptors, cytokines, neurotransmitters, hormones), through to the molecules we ingest (e.g., cholesterol, sodium, calcium, iron, lead, glucose, fats) and are exposed to (e.g., ROS, organic solvents, pesticides, silica) on a daily basis. Their commonality is that continued somatic exposure results in progressive dysfunction and shortened life span. Allelic variants may affect susceptibility to disease through direct and indirect pathways. Most DNA variants at most loci are likely to code for proteins that, on average, predispose somas to about equal reproductive success and life span. These averages are based on the species' minimal necessary life span (MNLS), previous evolutionary pressures, and current environmental contingencies. Others are likely to predispose to better or improved somatic survival and enhance "vitality", while still others predispose to poorer survival and increase "frailty" (Vaupel *et al.* 1979; Manton and Stallard 1984; Weiss 1989a). Still others may predispose to a reduced potential for reproductive success, but enhance somatic survival or increased reproductive success and reduced somatic survival. A number of such "longevity-enabling genes" have been proposed – e.g., APOE, angiotensin converting enzyme (ACE), human leucocyte antigen (HLA) Dr, and plasminogen activator inhibitor-1 (PAI-1) – by Barzilai and Shuldiner (2001). These all affect immune function and homeostasis and all have a high probability of predisposing individuals to CDCs. A number of mutants in mtDNA (Wallace *et al.* 1995; Tanaka *et al.* 1998; Benedicts *et al.* 2000a), along with presenilin-1 and -2 and the amyloid precursor protein (Clark 1999) may be added. The possible interactions among 30 000 coding loci, with as many as several thousand variations each, generally ensures that most genotypes produce phenotypes predisposed to about the average MRP and a soma built to outlast as far as possible the species' MNLS. However, how genetic propensities developed over evolutionary history are expressed depends strongly on current environmental circumstances.

Alleles associated with BP provide a physiological example of such phenomena. Those predisposing to low or moderate BP response to environmental perturbations may predispose to slower senescence and an increased vitality of carriers in cosmopolitan settings with constant stimulation, while variants leading to rapid or sustained response predispose to frailty, increased early life vascular disease, and shorter life span. These variants need have no direct associations with, or even secondary effects on, actual BP. Rather, pleiotropic, epistatic, and gene–environment interactions may occur through multiple pathways, as with the variety of proteins produced by the ACE, insulin, apolipoprotein, atrial natriuretic polypeptide (ANP), angiotensinogen, renin, and other loci influencing BP in humans (see Crews and Williams 1999). The ACE I/D (insertion/deletion) polymorphism further illustrates some of the molecular complications associated with senescence and life span. ACE I/D genotypes explain little variation in BP across or within populations (Barley *et al.* 1994). However, the number of ACD D alleles carried by a person is associated with greater morbidity and mortality during reproductive and late adulthood (Schuster *et al.* 1995). This association is probably due to untoward pleiotropic or epistatic effects of this allele elsewhere in the metabolic milieu. Protein products of both ACE alleles are equally capable of completing their supposed primary function: conversion of the largely inactive decapeptide angiotensin 1 after it is cleaved by renin to the active octapeptide angiotensin II (Erdos and Skidgel 1987). ACE functions in a variety of other pathways, cleaving a variety of decapeptides – for example, inactivating bradykinin by releasing a c-terminal dipeptide to form an inactive octapeptide (Erdos and Skidgel 1987). The D allele's association with stroke, MI, and nongevity may be related to its inactivation of bradykinin or pleiotropic affects elsewhere. D alleles are also associated with greater degeneration of myocytes *in vitro*, and appear to show a sex differential in their effects on survival, showing a stronger association with MI in women than in men (Schuster *et al.* 1995). To complicate matters, among late-life survivors, the D allele may increase in frequency, rather than decrease, as associations with mortality in mid-life and late-adulthood predict, suggesting that the effects of D alleles during late life may be reversed from those of mid-life.

Association of the APO E4 protein with late-onset Alzheimer's disease (senile dementia of the Alzheimer's type or SDAT) provides another complicated aspect of systemic dysfunction. Apolipoprotein E is a circulating plasma protein produced mainly by the liver and brain cells and is best known as a carrier protein for lipids. However, APO E proteins also participate in the repair of nerve tissues after injury, aid in aspects of dendritic remodeling, and may act as mitogenic factors (ADRA 1996). The *apo E*4* allele is also correlated with high total-c and LDL-c levels and CHD (Menzel *et al.* 1983). The *ApoE*4* differs from the *ApoE*2* and *E*3* alleles in its DNA base sequence. The APO E4 protein is the only one to show sequence similarity with a protein expressed only in brain tissue. These differences are likely to explain why APO E4 proteins become irreversibly bound to beta-amyloid in the neurological tangles characteristic of SDAT (Namba *et al.* 1991). APO E4's status as a marker for SDAT (Chartier-Harlin *et al.* 1994) may be a pleiotropic effect, having little relationship to its proposed primary function as a lipid transporter interacting with the LDL receptor. APOE protein types may also show direct effects on morbidity and life span. Among Finish centenarians, *ApoE*2* alleles are more frequent at older ages (8%, 18%, and 28% at ages 100–101, 102–103, and 104+, respectively); similarly, the frequency of *ApoE*4* alleles is much lower in centenarians (14%) than in middle-aged Finns (35–40%) (Chartier-Harlin *et al.* 1994). A sample of U.S. elders shows a similar trend, with *ApoE*2* frequencies increasing across the ages 75–79, 80–84, and 85+ among both men (12%, 14%, 17%) and women (13%, 11%, 15%), with those carrying the *ApoE*2/3* genotype show one-half and one-quater the mortality of those carrying the *E*3/3* and *E*3/4* genotypes (Corder *et al.* 1996). It is hypothesized that the APO E2 protein may enhance general neuroendocrine function, or may improve overall resistance to disease and senescence, in addition to influencing lipid metabolism senescence (Chartier-Harlin *et al.* 1994).

mtDNA variations have also been observed between the cells of younger and older samples. Among 37 Japanese centenarians, three single nucleotide mtDNA mutations (mt8414T, mt3010A, and mt5178A) occur at highly elevated frequencies compared with their rarity in 43 young Japanese (Tanaka *et al.* 1998); furthermore, those with mt5178C were more susceptible to CDCs throughout life than those with mt5178A (Tanaka *et al.* 1998). Among 212 centenarians, men showed a higher frequency of the J mtDNA haplotype than did 275 sex-matched controls (median age of 32) residing in the same region (De Benedictus *et al.* 2000b). Post-mitotic cells from older persons also show mtDNA mutants that are uncommon in the skeletal, heart, muscle, and brain cells of healthy middle-aged samples (Elson *et al.* 2001). Importantly, because of the random assortment of mitochondria into primary oocytes during intrauterine development, mutant varieties of mtDNA are incorporated into most human zygotes (Brown *et al.* 2001), endowing each individual with their own unique distribution and perhaps predispositions toward senescence. Such complexity of possible genetic promoters confounds understanding of the molecular biology of CDCs and senescence. Processes of senescence are not adaptations achieved through the actions of natural selection with specific alleles selected for their reproductive advantages. CDCs and senescence developed as byproducts of the need for somatic stability in an already sexually reproducing organism that provided greater parental investments in their young. Senescence results from selection for alleles that trade off reproductive and somatic survival benefits. This does not allow an immortal soma to develop due to evolutionary pressures for reproduction, nor can reproduction be infinite if the soma must survive to reproduce, although it may be quite prolific when the soma seldom survives reproduction.

Peptides made by an organism function and interact with multiple others across a variety of local internal milieus, cells, organs, and tissues. Primary functions of many are often difficult to determine, although such assumptions are frequent; ACE is a case in point – which of its multiple activities is its true major function, if there even is one, in humans (see Erdos and Skidgel 1987)? Many alleles at multiple loci must produce little difference in effect on their primary target; however, their pleiotropic, epistatic, and gene-environment associations may greatly alter other aspects of vitality/frailty and produce large differences in systemic dysfunction and senescence. Allele frequencies at various loci across cohorts within populations have been useful for identifying some that may influence senescence and life span. Among American-Samoans, age-specific selection may be occurring, with older persons showing more homozygosity at the ANP locus and more ACE heterozygosity than younger persons, although neither allele shows any real association with BP (Crews and Harper 1998). Proteins with limited variability in directly maintaining functional capacity and homeostasis in a particular system may also constitute or contribute to processes that increase overall frailty. Most insulin and insulin-receptor proteins appear to complete their primary function of maintaining glucose levels adequately. Seldom are such variants reported to be closely associated with hyperglycemia and associations tend to vary widely across cultures (Crews and Moore, submitted). The model developed here suggests that proteins not so directly responsible for maintenance of glucose levels, although still related to glucose metabolism, should show stronger associations with hyperglycemia, since their less direct involvement may allow them to vary more widely in structure and function (Gerber and Crews 1999).

In addition to inherited variability in DNA, both altered DNA and proteins occurring secondary to environmental stress (e.g., radiation, chemicals, viruses) and internal processes (e.g., ROS, replication errors, glycation) increase multiple morbid processes. Somatic mutations of DNA lead to structural, enzymatic, and housekeeping proteins that lack function or show altered functions. Additionally, internal cellular processes create oxidative stress and lead to alterations in function and loss of integrity that may produce a loss of needed cell numbers through apoptosis in organs and tissues or unregulated increases in cell numbers. A number of morbid endpoints – cancer, CVD, CHD, high BP, neuromuscular, neuroskeletal, and neuroendocrine disease of late adulthood and late life – are correlated closely with reduced activities of the antioxidants superoxide dismutase (SOD), catalase, and glutahione peroxidase (Jialal and Grundy 1992; Ames *et al.* 1993; Bowling and Beal 1995; de la Torree *et al.* 1996; Jun *et al.* 1996; Multhaup *et al.* 1997). Senescence proceeds more quickly in organisms that lack antioxidant defenses and in those that lose their defenses over time. Humans who survive to older ages generally maintain high antioxidant levels throughout late life.

Neurotransmitters, immunological factors, and hormones constitute an additional layer of molecular risks for, as well as protections against, degenerative outcomes. The neuroendocrine–immune system (discussed later in this chapter) includes multiple overlapping pathways that regulate homeostasis and defend the soma. The system may be either up or downregulated in response to environmental stimuli or internal stressors and influences everything from cellular metabolism to DNA expression, release of proteins from cells, organ activities, and the expression of antioxidant enzymes (Bolzan *et al.* 1995). Even during fetal development, aspects of the neuroendocrine–immune system are set and reset as they respond to environmental inputs, providing pathways connecting fetal exposures and stresses to late-life survival and disease. During human neurological development, about twice as many primary neurons are developed as actually ever establish themselves in dendritic/axionic connections. Patterns of neuronal integration ultimately established among surviving neurons are highly flexible depending on a variety of intrauterine factors, nutrient availability, maternal health, maternal and fetal stress, encounters with antigens, allergens, viruses, and bacteria, and genetic predispositions. Once established in response to the prevailing fetal and infant environment, neural connections remain and appear to be resistant to further environmental stimuli.

Multiple aspects of adult physiological responses to environmental stimuli are likely to result from adaptations during *in utero* and during infancy (developmental acclimatization) to stressful or abundant environments. Such responses apparently provide protections through adulthood from the original stressors but may increase risk for CDCs, early senescence, and shorter life span if a different stressful setting is encountered later. Associations of alterations and variation in multiple internal systems have been well documented. The influences of stress-related hormones (e.g., cortisol, adrenaline, noradrenaline, other glucocorticoids) on dysregulation of immune responses, damage to organ systems from stress responses, and increases in circulating stress hormones with increasing age are all proposed as possible mechanisms of senescence (Walford 1969, Finch and Rose 1995; Masoro 1996; McEwen 1998). A hyperactive hypothalamic–pituitary–adrenal axis (HPA) may be set early in life and then produce rapid and sustained stress responses throughout life. Another pathway for an increased risk of dysfunction is through constant wear-and-tear on cells and organs through excessive exposure to stress hormones; these potentiate increases in heart rate, BP, antioxidant activity, glycemia, respiration, and brain activity, increasing somatic wear-and-tear and the risks for CDCs and senescence. Stress hormones directly influence physiological and molecular processes including immune function, DNA transcription, release of proteins from cells, inflammatory responses, and cellular uptake of molecules (Finch and Rose 1995; Crews and Williams 1999).

Genetic variation seems to influence life span most at the extremes of age, over age 85, and during the first years of life, when multiple single-locus disorders and other genetic conditions occur. Between about ages 5 and 80, environmental and sociocultural–behavioral factors more actively determine who survives. However, after the vagaries of reproductive adulthood, genetics begin to exert their effects as those with inherent susceptibilities succumb more rapidly to the onslaught of their own somatic dysfunction. After age 40, heritability estimates for life span begin to increase and the siblings of centenarians are over three times more likely to achieve 90 years than are those of non-centenarians (Barzilai and Shuldiner 2001). Computer simulations also support a model that, at about age 100 years, alleles promoting slow senescence have an increased likelihood of representation (Toupance *et al.* 1998). As individual variations in allelic propensities become more important in late-life survival, not having frailty-enhancing alleles becomes as important as carrying those that enhance vitality (Perls 2001). After about age 85 or 100 years, multiple changes occur. Carriers of *ApoE*4* alleles are no longer disadvantaged in survival, and the ACE D allele tends to increase in frequency; concurrently, incidence and mortality rates for several CDCs particularly cancers, decline, or may stabilize, e.g., some cancers, type II diabetes, and high BP. This may reflect loss through death of susceptibles from the cohort or the fact that some degenerative processes may be limited by the progress of senescence itself.

Environmental risk factors

An organism's environment includes all it eats, drinks, and breathes, along with everything else from outside that finds its way inside or exists beyond their soma. Hypoxia, heat, cold, solar and human-made electromagnetic radiation, predators (both micro- and macro-life forms), crowding, family, life style, hygiene, nutrition, and intrauterine conditions are among the environmental stresses commonly limiting innate predispositions to life span. The environment is a complex web of interacting systems with multiple levels – and, thus, not an easy variable to measure. Environmental interactions influence all aspects of human life history, producing some obvious differences across populations. At high altitude, people grow slower; they also succumb in greater numbers and at earlier ages to respiratory diseases than those at low altitude. On Guam and the Marianas islands, until recent decades, large proportions of adults died of amyotropic lateral sclerosis (ALS), likely secondary to metals in native dietary staples. Environmental risks are often associated with human sociocultural and behavioral activities. Diets based on grain monoculture lead to vitamin deficiencies, lower RC, and earlier mortality. Compositions of soils in different locations may lead to mineral deficiencies (e.g., iodine, selenium, zinc, iron, magnesium) or allow minerals to build to toxic levels (e.g., aluminum, lead, arsenic). One bioanthropological model for examining major environmental stressors is to detail demographic, physiological, genetic, socioeconomic status (SES), sociocultural, and health variability in the affected population (Baker and Baker 1977). Such research designs continue to be helpful in examining the effects of additional stressors including pollution, undernutrition, poverty, toxic exposures, dietary contaminants, parents' exposures, and high-risk behaviors (Little and Haas 1989; Schell 1991; Goodman and Leatherman 1999).

In today's world, new environmental exposures, good or ill, frequently follow sociocultural developments and human-made environmental alterations (Goodman and Leatherman 1999). Inhalation exposures to smoke, soot, and toxic metals likely accompanied the use of fire for warmth, protection, cooking, and eventually smelting and mining. As humans aggregated into larger groups by adopting agriculture and animal husbandry, exposures to parasites and infectious organisms also increased, allowing viruses and crowd diseases (e.g., measles, chicken pox) to thrive (Fenner 1970). In recent decades, humans have added heavy metals, pesticides, hormones, and manufactured chemicals to the environment. Cancers and respiratory ailments along with the vascular diseases lead CDCs in their degree of environmental influence and loss of years of life in the short lived compared with the long lived. Many neoplasms and lung diseases occur secondary to occupational (e.g., radiation, asbestos, coal), life style (e.g., nitrates, alcohol, smoking), and geographic exposures (e.g., solar radiation, hypoxia).

Of all CDCs causing death, only cancers increased their representation among cosmopolitan populations during the twentieth century, while rates for all others fell (Olshansky and Carnes 1994). As a cause of death, cancer already may be close to its minimum in current human somas. Neoplastic transformation of cells is held in check by a wide-ranging system of controls, including apoptosis, telomeric shortening, antioxidants, and immune surveillance. All organisms must cope with a constant barrage of carcinogens in their air, food, and water, oncogenes, constant mutation, and alterations in cellular DNA. Rates of increase in mortality rates from cancer peak at about mid-life (ages 40–50) (Figure 4.1). Thereafter, they rise slowly, bottoming out and even droping during late life (ages 75+) (Weiss, 1989a). This suggests survivors to late life are less susceptible to carcinogenic stress and/or have fewer innate propensities to develop cancer. Inadequate thymus-dependent immune response in some cancer patients suggests that a variety of carcinogens may act by incapacitating the immune system, preventing cellular marking for destruction (Olshansky and Carnes 1994). In addition, neoplastic cells maintain active telomerase and telomeric length, and avoid apoptosis even when aneuploid. Apparently, it is those with greater thymus-dependent immunity who also maintain appropriate cellular apoptosis that manage to survive into late life and show lower cancer mortality. All animals and at least some plants are susceptible to neoplastic disease, a plesiomorphic condition. Over their long life spans, all mammals and primates must halt the insidious progression of cells toward the neoplastic state. However, such defenses must be traded off against the needs to reproduce, rear young, and maintain multiple additional and equally important systems. Given the amount of effort aimed at preventing uncontrolled cellular proliferation by human somas, and their general inability to do so for more than about a century, it is likely that susceptibilities to cancers represent a true underlying process of senescence. Cancer is not universal in all persons before death, but it is chronic, progressive, irreversible, and degenerative. Although a cancer may be removed and/or slowed, progression to neoplasm in other cells

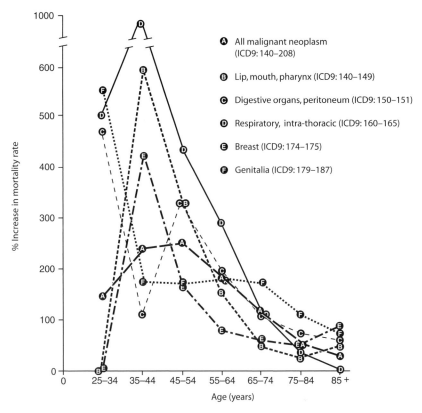

Figure 4.1 Declining rates of increase in mortality from neoplasms with age after mid-life (source: Murphy 2000, Table 11, p. 54).

may still occur such that persons may have several different types over their life span.

Some respiratory diseases that shorten life and perhaps speed up senescence arise from infectious exposures such as pneumonia and tuberculosis; some reflect life style choices, such as cigarette smoking and lung cancer. Other respiratory stressors come from industry, mining, energy, and processing systems, and the automobiles, air-conditioning units, and home-heating systems used by modern cosmopolitan societies. The smoke-filled huts, stale pathogen-filled air, hypoxia, and humidity endured by peoples in many non-cosmopolitan settings produce a different variety of respiratory stressors and diseases. Many incidences of respiratory conditions in modern settings (e.g., asbestosis, silicosis) are directly traceable to sociocultural processes. The commonality among these processes is that all harm the lung's cells, either filling them with contaminants, destroying them, or mutating their DNA and producing dysfunction. As with cancers, somatic susceptibilities to respiratory stressors seem to be quite ancient in mammals and to represent a compromise between function and cost. Better lungs and structures might improve life span, but lower RC. Among those surviving beyond age 75 years in the U.S.A., lung function and RC are high, while declining abilities to maintain blood oxygenation and loss of RC are associated with earlier mortality. Declining oxygenation of blood, loss of lung capacity, compliance, RC, and tidal volume are all predictive of mortality in both adults and elders. To the degree that measures of lung function are dependent on multiple physiological systems (e.g., heart, blood vessels, lungs, perfusion rates), they provide a systematic measure of functional capacity for elders during late life. Loss of lung function with increasing age is a component of senescence in human somas, and environmental stressors contribute to this loss.

Many postulate that human diets changed radically during the recent period compared with those throughout the later Pleistocene, leading to an abundance of calories and fats and subsequent vascular diseases (Eaton and Konner 1985; Eaton *et al.* 1988; Cohen 1989; Eaton and Nelson 1991; Armelagos 1991; Crews and Gerber 1994; Gerber and Crews 1999). Caloric overabundance in a large-bodied omnivorous hominid that until recently needed to store excess calories for leaner times is viewed as the cause for rampant obesity, vascular occlusions, high BP, hyperglycemia, and cardiovascular mortality in middle-aged humans today. All captive primates and mammals (e.g., macaques, chimpanzees, gorillas, horses, pigs) show obesity and atherosclerosis when not on a calorifically and nutritionally appropriate diet. Even wild-living chimpanzees show signs of arteriosclerosis in middle age, suggesting that at least to some degree vascular hardening with increasing age is a plesiomorphic trait among primates. Human energy and lipid transport, use, and storage systems were never designed to handle environments where such molecules were relatively abundant and constant. In today's settings, sociocultural alterations producing a new dietary environment have revealed this design flaw as higher lipid levels, coronary artery occlusion, and sudden death in those with more susceptible genotypes. Genotypes representing allelic variants that in previous environment would not be noticed by the scythe of natural selection and still, today, generally fell their carriers well after the period of MRP. As with environmentally induced functional losses elsewhere, those within the vascular system act synergistically with senescent processes occurring at other levels to promote CDCs and senescent alteration.

All human cultures determine what portions of their environment are appropriate foodstuffs and what are not, and when it is correct to eat them. This process is dependent on local availability, something most people who have ever lived never had any control over. Still today, most people control neither where they live, what foods are available, the cleanliness of their water supply, nor the quality of the air they breathe. Thus, they are dependent on what is locally available. Uses of non-food items – alcohol, hallucinogenic drugs, tobacco, and medicinals in traditional healing, religious, and recreational activities – are also environmentally and culturally determined. Environmental toxins from multiple sources may enter the food chain and build up in individuals, leading to premature death, but not necessarily senescence. Similarly essential nutrients may be unavailable to some populations and cause morbidity, but not death, before RS and fledging of offspring, e.g., iodine and goiter in New Guinea or minerals and ALS on Guam. Such environmental risks increase the risks for morbidity and perhaps early mortality, but do not satisfy additional criteria for a senescent process or function. However, they are likely to increase wear-and-tear on the soma and perhaps synergistically influence processes of senescence in some physiological systems. Late-life survivors (ages 75+) appear to have been less susceptible to multiple environmental stressors and insults during their reproductive (20–34), adult (35–49), and late adult years (50–64) and to have maintained greater function and RC into later life, as measured by lung capacity, immune function, and stress reactions.

Sociocultural/behavioral risk factors

There are many definitions of culture (Steward 1977); basically, it is composed of complex, often conflicting, constantly changing, and numerous overlapping sets of rules and beliefs that pattern peoples' current behaviors. Most sociocultural systems include general and specific rules allocating work, reproductive behavior, mate selection, food, and social roles. These often arbitrary patterns alter risks for CDCs and influence life span, thereby promoting or retarding senescent processes. Culturally sanctioned teenage pregnancy or multiple closely spaced pregnancies in marginal nutritional settings may drain body resources (pregnancy wasting), leaving women more susceptible to endogenous and exogenous causes of death, but may be necessary to provide for replacement-level RS. Conversely, late first pregnancy, no breast-feeding, and high-calorie diets increase the risks for neoplastic lesions of the female reproductive organs and breasts. Culturally determined age-graded reproductive norms, age at first mating, marriage, or pregnancy, desired number of offspring, along with workloads and places of work of reproductive-age men and women, and current demographic patterns profoundly affect women's health and life span. Survival is a tradeoff between current reproduction and future parental investment in the offspring produced. In marginal settings with high adult mortality, the "more is better" *r*-strategy may be more viable for women than counting on the survival necessary for a *K*-strategy of fewer higher quality offspring with long-term parental investment.

Culture and culturally determined beliefs also structure individual perceptions of health and well-being. For example, both self-perceived and observed health status are related to future survival (Borkan and Norris 1980; Schoenfeld *et al.* 1994). Even controlling for other risk factors (e.g., sex, recent hospitalization, number of CDCs, smoking, alcohol consumption, race, marital status, education), self-rated health was the strongest predictor of mortality in healthy 70–79-year-olds and those reporting poor/bad health were 19 times more likely to die than those in the best of health (Schoenfeld *et al.* 1994). That humans are capable of perceiving their own state of health to such a degree that self-measures are as closely related to survival as are commonly reported physiological and sociobehavioral risks factors is not puzzling. Self-perceived health is a psychosocial–biocultural self-construct. As a personal assessment, it is influenced by current and previous disease, disability, functional limitations, social support, current and previous psychological states, race, sex, gender, genetic predispositions, and degree of senescence (Johnson and Wolinski 1994; Harper 1999). Humans also have the unique capability of comparing themselves in their present state of function with themselves at early points in their life and with others they have known at the same stage of life, thus providing both temporal and comparative dimensions to self-perceptions of health, factors which are not determined by common clinical tests. Self-perceived well-being may also reflect underlying mental states such as depression (which is associated with immune suppression), reflect subclinical alterations in some physiological process, or allow subconscious perceptions of subclinical diseases to be reported (Schoenfeld *et al.* 1994). That self-perceived health predicts mortality in both high functioning and random samples of elders suggests that there may exist a natural progression from morbidity, to disability, to altered sense of well-being within individuals (Johnson and Wolinski 1994) or that self-perceptions of health are more sensitive to physiological homeostasis than are standard risk factors alone (Schoenfeld *et al.* 1994). This further suggests that humans perceive their overall health as a biocultural gestalt developed as part of their own life history, as they have attained MRP and adulthood within a particular ecological setting, and in comparison with their, and their cohort's, prime levels of function and well-being. Measures of self-perceived health may be the closest we have yet come to a practical and convenient way of assessing senescence. Assessing how a person feels about their own health may be a better indicator of senescence and biological (BA) age than are either chronological age or all the complicated ways of estimating BA from profiles of physiological characteristics yet developed.

Overnutrition illustrates how culture, behavior, environment, and genes interact to influence the pace of senescence and increase risk for CDCs. In combination with cultural innovations, many environments today provide excess calories and opportunities to over-ingest once scarce but now plentiful resources (Gerber and Crews 1999). Upon exposure to this overabundance, many overeat. Some show few adverse health consequences. Others rapidly become morbidly obese and succumb to CDCs early in adulthood (ages 40–54), while most show progressive increases in weight, BP, glycemia, and cholesterolemia, and an increased risk of death in late adulthood (55–69). Conversely, others show little inclination to overeat and show neither obesity nor increases in vascular risk factors. Nonetheless, it is only in environments with a surfeit of calories and adequate nutrition that most people today manage to survive into late life (70+). More than anything, this illustrates the value of adequate nutritional resources in slowing senescent damage and providing opportunities for late-life survival. One suggestion is that those whose alleles predispose them to rapid or sustained energy storage, slow metabolism of fats/lipids, maintenance of high glucose/lipid levels, or low basal metabolic rate (BMR) may develop nutrition-associated CDCs more rapidly throughout life and therefore die earlier than those without such predispositions when there is a surfeit of calories (Crews and Gerber 1994).

Poverty and SES provide another insight into how sociocultural and behavioral risks structure the biocultural dimension of human senescence and life span. Poverty (low SES) increases risks for mortality from conception through adulthood. Poorer pregnant women are less likely to carry conceptions to term and are themselves more likely to die during pregnancy and childbirth. Child survival is lower in the slums and barrios of the world's major cities than it is in upscale neighborhoods and suburbs. Among the poor, death rates during adolescence are often double those of the "best off". Differentials in life expectancy between the poorest and wealthiest segments approach 100% in some populations, while risks of death from infectious and porasitic causes decrease with increasing SES in all populations. Interestingly, those who do survive poverty and low SES untill their seventh decade are often healthier, less frail, and thereafter tend to survive longer than their high SES counterparts who showed better survival earlier in life. The maelstrom of poverty appears to select more heavily against those with inherent frailty at all ages, thereby increasing the representation of those with inherent vitality among survivors to later ages. Poverty does not restructure the processes of senescence; most poor people still die young. Poverty disadvantages everyone, but not equally. In general, the subset of the population with the highest somatic reserves, and lowest inherent risks for morbidity and rapid senescence, tends to survive the longest. Those who do survive to late life (70+), although impoverished throughout life, provide an opportunity to study truly slow-senescing high vitality (or just plain lucky) phenotypes. They also illustrate how sociocultural factors pattern biological processes in humans. Multiple aspects of cultural influence behavior and, ultimately, the pace of senescence. Particular religious beliefs may decrease participants' exposures to risks that promote degenerative somatic alterations (e.g., through fasting, dietary restraint, alcohol abstinence, celibacy), while others might increase such risks (e.g., feasting, endogamy, ritual tattooing, scaring, and drug and alcohol use). Because senescence represents such a broad array of biological processes acting jointly, synergistically, and often at odds with one another, to increase the probability of death with each passing moment after attainment of MRP, there likely exist infinite opportunities for sociocultural and behavioral factors to slow or hasten its overall rate and pattern.

Systems/integration

Representing three aspects of a multiply integrated neuroendocrine–immune system of controls and regulators, the neurological, endocrine, and immunological systems maintain somatic homeostasis. They are viewed as separate systems because their interconnections are not yet completely established and because of tradition. The hypothalamic–pituitary system regulates the neuroendocrine component of the brain, while the limbic axis of the cortex (the old brain) regulates autonomic action. Loci coding for enzyme regulators of hormone production by the endocrine system are located within the major histocompatibility complex (MHC)/HLA system, reflecting integration between the neuroimmunological and immunoendocrine components of the system (Finch and Rose 1995). Completing the picture of integration, psychosocial stress leads to altered immune function, illustrating connections between the endocrine and immunological systems (Kiecolt-Glasser *et al.* 1991). Traditionally discussed as three subsystems, dysfunction, alterations, and degeneration in neuroendocrine–immune function are likely initiators of senescence. Other somatic systems (e.g., cardiopulmonary, digestive–metabolic, sensory) all show age-related alterations. These were discussed earlier as CDCs and risk factors, and are reviewed extensively elsewhere (Arking 1998; Harper and Crews 2000).

Neurological systems

Functioning of the neurological system depends on the brain constantly integrating a mix of chemical and electrical messages. The human neurological system develops very rapidly during gestation, particularly during the last trimester. Continued rapid growth characterizes the first year of life, with cranial growth slowing gradually through the fifth year when most growth is complete (Bogin 1999). In general, cells of the central nervous system (CNS) do not show mitosis or cell replacement much after birth. Large numbers of neurons laid down during gestation and infancy remain unused and are never integrated into the CNS; these neurons are lost through apoptosis as the CNS matures during the first year of life. Cells that grow and develop retain the capacity to establish dendritic connections throughout life. Neuron numbers and dendrite connections to one another begin to decline following the attainment of MRP, leading to lower brain mass and neural dysfunction. Loss to intrinsic fragility and damage from external and internal stressors may produce mental defects ranging from loss of specific skills (e.g., speech, hearing, vision, motor capabilities, memory) to loss of more generalized functions (e.g., information processing or distribution) (Arking 1991, 1998).

Human brains lose both volume and weight with increasing age. Weight declines progressively from about age 55 to 80 years, averaging about 11% total loss over the adult ages (Davison 1987). Based on cross-sectional studies, human brain volume remains relatively stable, or enlarges slightly, from the third (ages 20–29) through to the sixth decade of life (ages 50–59); some individuals show volume expansion due to enlargement of both extracerebral and intracerebral spaces (Arking 1991). Decreases in neuronal numbers seem to occur mostly after age 60, with the most extensive losses, ranging from 25 to 45%, occurring in regions containing associative neurons of the cerebral cortex (neocortex, the new brain) where processes such as thought and memory are located (Duara *et al.* 1985). Synthesis of neurotransmitters within neurons also declines over the life span and may lead to progressive alterations in function and increases in behavioral alterations or disorders (Arking 1991). Reductions in activity characterize the neurotransmitters acetylcholine, dopamine, noradrenaline, adrenaline, serotonin, and gamma-amino-butyric acid (GABA) in older brains, affecting multiple and various areas of the brain differentially (Whitbourne 1985). In turn, variations in activity levels of these neurotransmitters are linked to mental health problems at all ages. Both mtDNA mutations and lipofuscin (an inert, yellowish pigmented waste product formed from cellular metabolism and mitochondrial debris) also accumulate in aging non-dividing neurons where they may contribute to malfunction and death (Elson *et al.* 2001). These alterations are likely to reflect multiple senescent processes, ROS, mutation accumulation, wear-and-tear, and loss of housekeeping functions common to all long-lived cells.

Human brains also show differential loss of function across hemispheres, with the right showing more loss with age, perhaps reflecting variable rates of senescence (Lapidott 1987; Mitrushina *et al.* 1995a,b). Comparisons of left (verbal) and right (imaginary-based) hemisphere single word associations show poorer right brain ability in elders (ages 60–73 years) than in younger adults (21–26 years) (Lapidott 1987). Similarly, among right-handed persons aged 60–64 without handicap or illness, right-handed (left-brain) performance tasks are superior, suggesting greater loss of right hemisphere function (Mitrushina *et al.* 1995a,b). Training, use, and activity, in a generally right-handed species, likely contribute greatly to the retention of left hemisphere abilities among humans into late life. Vascular and metabolic processes and dysfunction also influence neurological function. Both HT and diabetes are associated with cognitive declines during mid-life in the Atherosclerosis Risk in Communities (ARIC) study (ages 47–70 years) (Knopman *et al.* 2001). Original neuron numbers, dendritic connections, life style, use, training, education, and other CDCs all influence late-life cognitive abilities and may retard or enhance senescent processes in, and loss of, neurons. Still, whether slow or rapid, neuronal loss proceeds regardless of external influences, and represents a chronic, irreversible, progressive, and degenerative process.

Dementia

Dementia is a general category of neural malfunction. Symptoms include loss of aspects of higher cognitive functioning (e.g., memory, judgment, language capabilities) along with changes in personality (e.g., irritability, paranoia, hostility, aggressiveness). Organically, the dementias are a heterogeneous assortment of disorders of the CNS. All lead to progressive loss of time/space orientation, muscular control, long- and short-term memory, and general neurological dysfunction. This spectrum of neurological and cognitive impairments occurs so commonly among elders that neurodegeneration is likely to be an epiphenomenon of brain senescence (Finch and Seeman 1999). Neuronal loss, like loss of oocytes, appears to characterize all mammals that live sufficiently long. Modern human neurons, as do those of most primates, survive quite long and well. The somas they participate in function well beyond the minimum necessary life span to attain MRP, reproduce, and fledge offspring for any mammal. As a group, the dementias represent variable dysfunctional endpoints resulting from an array of neurological alterations common to all human brains, producing progressive loss of cognitive and motor function (Strong and Garruto 1994; Broe and Creasey 1995). Senile dementia, senile dementia of the Alzheimer's type (SDAT), and Parkinson's disease (PD) are described conditions included in the diagnostic category. Senile dementia is in many ways a catch-all classification; any neurological condition in the elderly with associated dementia that fits neither SDAT nor PD is attributed to the class. All persons who survive sufficiently long show progressive and irreversible neuronal damage and loss of function associated with dementia, and some more than others.

Over the average human life span, most human brains function adequately, although they vary widely in size, weight, number of neurons, and number of dendritic connections per cell. Most of this variation is set *in utero* and infancy, although connections continue to respond to culture, diet, life style, and post-natal events. As human brains age, individual differences in initial cell numbers, resiliency, RC, defensive systems, vascularization, proteins, and environmental stressors produce variation in function, reserves, and losses. Some individuals show aspects of dementia as early as their fifth decade of life; others show no overt evidence of late-onset disease until their eighth or ninth decade of life, and some may even survive to 120 years or more with no obvious dementia (Perls 1995). Both genetically transmitted familial forms and sporadic (non-familial), milder early- and late-onset forms of SDAT have been described (Arking 1991). Wide variability in the age of onset, rates, and degrees of neuronal and functional loss and amyloid deposition, and number of neurofibrillary tangles (NFTs), between patients suggests either a continuum of risk or multiple overlapping categories of loss and dysfunction. Multiple factors – alleles, gene–environment interactions, diet, nutrition, cultural factors, and perhaps even microorganisms – may influence the rates of cognitive loss and dementia in humans. Wallace (1992a,b) suggests some dementias are secondary to alterations in the proportions of mtDNA types in long-lived neurons. Other possibilities are that neurological alterations occur following exposures to oxidative damage, filament-producing microorganisms, or innate propensities to deposit minerals in neurons. Individuals with PD show reduced SOD and catalase activity (Bowling and Beal 1995; de la Torre *et al.* 1996) and mutations in SOD are found in a subset of individuals with familial ALS (Rosen *et al.* 1993). In addition, oxidative stress appears to contribute to some forms of Alzheimer's disease (Multhaup *et al.* 1997). Persons experiencing dementia at earlier ages are likely to be more susceptible to or incur more damage from these processes than do others who show late onset or no dementia. Those presenting with early dementia likely represent subgroups prone to rapid neurological degeneration due to increased genetic susceptibility and/or cultural/environmental factors predisposing to earlier symptomology.

SDAT increases with age and is a leading cause of morbidity among the elderly. SDAT affects about 5% of all persons aged 65+ and 20% of those over 85 (Arking 1991; Clark 1999). SDAT is characterized by a high incidence of NFTs in the anterior temporal lobe compared with "normal" individuals. Most affected are the pyramidal neurons of the associational areas of the brain, cells that are larger and have more interconnections with other cells than do other neurons (Clark 1999). Chromosomes 1, 14, 19, 21, and 22 have all been implicated as carrying risks for some familial forms, whereas the *Apoε*4* allele (*Apoε*4*, chromosome 19) is associated with some late-onset cases (Namba *et al.* 1991; Cotton 1994). Individuals homozygous for the *Apoε*4* allele have the highest risk of developing late-onset SDAT, while heterozygotes have a more moderate risk, and those without any *Apoε*4* alleles have the lowest risk. Those with two *Apoε*4* alleles show an age of onset before 70, while, among those with no *ε*4* alleles, onset averages over 85 years (Cotton 1994). The former show an 8-fold greater risk for SDAT, with their symptoms occurring almost a full decade earlier on average (68 years) than individuals carrying two *Apoε*3* alleles (75–85 years) (Cotton 1994; Perls 1995). Upon autopsy, the *ε*4* allele is significantly more frequent among 85+ year-olds with SDAT than those without (Tuomo *et al.* 1996). *β*-amyloid protein deposition in neurons also increases across *Apoe* genotypes, being lowest in *ε2/ε3* heterozygotes, intermediate in *ε3/ε3* homozygotes, and highest in *ε3/ε4* heterozygotes (the lone *ε4/ε4* genotype in the sample was diagnosed with SDAT before death) (Tuomo *et al.* 1996). Specific cells in the CNS (astrocytes and oligodendrocytes) produce ApoE protein, the major apoliopoprotein in cerebrospinal fluid, and ApoE immunoreactivity characterizes both the NFTs and amyloid deposits found in the senile plaques and cerebral arteries of SDAT patients (Namba *et al.* 1991; Cotton 1994). Current evidence suggests that ApoE proteins are tightly linked to these deposits and that ApoE4 protein is bound tighter and more frequently (Namba *et al.* 1991; Cotton 1994). Of all known factors, the *Apoε*4* allele confers the greatest risk for SDAT (Cotton 1994). In the brain, ApoE plays several major roles. Although the proteins each differ by only one amino acid from ApoE4, ApoE3 and ApoE2 differ in their activity levels. ApoE3 enhances branching of cultured neurites more than E4, and E4 binds more poorly to tau (the major component of NFTs) and other dendritic proteins, which may lead to destabilized microtubule structures and ultimate degeneration into tangles (Cotton 1994).

SDAT exemplifies the interplay of genes, environment, and life style expected of a senescent process. The *Apoε*4* allele is a documented risk factor for late-onset (after age 75) familial forms of Alzheimer's disease (Breitner *et al.* 1995). However, most cases of SDAT are sporadic (not familial), occurring in those without *Apoε*4* alleles. Further, most are of the early-onset type, suggesting non-genetic environmental or life style precursors (Breitner *et al.* 1995) e.g., that life style ameliorates risk. Diagnosis of familial late-onset *Apoε*4*-related SDAT occurs at significantly later ages in those with advanced educational degrees compared with those with only a high school education (Plassman *et al.* 1995). Women are at greater risk (adjusted relative risk of 1.54) than are men (Anderson *et al.* 1999). The incidence of SDAT rises more rapidly among European women (ages 65–69: 2/1000 person-years; 75–79: 11.8; 85–89: 44.3; 90+: 81.7) than men (0.3, 8.0, 38.6, respectively, falling to 24/1000 person-years at ages >90) (Anderson *et al.* 1999). In general, late-onset forms of SDAT do not appear to be associated with the same loci as are early-onset forms. Mutations in presenilin-1 and -2 may account for a majority of early-onset familial Alzheimer's disease, while the *Apoε*4* allele is observed in over 70% of late-onset cases (Clark 1999). Mutations in either presenilin-1 (PS-1, chromosome 14) or presenilin-2 (PS-2, chromosome 1) account for about 75%, while those of the amyloid precursor protein (APP, chromosome 21) account for 3–5% of early-onset (45–60 years) cases (Clark 1999). These mutations result in the production of a form of soluble amyloid protein that is slightly longer (41 or 42 amino acids) than the normal amyloid protein (40 amino acid); these longer forms condense more rapidly to produce amyloid plaques, which are the hallmark of SDAT (Clark 1999). Apolipoprotein, particularly ApoE4, also binds to intercellular amyloid, condensing at the specific sites that soluble amyloid molecules bind one with another to form the insoluble amyloid sheets found in plaques (Clark 1999).

Among healthy late-life survivors (average age of 93), only 14% carry one or more *Apoε*4* alleles; among those aged 85 years, 18%, and, among those younger than 65 years, 25% carry at least one *Apoε*4* allele (Perls 1995). Only 5.2% of centenarians, but 11.2% of controls, carry even one *Apoε*4* allele, while 12.8% and 6.8%, respectively, carry an ε*2 allele (Schächter *et al.* 1994). *Apoε*4* alleles appear to increase directly risks for hypercholesterolemia, stroke, CHD, and artherosclerosis during middle and later ages (Bockxmeer 1994). The reduced frequency of *Apoε*4* alleles in older cohorts suggests selective mortality of carriers from later adulthood throughout the remainder of life, a characteristic expected of a senescence-enhancing allele.

The underlying lesion(s) in PD remains unknown. Both genetic predispositions and environmental triggers seem to be involved. Among monozygotic twins, when the proband is diagnosed before age 50, the concordance rate is 1.0; for dizygotic twins, it is 0.167, yielding a relative risk (RR) of 6 (Tanner *et al.* 1999). The World War II Veterans Twins Registry data, the most comprehensive available, indicate that genetic factors do not increase the risk for PD after age 50, the typical case, with a monozygotic concordance of 0.155 and dizygotic concordance of 0.111 (Tanner *et al.* 1999). In England and Wales, idiopathic PD clusters by year of birth (1900–1930 cohorts; $p < 0.01$) and 4 of 6 years with clusters coincide with influenza pandemics (1892, 1918–19; 1929; $p < 0.005$; Mattock *et al.* 1988). PD births cluster in May, or March–June when divided into three equal periods, and the ratio of PD patients to non-PD controls correlates with observed influenza mortality ($r = 0.503$; $p < 0.005$) (Mattock *et al.* 1988; Torrey *et al.* 2000). Some influenza viruses are known to have neurotoxic effects. These may damage the fetal substantia nigra, leading to limited striatial neurochemical reserves in the neonate and a reduced nigral cell count; during later life, cellular involution or environmental neurotoxins may further decrease reserves, causing failure of the substantia nigra and PD (Mattock *et al.* 1988). While PD shows an excess of spring births, SDAT shows a non-significant winter excess (Torrey *et al.* 2000).

Neurodysregulation

Dysregulation of the HPA of the sympathetic nervous system may drive a variety of senescence processes. Finch and Seeman (1999) suggest that "... stress mechanisms interact at many levels with processes of aging ... ", while "... neuroendocrine reactivity might be related more generally to increased risks for disease and disability" (p. 83, 1999). Over time, multiple stressors lead to "... age-related ... resetting of the HPA axis poststimulation ... ", thereby leading to "... age-related increases in HPA dysregulation ... " (Finch and Seeman 1999, p. 83–4). Individual differences in the HPA may be set as early as the pre-natal environment and may even extend to the intrauterine environment provided by one's grandmother (Barker 1998; Cristafolo *et al.* 1999; Finch and Seeman 1999). The HPA orchestrates somatic homeostasis. Decreased or delayed responses to downregulating signals with age prolong somatic exposures to elevated glucocorticoids (GC) and to other GC-responsive hormones and proteins. Over the long term, chronic stress from high hormone levels could "... increase glucose/lipids, immunosuppression, and lead to increases (in) cardiovascular tone ... ", thereby causing "... mobilization of energy, and increased cv output and increased function in an emergency ... " (Finch and Seeman 1999, p. 83–4). Differences in initial and reserve capacity of HPA organs or their activity and resiliency may be set during pre-natal development. Data on middle-aged adults from a variety of European nations suggest that cortisol differences correlate significantly with both birth weight and abdominal girth at birth (Barker 1998), suggesting early environmental programming. Additional support for this model comes from the "... decreased HP sensitivity to negative feedback of GC and ... (the) ... loss of GC receptors in the hippocampus ... (that) ... cumulative exposure to GC potentiates ... " in rodents (Finch and Seeman 1999, p. 84). "... (I)ndividual differences in reactivity of the HPA appear stable over the life span and could lead to individual variation in neuroendocrine aging, disease, and disability" (Finch and Seeman 1999, p. 86).

The major GC in humans is cortisol, which increases immediately in response to stress. With increasing age, cortisol tends to remain elevated for longer periods of time before returning to its set-point. In cosmopolitan/modern and transitional culture settings, loss of traditional life ways, values, and rewards is thought to lead to chronic stress, chronic arousal of the autonomic nervous system, increased release of GCs, elevated BP, and, subsequently, somatic frailty. In humans "...astrocytes and neurons...(of the) hippocampus...the seat of declarative memory...show extreme sensitivity to GCs with aging...and GCs are associated with age-related neurodegeneration in the hippocampus..." (Finch and Seeman 1999, p. 84). In rodents, a specific protein, "Glial Fibrillary Acidic Protein (GFAP)...located in the cytoplasm of fibrous astrocytes...is now established as a general marker for aging in mammalian brains..." (Finch and Seeman 1999, p. 86). In humans, there is little change in initial cortisol response to ACTH (adrenal corticotropic hormone) or to hyperglycemia or corticotrophin-releasing hormone (CRH) with age; however, "...post-challenge recovery and resetting of HPA axis, suggests age-related prolongation of HPA response to ACTH and CRH" (Finch and Seeman 1999, p. 86).

Following menopause, women tend to show higher GC levels, social interaction and personal control seem to attenuate HPA responsiveness, and patients with certain CDCs (e.g., diabetes mellitus, hypertension, cancer, coronary artery disease, depression) show prolonged cortisol responses and less suppression of release by circulating GCs and glucose (Finch and Seeman 1999, p. 89). Decreases in HPA resiliency are not uniform across individuals; they are, however, life long, cumulative, influenced by life style, and may be irreversible (Finch and Seeman 1999, pp. 91–4). Such loss may also be an inevitable outcome of continued somatic response and reactivity to environmental stresses (Masoro 1996, McEwen 1998). As exposure to GC increases, HPA resiliency decreases, and risks for artherosclerosis, diabetes mellitus, cancer, and CHD increase. Elevated cortisol produces similar results in laboratory rodents and non-human primates. Elevated GCs also participate in the rapid post-reproductive decline of Pacific salmon following the release of their gametes (Finch 1994; Cristafolo *et al.* 1999; Finch and Seeman 1999). In humans, higher HPA and CNS activity are associated with personality type A (Finch and Seeman 1999). This may add to lifetime wear-and-tear on the neurological and cardiovascular systems of these types, and in part explain the marginal relationship of type A personality with CDCs and life span. Social stress and emotional loss may also increase HPA activity, contribute to loss of RC, and reduce resilience with age, thereby leading to losses in other systems. Loss of HPA resilience may partly explain observed links between socioemotional stress and immune dysfunction at all ages. Regardless of their etiology, high levels of GCs are a potent risk factor for early mortality and shortened life span. Dysregulation of the HPA and increased levels of circulating GCs during adult life conform to what is expected of a senescent process, suggesting that the sympathetic nervous system may potentate human neurological senescence.

Endocrine systems

The endocrine system monitors both the internal and external environment of the soma. In response to physiological alterations, it produces hormones that are important to all aspects of human survival and reproduction, the master regulators of life history (Finch and Rose 1995). Endocrine hormones modulate the timing, rates, and progression of conception/reproduction, growth and development, metabolism, responses to stressors, immune function, and body habitus during all phases of life. Endocrine hormones actively initiate and schedule life history events that lead to transitions across life stages, regulating all aspects. Endocrine hormones also monitor internal physiological processes such as serum glucose. Endocrine hormones are natural candidates for antagonistic pleiotropy, affect all physiological processes and fitness components, and are available for multiple epistatic relationships (Finch and Rose 1995, p. 36), characteristics expected of a system influencing senescence. That different hormone subsets are activated during each life history phase further suggests that hormones influence senescence, even if only by their relative absence (Finch and Rose 1995). Hormones act not only as modulators of cell and organ function (endocrine regulation) but also locally on nearby cells through diffusion into the extracellular spaces (paracrine), may affect the same cell from which they were released (autocrine), or may act within a single cell (intracrine) (Finch and Rose 1995, p. 4). Animals use their sensory and chemosensory arrays to monitor the environment and pass signals along their neural network to central processors in the brain. Responses to these signals rely on neuroendocrine connections to release a variety of hormones that adjust cellular activities and DNA expression in target cells and organs. Neuroendocrine functions are basic to life, respond to early environmental events, and structure long-term responses to the environment.

Multiple cell receptors and intracellular messengers, along with variable sensitivities of receptors and tissues, allow a wide latitude in target cell/organ responses to hormonal stimulation (Figure 4.2). At different life history stages, responses may be regulated by proportional and activity differences of multiple cell receptor types and intracellular cytokines (Finch and Rose 1995). Variable transcription factors, hormone processing peptides, receptor isoforms, and hormone-responsive intracellular elements permit wide inter- and intra-individual flexibility in endocrine regulation and response. This provides the neuroendocrine system with many avenues for programming species-specific life history and regulating species-specific senescent processes. All allelic variants influencing hormonal timing of developmental transitions potentially alter life history (Finch and Rose 1995). Loci involved in producing hormones, their receptors, and intracellular messengers likely have multiple alleles with

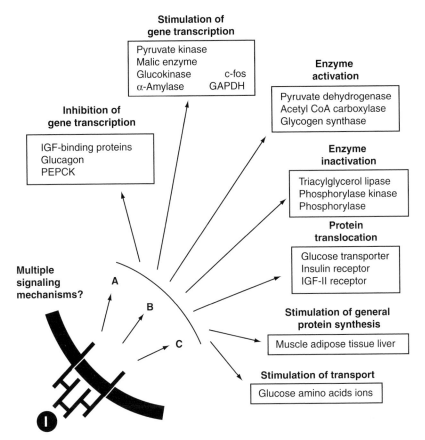

Figure 4.2 The intracellular response to insulin (reprinted with permission, Gerber and Crews, 1999, p. 154).

their own variable rates of transcription and activity and patterns of regulation. Additionally, all such loci may receive additional information and direction from life history complexes located elsewhere. Such a condition would then allow almost any "...gene to acquire importance in hormonal or neural mechanisms in life history evolution." (Finch and Rose 1995, p. 30).

Multiple and overlapping controls on hormone production, release, and activity allow wide variation in and rapid responses to selection pressures on life history characteristics (Finch and Rose 1995, p. 37). Alleles altering hormone signaling and enhancing life history (i.e., increasing inclusive fitness) are favored by natural selection, while those disrupting life history are not (Finch and Rose 1995). Most alleles at polymorphic loci should produce proteins that complete their primary function(s) with about equal reliability. Common alleles should not be strongly associated with reproductive success; they should be selectively neutral. Later-life pleiotropic effects of such "neutral" alleles should be highly variable and random (e.g., ACE, ApoE). Over time, natural selection weeds out those that limit RS, leaving a set of alleles that show about equal long-term reproductive benefits in the prevailing environment (i.e., the normal/average/common phenotypes). Multiple temporary and unstable compromises among alleles, variable environments, and sociocultural pressures produced quite variable patterns of growth, development, reproduction, and survival among *Homo* species over the evolutionary line leading to modern *H. sapiens*.

Loss of hormone receptors, messengers, cytokines, and processing proteins produces no or inappropriate signaling and activity throughout the soma. Accumulating over time, endocrine dysregulation and loss of hormonal control could pace senescent disintegration, just as growth and development of the system paced integration. Multiple overlapping components allow the endocrine system to show highly variable patterns of timing and activity across species (Finch and Rose 1995). Such variability is "... consistent with sets of genes that become assembled in such complexes as the MHC" (Finch and Rose 1995, p. 29). Systems such as the endocrine, neurological, MHC/HLA, and immunological systems, provide opportunities for even slight alterations in DNA to affect multiple phenotypic outcomes and to resonate across all phases of life history.

Immunological systems

The immune system is constantly attacking invading microbes, rogue self-cells, and non-self proteins. Immune function is directly controlled by the CNS through innervation of the hypothalamus (which controls the endocrine system), bone marrow (which produces red and white blood cells, and B cells), and thymus (which produces naive T cells and self-marker antigens) (Arking 1998). Most immune response tissues, including the spleen and lymph nodes, are densely innervated. In some circumstances, nerve endings and immune cells establish synapse-like connections (Sternberg 1997). Peripheral cytokines produced by immune cells also interact with the neuroendocrine system (HPA axis) to stimulate fever, physiological symptoms of sickness, and sleep patterns (Sternberg 1997).

The human immune system includes a variety of white blood cell types (leucocytes) differentiated physiologically by their activities and histologically by their variable staining patterns. Phagocytes (neutrophilic cells that ingest/engulf microbes, foreign proteins, and damaged cells) circulating in the blood are the first internal defense against pathogens. Other leucocytes attack and destroy invading proteins (eosinophils), transport heparin and histamine into wounded areas causing inflammation (basophils), or are held as undifferentiated lymphocytes available for recruitment and maturation (immature cells). B cells are produced by bone marrow. These lymphocytes mature in response to an immune challenge and produce antibodies to invading proteins. T cells are stimulatory cells of the immune system. They bring invading protein markers to immature B cells, thereby initiating an immune reaction. T cells also participate in cell-to-cell attacks that lyse and phagocytoze infected, cancerous, and rogue self-cells.

Our immune systems operate on at least three levels, non-specific, humoral, and cell mediated (Schlenker 1998). Non-specific immunity occurs in response to injury and infection when leucocytes act as part of the inflammatory response, and is dependent on lysosomal, phagocyte, and complement activity (Chandra 1992; Schlenker 1998). Lymphocytes and B cells produce immunoglobins and antibodies (humoral immunity) in response to specific foreign antigens, while cell-mediated immunity relies on T cells to identify/mark cells carrying foreign proteins (e.g., viruses, bacteria, prions) for apoptosis (Chandra 1992; Schlenker 1998). Older samples generally show a reduced or altered T cell function, lower antibody levels, a decline in bone marrow stem cells, and slower generation of fewer B cells when presented with an invading pathogen (Chandra 1992; Le Sourd 1995; Schlenker 1998). A progressive overall decline in cell-mediated immunity and T cell function with increasing age occurs. These alterations increase susceptibility to CDCs, specifically cancer and CVD, and increase risk of death with increasing age. They also are cumulative, progressive, and irreversible.

Numerous age-related alterations plague the immune system. These appear to reflect systemic dysregulation (LeSourd 1997) or breakdown (Imahori 1992), and many are likely to be senescent alterations (Walford 1984). Best known is involution of the thymus (composed largely of lymphoid tissue, located in the upper anterior chest or base of the neck, aids development of the immune system) at maturity, due mainly to atrophy of its cortex. The thymus develops during the pre-reproductive stage of most vertebrates, but disappears or becomes rudimentary in adults. In humans, it involutes following attainment of MRP during the early twenties. Despite thymic involution, total circulating lymphocytes remain relatively stable over the life span; however, proportions of various subpopulations of T lymphocytes alter with age (Arking 1991). One plausible explanation for thymic involution is that, with age, the body tends to increase its production of autoreactive antibodies (antibodies that attack self-cells). Thymic involution may, by limiting the number of self-marker antigens, reduce the probability of producing autoreactive antibodies and prevent late-occurring somatic mutations in self-marker loci from being transcribed. An age-related increased prevalence of autoimmune disease (when autoreactive antibodies attack the soma, e.g., multiple sclerosis, systemic lupus erythematosus) in humans supports this hypothesis. Still, most individuals do not show autoimmune disease, perhaps because they lack autoimmune related HLA specificities. Thymic involution also limits the total number of naive T cells available for initiation by antigens. This leads to a decreased ability to repel new antigens with increasing age, although the ability to respond to previously encountered antigens remains (memory T cells). Overall, the immune response becomes sluggish and less effective as one ages, and loss of immune competence seems to keep pace with general senescence.

Multiple aspects of the human immune system alter with increasing age (Mayer 1994). Immunoglobulins G (IgG), IgA, and IgM and the ABO antibodies (circulating antibodies always present in the blood) respond to invading antigens by coating the surfaces of microbes and marking cells for phagocytosis. Among those who survive to ages over 70, there appear to be higher circulating levels of IgG relative to IgM in plasma (Buckley and Roseman 1976), while other reports suggest that IgG concentration is decreased, while IgA increases (Chandra 1992). It is also possible that individuals with O blood type of the ABO system, as opposed to A, B, or AB, may be more common among elders, at least in more cosmopolitan settings. Interleukin-2 production also decreases with increasing age (Lakatta 1990; Mayer 1994). A reduced ability to eliminate extrinsic and monitor intrinsic cellular transformations increases the risk of death with increasing age. Still, it is not likely that the immune system regulates senescence in any broad sense, such as suggested for the neurological and endocrine components. Rather, genetic variability within the immune system likely provides some individuals with more efficient, versatile, or longer-lasting systems allowing them to evade wear-and-tear and death from microorganisms and cancerous cells and express their intrinsic somatic vitality in other physiological systems.

The HLA system

The HLA system, the MHC in other organisms, regulates histocompatibility, self-identifications, and resistance to specific infectious organisms. Since the 1970s, numerous studies of different ethnic populations have examined HLA phenotypes in elders and compared frequencies with those of middle-aged and younger samples. Some alleles and haplotypes at HLA loci show clear associations with autoimmunity and pathogen resistance. Some are related to disease processes that reduce survival to old age, showing post-natal or age-specific selection pressures. Others appear to exert sex-specific effects on longevity. A wide variety of HLA antigens have been identified as possible markers for longevity across different populations, likely reflecting environmental, ethnic, historical, and ecological variability (Macurová *et al.* 1975; Greenberg and Yunis 1978; Yarnell *et al.* 1979; Proust *et al.* 1982; Thompson *et al.* 1984; Takata *et al.* 1987; Yunis and Salezar 1993; Papasteriades *et al.* 1997). Because of its multiple pleiotropic effects and influences on all phases of life history and life span, the MHC system may be a life history gene complex (Finch and Rose 1995). However, most alleles/haplotypes appear to be unrelated to longevity and no strong Methuselah haplotypes have been identified.

Multiple examples of HLA associations with longevity exist. In Greek elders (ages 75–104 years) and middle-aged controls (ages 18–65 years), the B16 and Dr7 antigens were observed at significantly higher frequencies in the aged, while the B15 and Dr4 antigens and B8Dr3 haplotype were seen at lower frequencies (Papasteriades *et al.* 1997). In addition, sex-specific differences were noted, with aged men showing increased frequencies of the B16 and B21, and decreased Dr7, B8, and B15, alleles. Papasteriades *et al.* (1997) suggest that in women the HLA-Dr7 antigen, and in men the HLA-B16, are markers for longevity in Greece. The HLA-A1-B8 linkage group appears to carry a survival disadvantage for women, but not men, over the age of 70 years, also suggesting sex-specific effects (Greenberg and Yunis 1978). The HLA-B8 allele is associated with several autoimmune disorders (e.g., type I diabetes, Addison's and Graves's diseases, myasthenia gravis) and occurs at a reduced frequency among older members of several populations (Greenberg and Yunis 1978). Among Japanese centenarians and nonagenarians, the HLA-Dr1 phenotype is seen at an increased frequency (Takata *et al.* 1987), while in the Netherlands Dr5 shows most difference in frequency between elders and non-elders, more so in women (Lagaay *et al.* 1991). Elsewhere, the HLA-B40 antigen is more frequent among elders, while the A1 and B27 are less frequent (Macurová *et al.* 1975; Yarnell *et al.* 1979), a finding not replicated in other samples (Hansen *et al.* 1977; Lagaay *et al.* 1991). Both the HLA-Dr 3 phenotype and the A1B8Cw7Dr3 haplotype are over-represented in very long-lived German men (Proust *et al.* 1982), while the A29, B27, and B35 antigens are more frequent in centenarians (Thompson *et al.* 1984). The increased frequency of haplotypes carrying A1B8 antigens reported by Proust *et al.* (1982) contrasts with Papasteriades *et al.*'s (1997) report that A1B8 haplotype is less frequent among Greek elders and of Macurová *et al.* (1975) that the A1B8 haplotype is less frequent in elder samples. Elsewhere, the HLA-B8 genotype declined with increasing age among women, but not men, while the Dr7, Dr11, and Dr13 alleles were more frequent and homozygosity for HLA alleles less frequent among the long lived (Ivanova *et al.* 1998).

Results from different studies of the HLA system and long life are seldom comparable. Most examine a variety of antigens and few show wide overlap. For those that are comparable, results are often contradictory across samples. Conflicting results suggest that multiple factors structure HLA distributions among different ethnic groups. Both gene–environment interactions and patterns of age-specific selection differ across samples, and generational birth cohorts may differ in haplotype distributions. Most studies of HLA and age have used classic two-step micro-lymphocytotoxic techniques that are not allele specific. The same immunoreactive phenotype may result from a variety of alleles. Different alleles segregating in different ethnic groups may show the same immunoreaction, but respond differentially to selective pressures. Development and application of allele-specific probes will further clarify HLA associations with life span (Severson *et al.* 1997; Severson 2001).

HLA allelic variability also shows direct associations with human life span. Specific autoimmune conditions are clearly associated with particular antigens (e.g. B8, B27, Dr3, Dr4) and these alleles are observed less frequently with increasing age regardless of associated haplotypes. Other phenotypes (e.g., B16, Dr7, B21, Dr1, Dr5, B40, A29, B35) seem to enhance longevity in some general fashion beyond a reduced risk for autoimmune diseases (Greenberg and Yunis 1978). These may provide broad resistance to deteriorative processes or show hyperhumoral reactivity, providing high host protection from infectious or neoplastic disease (Greenberg and Yunis 1978). Several HLA antigens show higher frequencies in either older men or women, but not both. This observation holds across ethnic groups and ecological settings, although the particular antigen may differ, supporting the hypothesis that associations of HLA alleles with retarded senescence are sex specific. Close associations of the HLA system with survival was to be expected, but the sex specificity of both autoimmune diseases and age-related selection against HLA alleles was not. Some HLA linkage groups in association with female hormones that alter immune regulation may produce a hyperimmune response at later ages and this may then predispose to the development of autoantibodies (Greenberg and Yunis 1978). More recent research suggests that HLA alleles influence aspects of reproduction, including mate selection and the frequency of spontaneous abortion. HLA haplotypes may influence the responsiveness of cells to hormones or HLA loci may be hormonally or immunologically modulated (Finch and Rose 1995). For example, expression of class I and II HLA loci is stimulated in many cell types by interferon-γ (Finch and Rose 1995). In rodent models, different MHC haplotypes show variable life spans as in humans, and the expression of SOD appears to be regulated by factors within the MHC system (Finch and Rose 1995; Cristofalo *et al.* 1999). The HLA system is likely to act in concert with the neuroendocrine system to regulate multiple aspects of human life history (Finch and Rose 1995).

Immunological senescence

Associations of increasing age with decreasing immune function and increased autoimmune diseases have been taken as support for an immunological theory of senescence (Walford 1984). An immunological component to senescence originally stemmed from the somatic mutation and error accumulation theories of senescence. Mutations were thought to convert self-marker antigens to "not-self" or alter immune cells to recognize somatic cells as "not-self". According to the immunological theory, thymic involution prevents increases of T lymphocytes with autoimmune properties as the thymus ages, accumulates somatic mutations, and produces non-self-recognizing clones. The thymic cortex turns cells over more rapidly than any other bodily organ. However, not more than 1% of all thymocytes survive to enter the extrathymic circulation; in addition, serious body stress (e.g., trauma, massive infection, starvation) accelerates thymic atrophy and decreases available thymocytes (Burnet 1970a,b). Proponents of the immunological theory suggest that exhaustion of the thymus-dependent immune system due to limited cell proliferation (the Hayflick limit) is a key factor in immune-driven somatic senescence (Burnet 1970a,b). Immune surveillance and restricting cellular anomalies arising secondary to somatic mutations are the main functions of thymus-dependent immunity; loss of this system should increase risks for cancer and autoimmunity similar to the situation which is observed during late life (Burnet 1970a,b).

Finch and Rose (1995, p. 1) reviewed evidence that: "the MHC and other complex loci may be considered as life history gene complexes, with pleiotropic influences throughout the life span". They (1995, p. 20) define a "life history gene complex" as a set of: "(1) closely linked genes that directly influence fitness components . . . (2) with some genes that show extensive polymorphisms; and (3) (with) genes in the complex that influence the scheduling of activities at one or more stages of life history and are highly pleiotropic." The HLA system shows large amounts of pleiotropy (similar to neuroendocrine hormones), acts at many stages during individual life history, and shows the highest level of polymorphism of any known set of loci (Finch and Rose 1995). The HLA system influences everything from reproduction through survival, including immune response, hormone receptor function, and metabolism of toxins and carcinogens (Finch and Rose 1995). The HLA system is also evolutionarily stable, including trans-species haplotypes shared across primate lineages since before the prosimian and anthropoid branches diverged 85 million years ago (Satta *et al.* 1994; Finch and Rose 1995).

Although immunosenescence appears to be a fact of life in humans, many components of the immune system do not decline in function in a post-maturational age-related fashion, and some may even increase (Mayer 1994). Still, when compared with younger cohorts, older adults are more susceptible to a variety of infections and seem to suffer more severe consequences (Mayer 1994). Rather than loss of ability to respond, this suggests loss of either RC to mount an adequate immune response or slowed response due to lowered surveillance, detection, or production capabilities. Numerous aspects of the immune system undergo alterations with increasing age (Walford 1984; Chandra 1992; Hirokawa 1992; Imahori 1992; LeSourd 1997; Schlenker 1998). However, most are far from universal and many may be reversed with alterations in nutrition and activity. In addition, age-related immune dysfunction may occur secondary to declines in other physiological areas – e.g., age-related dysfunction or loss of hormone receptors (Arking 1998). Immune dysfunction results in multiple stressors, and increased infectious disease, cancer, autoimmune disorders, allergies, and increased healing time (Chandra 1992; LeSourd 1995; Schlenker 1998), all of which may influence life span and senescence.

The immune system does not act in isolation; it is but one part of a complex psychoneuroendocrine- immune system. Alterations in either hormone or neurotransmitter functions, whether long or short term, also alter immune function. Life events and stress promote the sustained release of stress hormones and are associated with reductions in circulating antibodies, slowed wound healing, and more episodes of infectious disease (Thomas *et al.* 1985; Kiecolt-Glasser *et al.* 1987; Kaplan 1991; McIntosh *et al.* 1993). Psychological factors directly affect immune function, in addition to the major role of GCs. For example, depression (dysphoria) is associated with immune suppression, while greater social support is associated with greater total lymphocyte counts (Thomas *et al.* 1985) and enhanced immune response (McIntosh *et al.* 1993). Social support is also significantly correlated with total lymphocyte counts, and more strongly so in women than men; thus, social support may not only buffer individuals from physiological responses to stress, but women may benefit more than do men (Thomas *et al.* 1985). Data on life events and immune function also show differential influences on men and women. Among elderly women, but not men (aged 58 and older), disruptive family stress or legal problems are associated with higher lymphocyte counts, while, for men, only recent sexual dysfunction is associated with reduced counts (McIntosh *et al.* 1993). Apparently, in addition to defense, the immune system functions as a regulatory system interacting with the psychoneuroendocrine system (Hirokaw 1992).

Older adults are also more likely than younger ones to be under- or malnourished and to have specific deficiencies in protein, iron, thiamine, and vitamins C and D (Tucker 1995). These conditions may increase susceptibility to infections and parasitic conditions, reduce immunocompetence, and lead to slowed or lost immune function. For example, immune function in elderly samples improves with vitamin, mineral, calorie, or protein supplements (Chandra 1992; LeSourd *et al.* 1998). Relatively minor nutritional differences may yield significant increases in life span; intake of vitamin C by older National Health and Nutrition Examination Survey (NHANES) participants, for example, is associated with a 6-year greater average life span of men, but only 2 years for women (Bendich 1995). Men may be at greater risk for the ill effects of malnutrition and subsequent lower immune function than women as they are to many environmental impacts throughout life (Stinson 1985; Stini 1994). Although immune function declines in most humans with advancing age, immune losses are not necessarily innate biological senescence. Immune dysfunction/loss results from multiple psychosocial/cultural, nutritional, and environmental factors, and these overlay innate differences in cell numbers and organ capacity and degenerative alterations occurring throughout life. Whatever the cause, declining immune competence is cumulative and progressive, increases susceptibilities to infectious and CDCs, and reduces life span among elders (Hirokawa 1992).

Even though immune function declines as humans age, data do not support models suggesting that this decline is the underlying cause of senescence. For example, although autoantibodies increase with age in humans and other mammals, few elders actually suffer from any overt autoimmune disease (Greenberg and Yunis 1978). Also, the decline in cell-mediated immune function often reported with increased age is not due to a general decrease in T lymphocytes (Greenberg and Yunis 1978; Arking 1991; Cristofolo *et al.* 1999). Antibody responses depending on T cell generated helper cells are, however, slowed and newly generated antibodies show less antigen affinity in older samples (Chandra 1992). These findings may reflect selective loss of specific cell types – for example, of T-helper cells – due to processes outside the immune system, as is seen with immunodeficiency viruses, or errors in antibody-generating mechanisms. Another problem for the immune model of senescence is that age-specific cancer mortality curves do not conform to age-related patterns of loss in cell-mediated immune function or humoral responses to T-dependent antigens (Greenberg and Yunis 1978; Weiss 1989a). Even allowing for lag periods in cancer mortality, which would follow if loss of thymic function decreased immunosurveillance for abnormal cells and increased cancer risk, little association is observed. Elders with defective cell-mediated immunity do show decreased life expectancy, as compared with those with normal T-cell function (Greenberg and Yunis 1978), and poor resistance to infections at older ages is associated with a decreased proportion of naive T cells (Cristofalo *et al.* 1999). Thus, one of the primary age-related alterations of the immune system appears to be loss of T-cell functional capacity (Greenberg and Yunis 1978; Cristofalo *et al.* 1999). This may be associated with variable HLA expression, which is necessary for T-cell activation. This provides a mechanism by which certain HLA phenotypes, those that continue to promote T-cell activation, could become more common among survivors to old age and thereby promote variable patterns of life history, senescence, and life span.

Variation in human life span

Patterns of morbidity, mortality, and longevity vary across all populations, from nation to nation, between cities within nations, and between subgroupings within national and municipal boundaries (Tables 1.3, 2.1, 2.2). Traditional and convenient methods for grouping populations within these various levels of sociopolitical organization generally have relied on age, sex, and race. Numerous national surveys and censuses in the U.S.A. show lower life expectancy among subgroups labeled "African-American/Black" as compared with subgroups commonly labeled "European-American Caucasian/White" – poorly defined terms (Crews and Bindon 1991; Goodman and Leatherman 1999; Silva and Crews unpublished data; recall here and elsewhere that population/ethnic/racial terms used in other publications are enclosed in quotes. In the U.S.A., lower survival rates for African-Americans are observed at most ages, from infancy, through adulthood, and into old age. African-Americans also suffer higher morbidity and mortality rates for all major causes of death (Rogers 1992; Perls 1995; Rakowski and Pearlman 1995; Sorlie *et al.* 1995). This across the board difference in mortality rates likely follows from multiple factors, including poverty, discrimination, and lack of adequate health care, cultural/behavioral propensities, and genetic variation.

Based on early data, a major contributor to such differential survival was thought to be unequal access to resources and care. Those who tended to die earlier were often disadvantaged and had less access to health care (Perls 1995). However, even in samples selected to control for economic and educational differences, "African-Americans" still experience greater morbidity from major CDCs (NIDDM, stroke, hypertension, obesity) (Rogers 1992; Sorlie *et al.* 1995). In the U.S.A., at all ages up to about 70 years, African-Americans continue to experience more frequent mortality from chronic and infectious diseases, maternal and perinatal conditions, and trauma (accidents/homicide/suicide), along with having elevated levels of most of the established risk factors (e.g., serum cholesterol, triglycerides, LDL, uric acid, glucose, weight, and BP) for CDCs than do European-Americans. Socioeconomic factors alone do not lead to higher morbidity and mortality among this subgroup of the U.S. population. Rather, a web of causality linking genes, environment, and culture produces differential mortality among various subgroups and age groups. Although socioeconomic factors influence survival among

African-Americans, they fail to explain completely morbidity and mortality differentials.

At older ages, African-Americans show better survivorship than do European-Americans. The age at which this "mortality cross-over" occurs has increased in recent decades (Perls 1995). In the early twentieth century, it occurred at about 50 or 60 years; currently, it occurs at about 85 years. A similar mortality cross-over occurs between men and women at about age 85 (Perls 1995). At all ages below 80 years, men show poorer survival than women. Conversely, men who survive to 90 years tend to be less affected by Alzheimer's disease and show better average mental and physical health than same-age women (Perls 1995). Membership in any "minority" subgrouping of the U.S. census is associated with a greater risk of morbidity and mortality at all ages from conception to late life. Those capable of surviving the multiple risks associated with "minority" status probably tend to have greater innate vitality. Although various combinations of genetic, environmental, and cultural phenomona determine this innate vitality, genetic factors appear to be the major influence (Schoenfeld *et al.* 1994; Sorlie *et al.* 1995; Perls 2001). Those without winning gene, environment, and cultural combinations are frailer; these individuals generally die younger, leaving those with greater vitality (healthier) more represented at later ages.

Compression of the cross-over between African-American and European-American subgroups in the U.S.A. to ages beyond 85 years occurred in the late, compared with 50 or 60 years in earlier, decades of the twentieth century. Gene–environment–cultural interactions that structured previous mortality in these groups were altered in recent decades. This has allowed more members of both groups to achieve more of their potential life spans. Socioeconomic and environmental factors may now be more equal, allowing more African-Americans to survive to late life and express their innate vitality. Still, at all ages up to at least age 65, mortality rates among minority subgroups (including native Americans, Pacific islanders, Hispanic Americans, and African-Americans) in the U.S.A. remain above those of the majority comparison groups. In addition, regardless of race or ethnicity, poorer, less educated, and persons in the lowest occupational classifications, along with the unemployed – conditions that affect proportionally more minority individuals – are at greater risk of morbidity and death than those who are better advantaged (Sorlie *et al.* 1995). In the U.S.A. today, the majority of those over the age of 65 are classified as "white" (86%); the remainder are classified as being of various minorities (Rakowski and Pearlman 1995). By 2030, about 20% of the older group is expected to be "non-white", due to more rapid growth anticipated among minority elders in the next 30 years (e.g., "Hispanics" 395%, "blacks" 247%, "whites" 92%) (Rakowski and Pearlman 1995).

Race, ethnicity, culture, and senescence

Race is often viewed as among the most important categories in health, late-life survival, and senescence (Brach and Fraserirector 2000; Laveist *et al.* 2000). To others, race is an overused and poorly defined anachronism from another time and place (Montagu 1962; Marks 1994; Gould 1996; Williams 1998; Caspari *et al.* 2003). Culture and ethnicity are increasingly introduced into discussions of health, aging, senescence, mortality, and life span to replace the long-disparaged term "race". Unfortunately, very often neither ethnicity nor culture are used consistently within or across disciplines, nor are they consistently defined (Crews and Bindon 1991; Silva and Crews unpublished paper). Race is a complex and poorly defined combination of biological, cultural, socioeconomic, political, and legal factors. However, pragmatically, populations must be tabulated, even if in an arbitrary fashion, to compile health statistics, morbidity/mortality data, and complete congressionally mandated sociopolitical analyses. An alternative view is that "race" be used only to connote biology, since it is equivalent to a subspecies in formal taxonomic terms (Montagu 1962; Marks 1994; Gould 1996). Others have suggested that "ethnicity" be used for population tabulations and research since it represents a conglomerate of biology, behavior, ecology, culture, historical antecedents, ideology, and life ways that one acquires by virtue of being born into or living in a particular cultural/ecological setting (Montagu 1962; Crews and Bindon 1991; Silva and Crews unpublished paper). Modern humans represent a single subspecies (race), *H. sapiens sapiens*. They, along with Neanderthals (*H. sapien neanderthalensis*), are the most recent two biological races (subspecies) of H. sapiens to have existed. Although they are likely to have had distinct patterns of growth, development, maturation, and life history and ultimately went extinct as a subspecies, Neanderthals are likely to have contributed alleles to the gene pool of extant modern humans (Wolpoff 1998; Templeton 2002).

Some recent commentaries suggest that "race" categories are biologically meaningless and of little value in understanding human variation and adaptability (Montagu 1962; Crews and Bindon 1991; Marks 1994; Gould 1996). However, patterned external and internal differences between populations also reflect to some degree responses to variable ecological and cultural circumstances, random drift in small breeding isolates, sexual selection, and cultural influences on mate choice that have structured human physiological responses (Baker 1982; Gerber and Crews 1999). Clinical medicine provides many examples where "race" or "ethnicity" is important in determining treatment modalities, responses to pharmaceuticals, and adverse reactions (Gerber and Crews 1999). Specific types of anti-hypertensives control BP more effectively in "blacks" than they do in "whites", likely reflecting innate factors that although generating a similar phenotypic outcome are themselves variable. Similarly, in some "Asian" populations, cholesterol-reducing drugs must be administered at a 10-fold higher level to achieve the same "effectiveness" as in "U.S. white" controls (reviewed in Gerber and Crews 1999). A polymorphism of the alcohol dehydrogenase locus (ADH, the major enzyme for metabolizing ethyl alcohol) illustrates how genetic factors may produce ethnic variability in drug response. ADH comes in two major forms, fast and slow, resulting in homozygous fast and slow metabolizers and an intermediate heterozygote phenotype. This polymorphism shows variable distributions across populations. Asian-descended samples generally average about a 0.85 frequency of the "slow" allele, while European-derived samples show about a 0.85 frequency of the "fast" allele. For many Asians, once ingested, alcohol is retained longer in and builds to a higher proportion even with fewer grams ingested. Variable allele frequencies are reported for alleles at all loci yet examined across all human populations. Often, an allele found in one population at polymorphic frequency is absent from unrelated populations. Such variation reflects population history, including local environment, selection pressures, and cultural competencies, along with multiple random factors. Allelic variation at multiple such polymorphic loci structures disease risks, senescence, and life span across populations. This allows etiologies for multiple CDCs to differ in subtle ways across human groups and alters the effectiveness of clinical treatment modalities.

Epidemiological and demographic transitions

Although disciplines may emphasize either "epidemiological" (e.g., epidemiology, public health, medicine) or "demographic" (e.g., demography, anthropology, geography) transition theory, these constructs are widely applied in sociocultural and biomedical research (Omran 1971; Tietlebaum 1975; Omran, 1983). The epidemiological transition (transformation of leading causes of morbidity and mortality from infectious/parasitic to chronic degenerative following adoption of a suite of public health behaviors, improved sanitation, hygiene, medical care, and nutrition) occurs in three phases. Pre-transition is characterized by high death rates, particularly during childhood from infectious/parasitic diseases. Changes in medical technology and public health characterize the transition phase. These lead to fewer epidemics, reduced infant/childhood mortality, improved reproductive health, and longer survival. Post-transition is achieved when deaths from infectious/parasitic diseases are replaced by CDCs as the major cause of death and life expectancy increases (Omran 1971, 1983; Crews 1988). Epidemiological changes underlie the demographic transition. Pre-transition is characterized by high fertility and mortality rates, particularly during infancy and childhood and is related to reproduction in women. These generally balanced one another out, producing moderate population growth. Transition in demographic rates follows when improved nutrition, sanitation, and hygiene are adopted, improving health in all age groups. Subsequently, more of those born survive to reproduce, resulting in rapid population growth as high fertility prevails. At this time, a population's average age decreases, younger cohorts survive better than previous ones, and fertility remains well above replacement level. Following about 30 years of improved survival and increasing proportions of children to support, birth rates decline, population growth slows, the population ages (increased average age of the population), and a relatively stable population distribution is established. Forces reducing earlier pro-natalist tendencies are thought to include high ratios of dependants to working-age adults, lack of land, and reduced political, sociocultural, and religious support for large families. Transition theorists suggest that many European and European-derived populations experienced these alterations during the nineteenth and twentieth centuries, particularly following World War II.

Those experiencing epidemiological transitions during the latter part of the twentieth century have followed different demographic paths (Tietlebaum 1975; Crews 1988). Mortality rates declined neither as rapidly nor did they affect as broad a cross-section of these populations. Neither are infectious causes of death as well controlled in many of today's transitional populations, in part because of their tropical and semi-tropical environments (Crews 1988); nor have birth rates fallen as precipitously. These variations likely reflect in part pre-existing differences in initial fertility and mortality rates, religiosity, pro-natalism, marginalization, and poverty. In addition, for recently transitioning populations, mortality declines are precipitated through outside forces, resulting in more rapid declines than those observed earlier. Alterations in multiple sociocultural, political, and economic patterns (e.g., a world economy, media-sponsored lifestyles, and extractive capitalism) accompany the transition today. Although larger proportions of each cohort are surviving to older ages, population aging may be slow as younger age groups continue to expand. In some populations, the elderly (ages 65+) decrease as a proportion, while increasing in total numbers. Among populations who experienced transitions earlier in the twentieth century, over 95% of female newborns survive to adulthood (arbitrarily set at 16 years), a majority reproduces, and most women survive past menopause (>75%). Epidemiological and demographic transitions have led to a world population of over 6 billion and rapid growth in the number of persons aged 65+, producing national populations where as many as 20% are over age 65 (e.g. U.S.A., 12.6% in 1990, with 22% predicted for 2000, see Table 4.1).

Population aging increases dependency ratios (number of children and elders/working aged adults), a crude measure of economic dependency, and

Table 4.1 *Proportion of persons born surviving to select ages: U.S.A. 1900–02, 1949–50, 1998*

	Men						Women					
	1900–02		1949–50		1998		1900–02		1949–50		1998	
Age	White	Black	White	Black	White	Black	White	Black	White	Black	White	Black
1	86.7	74.7	96.9	94.9	99.4	98.4	88.9	78.5	97.6	95.9	99.5	98.7
5	80.9	64.4	96.4	93.9	99.2	92.2	93.4	68.1	97.2	95.1	99.5	98.5
20	76.4	56.7	95.1	91.9	98.6	97.1	79.0	59.0	96.4	93.5	99.0	98.1
40	65.0	43.0	91.2	82.8	95.7	91.4	67.9	46.2	94.1	86.5	97.7	95.3
60	46.4	24.2	73.2	55.5	85.6	72.2	50.8	27.5	83.3	61.8	91.6	83.7
80	12.3	4.8	24.0	16.9	43.5	28.4	15.4	6.7	38.0	23.7	59.8	46.0
90	0.3	0.6	4.2	4.6	12.6	7.5	2.3	1.5	08.7	8.6	24.8	17.3
100	0.02	0.02	0.1	0.2	0.7	0.6	0.04	0.1	0.3	0.4	2.4	2.0

Source: Anderson 2001, Table 10, pp. 25–28.

Table 4.2 *Changing life expectancy in the U.S.A.
during the twentieth century*

	Men		Women	
Year	White	Black	White	Black
1900	46.6	32.5	48.7	33.5
1910	48.6	33.8	52.0	37.5
1920	54.4	45.5	55.6	45.2
1930	59.7	47.3	63.5	49.2
1940	62.1	51.5	66.6	54.9
1950	66.6	59.1	72.2	62.9
1960	67.4	61.1	74.1	66.3
1970	68.0	60.0	75.6	68.3
1975	69.5	62.4	77.3	71.3
1980	70.7	63.8	78.1	72.5
1985	71.8	65.0	78.7	73.4
1990	72.7	64.5	79.4	73.6
1995	73.4	65.2	79.6	73.9
1998	74.5	67.6	80.0	74.8

Source: Anderson 2001, Table 12, p. 22.

increases the average population age. In subsistence agricultural systems, dependency ratios are relatively low. In the U.S.A., around the 1950s when the
Social Security System was first established, the ratio of those aged 65+, and
thus eligible for support, to "workers" was about 22:1. Today, the "worker"
to age 65+ ratio is closer to 3:1, and decreasing rapidly to what is expected
to be about 2:1 in 2020 A.D. Such demographic trends were a surprise to
those who forecasted future life expectancies based on 1950 demographic data
(Table 4.2). Rapid improvements in nutrition, education, and health technology
since the 1950s have led to more people surviving longer and have produced
a worldwide crisis in financing retirement, social security, health care, and
housing for elders. Populations hardest hit are those still experiencing transition while beset with additional sociopolitical problems (e.g. human immunodeficiency virus/acquired immunodeficiency syndrome and orphans, childhood
poverty/alcoholism/drug abuse, ethnic/religious/political violence, and natural
disasters). The burgeoning number of elders (12–20%) is overtaxing resources
even in nations with stable socioeconomic and sociopolitical systems; none
was prepared for the rapid increases in life expectancy and health that have
characterized those aged 55+ during the last half of the twentieth century. Increases in life expectancy at ages 65+, although small in magnitude, have been
proportionately large in recent decades. These extra years of life place more
people at risk for age-related CDCs. In the latter part of the twentieth century,

morbidity and mortality from CVDs (incidence) decreased as life expectancy increased and more persons lived longer with one or more CDCs (prevalence). Improved survival post-CDC may produce greater debilitation among late-life survivors, increase frailty, and increase health care costs.

Today's epidemic of CDCs is an artifact of humanity's successes in eliminating most maternal, infant, and childhood mortality, in halting or at least slowing many infectious and parasitic diseases, and in bringing adequate nutrition to much of the world's population (Cassel 1976, 2001). Bioculturally determined alterations in life ways, however, have provided opportunities for more individuals to express any pre-existing genetic potential for late-life survival by constantly altering gene–environment interactions. Stated thusly, recent increases in average, and perhaps maximum, human life spans seem unlikely to be due to any newly arisen alleles. Rather, by providing opportunities for late-life survival, biocultural processes have revealed innate propensities to both CDCs/frailty (nongevity) and health/vitality (longevity). Most CDCs develop not for any one reason, but rather are the outcome of a web of causality including genes, culture, and environment. Major factors in today's epidemic of CDCs are that humans lack prior exposure to extended life and have never before needed to eliminate any genetic propensities to late-life CDCs. CDCs are the legacy of our species' successful biocultural adaptations.

For some, long life has become a double-edged sword. Through biocultural mechanisms, humans are creating populations that include many Tithonus-like individuals (Tothonus (Roman)/Tithonus (Greek), the unfortunate young mortal who fell in love with Aurora (Roman)/Eos (Greek) Goddess of the Morning and was granted eternal life by Jupiter/Zeus but not eternal youth; when Tithonus became decrepit, Zeus turned him into a grasshopper (or cicada), a classical image of old age), intermingled with those who manage to retain vigor and health (if not youth) well into their tenth decade of life. Some suggest that this will lead to an increasing burden on the health care system due to high morbidity and debilitation (Rosenwaike 1985; Stini 1990; Rakowski and Pearlman 1995), while others suggest that these elderly will be healthier and less in need of health care services than currently predicted (Alterm and Riley 1989; Perls 1995; Manton *et al.* 1997; Manton and Gu 2001; Perls 2001).

Debilitation/functional loss

Along with CDCs, survivors to late life (ages 75+) also experience debilitation and functional loss to a greater or lesser degree (Johnson and Wolinsky 1994). Humankind's continued care of elders with disabilities that lead rapidly to death in other mammals is a hallmark of their biocultural adaptability. Few

Table 4.3 *Activities of daily living (ADLs) and instrumental activities of daily living (IADLs)*

Basic ADLs	Instrumental ADLs
Bathing	Using telephone
Dressing	Grooming
Transferring (moving from chair to bed or chair to toilet and vice versa)	Laundry
	Shopping
Toileting	Housework
Feeding	Taking medicine
Ambulating	Managing money

animals continue to survive once they are unable to feed themselves or ambulate. Humans survive decades with major and minor disabilities and debilitations. Although distinct from illness/morbidity, they often occur as sequelae to disease. Debilitation refers to the loss of ability to complete necessary tasks on a regular basis. Debilitation follows from both intrinsic (e.g., mutation, loss of organelles, cells, and sub-units of organs, collagen cross-linking, errors in DNA, RNA, and proteins) and extrinsic processes (e.g., radiation and chemical exposures, wear-and-tear, injury and accidents, predation by bacteria, viruses, fungi, and prions, and cultural activities). Some disabling losses result as the aftermath of CDCs, while others do not and just show continuing declines over several decades. Disability also leads to a progressive deterioration in function and an increased susceptibility to mortality, a pattern suggesting that debilitation and functional loss are processes of senescence (Johnson and Wolinsky 1994).

Declining abilities are often assessed by determining a person's ability/inability to perform various tasks considered to be essential for self-maintenance and independent living (Katz *et al.* 1963; Harper and Crews 2000). Collectively termed activities of daily living (ADLs) (Table 4.3), these performance tasks include basic skills such as feeding, dressing, and bathing one's self, along with abilities to use sanitary facilities and to walk. Those debilitated sufficiently that they are unable to complete two or more ADLs generally show elevated risks for institutionalization, higher need for health care, and increased risk of death. In addition to ADLs, researchers have identified other sets of activities necessary to function well in cosmopolitan societies, Instrumental Activities of Daily Living (IADLs) (Table 4.3). More removed from physical and self-care than ADLs, IADLs include aspects of cognitive functioning and planning, such as money management and the timing of ingesting prescribed medications (Harper and Crews 2000). In general, those failing to complete one or more of their ADLs and IADLs report poorer health, increased infections, and fewer social contacts; they also show greater risk for mortality, functional decline, and institutionalization whether elderly, middle-aged, or young than those who complete all (Baltes *et al.* 1990; Guralnik *et al.* 1994; Mulrow *et al.* 1994). As a third dimension to measures of functional limitations and assessments of debilitation, basic, household, and "advanced" ADLs have been proposed (Johnson and Wolinsky 1994). In this conceptualization, "basic" is equivalent to traditional ADLs and "household" is equivalent to IADLs. However, "advanced" ADLs target cognitive abilities and mental functioning dependent on higher cognitive levels – for example, planning activities or balancing a cheque book (Johnson and Wolinsky 1994).

In 1984, among those aged 65–74, 75–84, and 85+ in the U.S.A., 15%, 20%, and 40% of men, along with 18%, 31%, and 53% of women, self-reported difficulties completing at least one ADL on the National Health Interview Survey (National Center for Health Statistics 1987; Brock *et al.* 1990). However, data from 1987 suggest significantly lower rates of loss of at least one ADL: 65–69-year-old women, 6.5% and men, 5%; 85+ women, 38% and men, 26% (Rakowski and Pearlman 1995). In addition, between 1982 and 1994, the prevalence of chronic disability among U.S. elders (ages 65 and older), as reported in National Long-Term Care Surveys, declined from 24.9% in 1982 to 21.3% (-3.6%) in 1994 ($p < 0.0001$), with a greater rate of decline between 1989 and 1994 (-1.7%, in 5 years) than between 1982 and 1989 (-1.9%, in 7 years) (Manton *et al.* 1997). Elsewhere, it is reported that as many as 80% of elders (ages 80+) require some aid with cooking, shopping, or other ADLs/IADLS, and that 50% of those over age 65 have osteoarthritis, one-half of whom are disabled (Cassel 2001). These latter numbers seem out of line with national surveys and may represent only specific subgroups of the older population. For example, in the most recent representative survey available, men showed greater disability than women from ages 65 to 84 (65–69:5% versus 4%; 70–74: 10% versus 7%; 75–79: 17% versus 12%; 80–89: 23% versus 19%), but almost equal disability at ages 85+ (37% versus 35%), although men still remained slightly above women (Federal Interagency Forum on Aging Related Statistics 2000). These rates of disability are well below either 25% at age 65, or 80% at ages over 80, reported elsewhere, and suggest that assessments of disability among elders may need to pay more attention to sampling strategies.

Overall, available data suggest (as do mortality data reviewed later) that currently aging cohorts are both healthier and less frail than those who were the same age in previous decades. Such data support the idea that both morbidity and subsequent disability are being put off to the later stages of life (Cassel 2001). In the LSOA (Longitudinal Study of Aging, a 6-year follow-up of the 1984 National Heath Interview Survey (NHIS)), at older ages women reported a higher frequency of disability than did men (Cassel 2001). This finding is misleading to some extent, since a much larger percentage of men are already dead, a permanent form of disability. Because of their greater frailty and mortality at earlier ages, fewer men are alive to report disability than women, and older men represent the hardiest of males of their cohort.

All through life, men are more susceptible to death from a variety of morbid conditions, whereas women with the same conditions tend to survive more frequently, but they do so with greater functional disability (Johnson and Wolinsky 1994). The inherent frailty of men leads to proportionally fewer, but to some degree more robust, late-life survivors, so much so that men aged 90 and over show less disability and better survival than do women (Perls 1995). Even at older ages, as at younger ages, men may be more susceptible to environmental and physiological stress than are women. In addition, these national survey datasets include only non-institutionalized individuals, excluding a group who likely has the greatest disability. Self-report data also have low reliability, suggesting they likely provide but a lower boundary for debilitation among U.S. elders. Finally, any underestimates may increase with increasing age of the cohort.

Although most elders in the U.S.A. today likely are healthier, more active, and financially better off than were those of previous cohorts (Perls 1995; Cassel 2001), the sheer numbers of today's elders leads to a larger than ever before number in need of health care and services. In a very general sense, at least in the U.S.A. and other nations with well-developed systems of medical care and fairly universal access to that care, mortality rates along with rates of debilitation have declined at all adult ages. Today's elders are experiencing mortality and debilitation rates that their great-grandparents may have experienced as much as a decade earlier in life. This phenomenon can be illustrated with age-specific mortality rates from the twentieth century. When coupled with data on the proportion surviving to older ages over the twentieth century (Table 1.3), they illustrate why the numbers, if not the proportions, of elders in need of health care and ancillary services (e.g., homemaking, bathing, shopping, transportation) are rising so rapidly. Simply put, large post-war birth cohorts (1946–1964) are now aging. They are healthier, wealthier, and more active than previous elderly cohorts, and they have experienced much less attrition due to intrinsic and extrinsic factors. However, as they age through their 60s, 70s, and 80s, they may be expected to suffer debilitation patterns similar to those that were suffered by earlier birth cohorts at somewhat earlier ages. Baby-boomers have not escaped the CDCs and disabilities of their forebearers – they have only postponed them. A retirement community in central Ohio, U.S.A., provides one simple illustration. Over 20 years of existence (1977–1997), the average age of entry increased from 62 to 75 years (Crews and Harper 1997). All residents entered voluntarily and must have been able to complete sufficient

ADLs to live independently or with a spouse upon entry. Given these criteria, the average age at entry rose by 13 years over the first two decades of operation. Decreased debilitation and morbidity in later cohorts may be contributing to this trend to older average age of entry since residents tend to enter the community when their self-perception is that they soon will not be able to maintain their own residences and independent life styles.

Debilitation and loss of function may also influence self-perceptions of health and well-being, "subjective health," discussed earlier as a major independent risk factor for morbidity and mortality (Mossey and Shapiro 1982; Idler and Kasl 1991; Schoenfeld *et al.* 1994). As yet, the bases for strong associations of self-ratings of health with subsequent mortality remain speculative. Direct influences of stress or psychosocial malfunctioning (e.g. depression) on the HPA axis could lead to constant arousal, depression of immune response, and lower host resistance to environmental insults, or self-monitoring of one's physical condition may reveal subclinical alterations in physiological homeostasis that over the short term have detrimental effects on survival (Schoenfeld *et al.* 1994). Another possibility is that self-perceptions of health represent a cumulative evaluation (gestalt) of current psychosocial/cultural/physical well-being, based on one's personal life history and previous levels of function. If subjective well-being represents a comparison of one's current position, function, and level of debilitation to one's own functioning during earlier life phases, subjective health may reflect cumulative aspects of personal life history. As such, it may be a sensitive indicator of one's relative trajectory with respect to one's period of maximum function. "Subjective" or "perceived" health is a uniquely human variable/risk factor. No other animal can tell researchers how it feels or judge its health relative to conspecifics or itself at an earlier age.

Perceived health seems to be tightly connected to biological health, disease, and functional loss. Johnson and Wolinsky (1994) examined the structure of self-reported health as determined by age, education, disease, disability, and ability to complete ADLs in a nationally representative sample of older Americans. The impact of disease on perceived health appeared to be mediated mainly through disability and functional limitations, associations that differed by gender and race in older American adults. Basic (traditional) ADLs all showed differential validity by gender and race for self-perceptions of health. For example, household ADLs did not affect perceived health of men, while advanced ADLs influenced perceived health in "white" men, but not "white" women or "blacks". Further, lower body disability for men, but upper body disability for women, was significantly related to perceived health (Johnson and Wolinsky 1994). Such results indicate multiple connections among biology, physiology, sociocultural attributes, illness, disability, cognition, and self-perceptions of well-being and health.

Chapter synopsis

Among Earth's many species, humankind's biocultural responses to environmental variability stand apart. These biocultural responses have enabled only humans and the species they protect (e.g., mice, rats, cats, dogs, horses, and cows) to colonize and reproduce on a worldwide basis. Humans inhabit all areas of the world, and have done so for millennia. Well before the development of settled agriculture, humans penetrated and occupied all the continents while still remaining a single interbreeding species composed of local populations, with superficial variations in external features likely to have been developed in response to local environmental stressors (e.g., heat, solar radiation, hypoxia, cold) and cultural preferences (e.g., mate choice, body enhancements, dress), and various local subgroupings which remained more or less connected to one another through mate exchange. Observed differences in allele frequencies are likely to reflect local selection pressures, random genetic drift in population isolates, and mutation. Alleles that were more fit for life and reproduction in one setting may have been of only local adaptive value and may often show geographic clines. However, any mutant allele useful throughout the species range arising upon occasion in such groups likely would, given the branching lattice model for human mating patterns (Lasker and Crews 1996), eventually spread between neighboring demes and ultimately throughout the species gene pool. Humankind's peculiar adaptations and mating patterns have assured that most alleles with positive benefits on survival and reproduction are widely distributed throughout the species (Lasker and Crews 1996). Further, sociocultural processes and new developments continue to restructure all available genetic variation and alter gene–environment relationships. This provides ongoing opportunities for late-life survivors to contribute to the welfare and reproductive success of their descendants and extended kindred, thereby further enhancing the representation of any allelic propensities to senescence slowing in future generations.

5   *Human life span and life extension*

Life span

All sexually reproducing species, in their genomes, carry numerous allelic variants predisposing to slow senescence and extended longevity (vitality). All also include numerous variants predisposing to more rapid senescence and shorter life (frailty). Still, the majority of our available alleles are likely to have little direct influence on life span. In wild populations, senescence-promoting (frailty) and senescence-slowing (vitality) variants are likely to be in balance. Alleles predisposing to senescence slowing, and those that shorten life too greatly (high fraility), generally are outcompeted by those predisposing to more modal life spans and reproductive success. Over evolutionary time, genetic variants promoting longevity and nongevity likely have, for most wild populations, been "non-essential" with respect to natural selection and reproductive success (RS). Allelic frequencies for such variants will drift along randomly, except when very early senescence-promoting types reduce RS sufficiently to be selected against. Life extension experiments on laboratory models (calorie restriction, late reproduction, temperature modulation, selective breeding, and transgenic manipulation) expose this underlying DNA variation. In some cases, by revealing genetically determined phenotypic plasticity in response to harsh environments, in others, by selecting more long-lived phenotypes with underlying alleles predisposing carriers to late-life reproduction and survival (traits not conducive to high RS in the wild) to produce later generations. Similarly, reports showing longevity-enhancing or -retarding DNA/protein variants with increased/decreased expression of specific proteins (superoxide dismutasel (SOD), catalase, and the like) and linkage of allelic variants at specific loci to life extension reveal simple mechanisms with large influences on rodent, insect, and worm life spans. Still, long life is not a common feature of non-laboratory life for any of these models, suggesting, that this variation is non-essential with respect to RS/natural selection in the wild.

Among humans (and other K-selected species), simple processes (similar to those used to extend rodent and invertebrate life spans) are likely already contributing to their higher life spans. The exact same loci and systems do not influence life span in humans. Rather, early evolutionary alterations among phyla adopting a K-selection strategy may have incorporated many systems used to enhance life span among today's laboratory models. Today, in humans and other long-lived species, simple mechanisms that extend rodent life spans may have different functions. In rodents, longevity and reproduction are interlocked in a reciprocal negative relationship – any increase in one reduces the other. Long-lived strains have later-reproducing individuals, while short-lived ones reproduce earlier and more often. Over several million years, humans have relied on culture for survival. Cultural evolution probably was somewhat haphazard, with multiple long hiatuses. However, over the past several 100 000 years, developments have proceeded more rapidly. This coincides with rapid expansion of the neocortex, particularly the language centers of the brain, and reduction of dentition. This period is also believed to have seen a 3-fold increase in human life expectancy. Although much of this increase has been ascribed to culture, it seems likely that positive fitness has also characterized some variants predisposing to senescence slowing.

Hominids were bipedal well before the final episodes of brain expansion that today allows the characterization of *Homo sapiens* as large brained. What pushed this lineage to expand an already large brain with highly dependent neonates even further and resulted in extended child dependency through the fifth to tenth years, and extremely altricial infants? Humans, through their development of culture, created an environment similar to that provided for long-lived fruitflies and rodents. Segregating into breeding populations allows alleles predisposing to longevity or nongevity randomly to attain high frequencies (e.g., founder effect and inbreeding) and to respond to local environmental/cultural pressures. This process allows multiple vitality-enhancing alleles to remain in the gene pool. Cultural processes further led to greater access to reproductive-age women by older men (in the beginning of cultural development, a 30–39-year-old likely was a rarity). Reproduction at these ages may have been rare early in hominid evolution. However, once sociocultural power differentials became entrenched, older men may have experienced greater reproductive success than ever before. They may have gained access to fertile young women through cultural proscriptions and power relations and held sway over local group(s) for considerable time, as seen among the Yanomami (Chagnon 1968). Once elders had greater opportunities to achieve RS, practice nepotism toward those carrying alleles identical to theirs by descent, and enhance the social, economic, and reproductive opportunities of their descendants, this produced additional selective pressures against senescence-promoting predispositions.

This is not to say that human life expectancy and maximum life span may not be extended further, or, for that matter, extended indefinitely. Rather, it is a warning not to expect that life extension of laboratory models will have direct applications to humans. Not all options available in laboratory settings are likely to be available, or practical, for extending life span among humans in real-life settings. A wide variety of alleles/proteins are likely to predispose to senescence slowing in different organisms. Numerous senescence-promoting counterparts are also likely to exist. These interact in complex physiological webs to produce senescence and increase risks for age-related disease within and across all species. Among humans, a wider range of the former may have already become fixed or at least be at higher frequencies than observed in organisms for whom life expectancy has not already been influenced by sociocultural forces. Undoubtedly, many alleles and genotypes are, given the appropriate environmental circumstances, capable of enhancing individual life. Unfortunately, most are surrounded by alleles with average or median life-sustaining capabilities.

Weisman's (1891) conclusions that "... as a rule, life does not outlast reproduction except in species that tend their young ... (and that there is a) need for life to be as short as possible ... (whilst long life is a) luxury without an advantage ... (unless) lengthening life is connected with an increase in the duration of reproduction ... " bear directly on human longevity. Humans tend their young not only longer than any other species once born, they often do so for the remainder of their lives. Human offspring are never fully fledged. During human evolution, the duration of "reproductive investment" has been greatly extended such that today investments extend across multiple generations (commonly three, often four, and sometimes five or more). Among humankind, long life is not a luxury without an advantage. Late life is a period for continued behavioral, biological, and sociocultural investment in offspring and descendants. Availability of this additional period of investment and socioculturally patterned behavior, places new and never before encountered selective pressures on human life history regulators. Even investments that improve fitness by just 1% or less per generation may, over evolutionary time, exert substantial selective pressure against senescent processes. Such a small differential may not be able to push loss of reproductive capacity (i.e., menopause), but, a 0.5% or 0.25% selective pressure against senescence-promoting hormones could gradually alter patterns of endocrine regulation of life history and enhance vitality. Such pressures likely have led to humankind's current optimal evolutionarily stable strategy (ESS) of extended late-life survival. During this period women are non-reproductive. However, men may continue to invest in RS, while women who accumulate inclusive fitness (IF) through parental and grandparental investment may increase their total fitness.

Many physical measures are highly correlated with inter-individual and inter-species variations in longevity. The most noted of these are metabolic rate (Sacher 1975, 1980; Hoffman 1983), brain and body size (Cutler 1980; Hoffman 1984), encephalization (Weiss 1981), growth and development (Pfeiffer 1982; Smith B.H. 1993; Smith and Tompkins 1995; Bogin and Smith 1999), reproductive strategies (Mayer 1982; Rose 1991; Turke 1997), allelic variants at multiple loci (Schäcter *et al.* 1994; Crews and Harper 1998), and levels of reactive oxygen species (ROS) (Richter 1995). The extended period of human "post-reproductive" survival (over 72 years in the longest-lived women) has been of particular interest to gerontological theorists (Cutler 1980; Weiss 1981; Hoffman 1984). One idea is that increased IF accrued to post-menopausal women through their investments to increase the fitness of their descendants and relatives, the grandmother hypothesis (GMH). Another idea is that IF increased for all older individuals when humankind came to rely on learning, culture, social reciprocity, and oral history for its survival and reproduction. Another idea is that no or little pressure for late-life survival occurred over the majority of human evolution. In this model, human longevity today is simply a byproduct of modern cultural life, without any particular evolutionary (genetic) basis (Weiss 1981). For all sexually reproducing species, life history varies such that age-specific reproductive schedules determine rates of senescence (i.e., late reproduction favors longer life and early reproduction favors short life).

Humankind's current life history suggests several long-term evolutionary trends associated with longevity/senescence. Foremost among these is pedomorphism, the slowing and extension of developmental phases through life's second decad which are more than twice those of other apes (as discussed earlier). Another is long life itself, particularly late life (for most women post-reproductive) survival, to twice the maximum life span of existent apes. These are closely interconnected trends. If your offspring does not need you to grow and develop, parental investment is complete before birth (salmon, insects) and you may die once you reproduce. If your offspring needs 4–6 weeks of care, the need to survive after reproduction increases apace. When offspring take years to develop and require constant parental care for a decade or two, parental survival is a necessity. Slowing of development, increased the need for parental care, and somatic survival requires greater somatic investment and proportionally longer life spans. Among humans, long life appears to be associated with greater RS and IF.

Humans are also among the latest reproducing of mammals. Thus, longer life is necessitated by longer time to maturity and extended growth. Humankind's basic life history adaptation includes immature neonates, pedomorphism, slow development, late maturation, extended parental investment, and biocultural influences on fitness, reproduction, and life span that promote late-life survival. Culture may directly (e.g., by ingesting toxic substances) or indirectly (e.g., through residing in a low iodine environment) alter these underlying biological processes.

Culture may ameliorate (e.g., drugs that reduce blood pressure) or enhance (e.g., suntanning) the biological effects of specific environmental stressors and thereby produce additional selective pressures. Current life spans in different human populations are the product of culture interacting with biology and environment. As culture has influenced survival and reproduction over evolutionary time, it has also patterned much of human biological variation. Over the short run (the past 200 years), modern human culture has improved survival at all ages for most people and reduced opportunities for natural selection to act through early mortality. These alterations have also provided additional opportunities for grandparents and great-grandparents to contribute to their kindred's welfare and overall fitness and their own IF. To the extent that longevity has a genetic basis, such biocultural processes continue to produce selection biases against shorter life spans (senescence-promoting alleles). Whether there are additional prospects for furthering human life extension is unclear. But, before examining that possibility, it may help to examine briefly studies of life extension in non-human models. This will help to detemine whether such designs can be extended to humans. These studies also may help to answer the question of whether there are limits to human longevity.

Non-human animal models

With short life spans, rapid reproduction, and inexpensive upkeep, invertebrates (e.g., flies and worms) are standard laboratory animals for research on senescence. Laboratory rodents, although more expensive to keep and longer lived, are the standard mammal model. Age-related alterations in function, both senescent and non-senescent, characterize the entire animal kingdom. No species totally escapes senescence. Often, processes seem so similar across species that they appear identical. However, such superficial similarities provide insufficient information on senescence to generalize about mechanisms across phylogenetically different organisms. Very different underlying biological mechanisms and physiological alterations may produce very similar processes and endpoints (e.g., organ dysfunction, frailty, stroke, dementia, and death). For the multitude of senescent mechanisms to be identical even between two individuals within one species is highly improbable; for the mechanisms to be identical in any two species without some close phylogenetic ties is unlikely. Due to the long-term influences of environment on selection for earlier life history events, mechanisms and processes of senescence are likely to vary widely even within single genera (see Finch and Rose 1995). At present, we know too little about the underlying mechanisms of senescence or the physiological and biochemical alterations with age in any species to generalize mechanisms across phyla. This general lack of knowledge is why study of non-human animals is a necessary step in determining possible mechanisms of senescence in humans. Furthermore, data from as many organisms as possible are needed to test the biological hypotheses generated by evolutionary theory.

Caenorhabditis elegans

The entire *C. elegans* genome, along with the site of activity for each locus, has been identified (The *C. elegans* Sequencing Consortium 1998), and the genetic basis for nematode longevity has been studied extensively. Current research centers around several mutant alleles, in two different pathways identified as "longevity assurance" genes that appear to determine *C. elegans's* life span (Lakowski and Hekimi 1996). The first observed was *age-1*. Mutated *age-1* alleles extend both the mean and maximum life span of *C. elegans* in the laboratory, while reducing *C. elegans*' fertility significantly below that of the wild type (Friedman and Johnson 1988). Other loci include mutant alleles (*daf-2/e1370*) associated with as much as a 100% increase in the mean life span of roundworms (VanVoorhies 1992), and a series of clock genes, one of which (*clk-1/e2519*), when carried in combination with *daf-2/e1370*, extends life by almost five times that of wild types (Lakowski and Hekimi 1996). Although the phylogenetic distance separating roundworms and humans is wide, sequence homologs for both the *age-1* and *daf-2* alleles occur in the human genome (the human insulin/insulin-like growth factor I receptor is the *daf-2* homolog). Alleles at such homologous loci are unlikely to have similar associations with human longevity, where they are interacting with many more loci, cells, and physiological processes. Still, identifying loci and proteins that alter senescent processes and extend longevity in organisms carrying less DNA (and having fewer physiological systems and cells) may provide valuable insights into candidate loci/mechanisms in more complex organisms. However, similar influences of these homologs on either senescence resistance or proneness in humans should not be expected.

Drosophila

The success of artificial selection for longevity in *Drosophila* species is generally well known. Numerous research programs (see Luckinbill *et al.* 1984; Rose, 1984; Pletcher and Curtsinger 1998; Service 2000a, b; Charlesworth 2001) demonstrate significant increases in mean and maximum life spans of fruitflies following a few generations of selection for either long-lived or late-reproducing phenotypes. Since environmental factors remain constant for subsequent generations, these results suggest the existence of a large reservoir of allelic variation predisposing to longevity. Even after several generations of such breeding, additional variation exists, since continued experimentation produces even more long-lived populations. As yet, the limits, if such exist, to *Drosophila* longevity have not been observed (see Curtsinger *et al.* 1992). Many studies of *Drosophila* concentrate on breeding artificially selected long-lived varieties and then looking for linkage to specific bits of DNA to identify candidate "longevity-assurance" alleles.

In addition to illustrating the extent of possible genetic variation for longevity and late-life vitality in flies, artificial selection data from *Drosophila* support the evolutionary theory linking reproductive behavior and senescence. Artificial selection for longer-lived individuals is associated with a decreased reproductive efficiency at younger ages. Long-lived individuals show delayed senescence, produce fewer eggs, have smaller ovaries as young adults, and show peak egg production later in life than short-lived individuals (Luckinbill *et al.* 1984; Rose 1984). To control for possible confounders, Zwaan *et al.* (1995) selected directly for longevity in *Drosophila*. They found that the production of progeny by long-lived strains was below that of short-lived strains. In addition to having lower reproductive output, longer-lived *Drosophila* are more resistant to environmental stress, including exposures to starvation, desiccation, ethanol, and paraquat (Service *et al.* 1985; Service 1987; Arking *et al.* 1991). They also exhibit increased SOD and catalase activity (Arking and Dudas 1989; Dudas and Arking 1995) possibly providing protection against ROS (Dudas and Arking 1995). Data from *Drosophila* breeding experiments are also consistent with the antagonistic pleiotropy model for genetic control of senescence and tradeoffs between reproduction and senescence. Results also accord well with the disposable soma theory of aging; that is, senescence results when somatic cells and tissues are no longer able to protect or repair themselves when damaged and longevity is associated with an increased competency of defense mechanisms.

Drosophila also provide clear evidence that environmental factors mediate genetic influences on longevity. *Drosophila* selected for late-life fitness will fail to express their extended longevity when the larval density at which they are raised falls too low (Clare and Luckinbill 1985; Arking 1988; Buck *et al.* 1993; Khazaeli *et al.* 1996). Although the mechanisms by which this gene–environment interaction works are unclear, enhanced transcription of antioxidants characterizes lines with extended longevity (Dudas and Arking 1995). In addition, female medflies show greater variation in mortality in response to an environmental change (altered diet) than do males; such differential environmental sensitivity may also underlie mortality differentials following irradiation (Carey *et al.* 2001). These data reveal two important phenomena. First, environmental responses occurring early during larval development may influence the transcription of DNA in later life. Second, and perhaps more interesting, once set, such responses apparently remain set to the environment encountered during the earliest phases of development as they apparently do among human fetuses exposed to low nutrients. These data suggest that the gestational programming model proposed for human infants may have a counterpart in fruitfly larvae.

Rodents

Since the early calorie-restriction experiments of McCay *et al.* (1935), rodents have become "the animal model" for laboratory research on the "biology of aging" in mammals. Often, data from such studies are extrapolated uncritically to humans and other organisms. Such generalizations may be misleading. Still, rodents are the best-studied mammal and they have revealed much about the mechanisms and processes of senescence. Laboratory rodents mature rapidly and reproduce early. Wild types have relatively short mean and maximum life spans. Short generation times and life spans (relative to humans) allow rodent breeding experiments to provide fairly rapid results. These have shown the ease with which life span may be extended and inbred strains susceptible to specific chronic degenerative conditions (CDCs), or types showing accelerated senescence, developed. Because of their longer life spans, slower breeding, more complex social behavior, and costly upkeep, dogs, cats, and non-human primates are less acceptable for laboratory research.

Voluminous research has shown that it is relatively simple to increase both mean and maximum life spans of rodents using a variety of simple mechanisms – calorie restriction, selective breeding, castration, reduced activity, or reduced ambient temperature (Johnson *et al.* 1995). Mice or rats receiving a complete diet, including all essential nutrients in the proper amounts and proportions, that is reduced in calories by 30–40% of their non-restricted littermates show longer mean and maximum life spans (McCay *et al.* 1935; Masoro 1995). They also generally show extended periods of growth and development, later sexual maturation and reproduction, later onset of multiple CDCs, smaller body sizes, and produce fewer offspring over their life spans than do littermates provided *ad libitum* calories. These extended average and maximum life spans occur whether calories are reduced by decreasing fats or carbohydrates, but not when calories are reduced via protein restriction. Calorie restriction is the best illustration of an environmental influence on senescence and life span; it is also the one environmental manipulation that both reduces the rate of mortality and extends mean and maximum life spans (Finch 1994; Roth *et al.* 1995). Restriction of calories by 30–40% of *ad libitum* fed animals generally extends the mean and maximum life span by about 10–30% (Weindrunch and Walford 1988). Reduced caloric intake is associated with fewer tumors (Weindrunch 1989), less atherosclerotic damage, fewer autoimmune lesions (Rose 1991), and less oxidative damage secondary to the generation of fewer ROS and greater antioxidant activity (Yu 1993).

Although mammals, rodents are phylogenetically far removed from humans, with a very different life history. They mature in 35–50 days, live only about 4 years, are small bodied with large litter weights, and practice an *r*-selection strategy (Austad 1997). Calorie-restricted rodents show altered patterns of reproductive hormone release, timing of maturation, and reproductive output (compared with non-restricted individuals); the physiological processes to alter such parameters may not be available to *K*-selected species with a very different life history (Austad 1997). However, although dietary restriction in rodents may have its counterpart in humans, study of calorie restriction among non-human primates may provide a better model and reveal how restriction affects members of our own order, which branched from other mammalian orders over 65 million years ago.

Selective breeding for late-life reproduction can also be used to produce a rodent population that breeds later while also showing longer mean and maximum life span than the original population. Successive generations tend to show slower growth and development and later onset of reproduction than other populations that breed at younger ages. The extension of life span resulting from selective breeding for late reproduction is about the same as that observed for calorie restriction. Castration is also accompanied by extended average and maximum life spans for both male and female rodents. These latter two results support suggestions that reduced investment in early-life reproduction is associated with longer life, even when investment in reproduction is artificially induced in newborns. Selective breeding for reproduction at later ages, castration, and calorie-restriction studies illustrate evolutionarily balanced tradeoffs between reproduction and survival. Selective breeding also reveals the presence of allelic variants that predispose to senescence slowing.These apparently are outcompeted by those predisposing to greater early RS in wild settings.

The Senescence Accelerated Mouse (SAM) was introduced to rodent research in the latter decades of the twentieth century. SAM models have been bred that are either senescence prone (SAMP) or senescence resistant (SAMR) (Takeda *et al.* 1991; Yegorov *et al.* 2001). Interestingly, the senescent-prone varieties, in addition to exhibiting early senescence, also tend to succumb to specific CDCs (degenerative joint disease, osteoporosis, and cataracts) within specific lineages (Takeda *et al.* 1994; Yegorov *et al.* 2001). In these models, the processes underlying early senescence appear to be linked to accelerated pathological processes, supporting the conceptualization of age-related pathology/disease and senescence as part of a continuum. Understanding the genetic basis of these specific CDCs in SAMP models may reveal in part the molecular basis for senescence in that particular system. Examinations of reproduction and longevity among senescent-prone varieties have not provided conclusive evidence as to any tradeoffs. Compared with senescence-resistant forms, senescence-prone types may show either reduced fertility or no difference in fertility (Miyamoto *et al.* 1994a, b). If RS is equal between the two types, loci involved in accelerated senescence in this model may not exhibit antagonistic pleiotropy. Equal reproductive success also suggests no tradeoffs between somatic survival and reproductive investment in the senescence-resistant forms. Apparently, the involved loci/alleles act directly on disease processes that retard or accelerate some senescent process (osteoporosis, cancer, degenerative joint disease, or atherosclerosis), but are unrelated to early-life fertility. One suggestion is that at least some of the observed longevity-lowering effect is due to lowering of the age at which detrimental alleles express themselves. This supports the mutation accumulation model. This model suggests that genetic modifiers postponing the ill effects of early-acting detrimental alleles until later in life will accumulate in wild populations. These SAM experiments suggest that breeding for either senescence-prone or -resistant types may decrease or increase the frequency of delaying alleles without altering RS. They also illustrate the disconnection between genetic variation for life span and RS, a disconnection that is likely to characterize most wild populations.

Non-human primates

Longitudinal studies of non-human primates as they grow old and senesce are rare due both to their lengthy lives and the recency of biological research on senescence. Unfortunately, due to natural mortality and predation, studies of animals in the wild are unlikely to include sufficient numbers of old and very old individuals for representative research on aging and senescence. Only controlled colonies may provide adequate data to examine questions of life span, longevity, and reproductive behavior. Even then, only groups that have been observed for several generations may provide the type of historical data needed to aid in understanding biological senescence (Corr 2000). In addition, as with RS, dominance hierarchies, and activity/proximity in non-human primates, it is unlikely that a single year or even several years of cross-sectional research will provide clear data on biological senescence. Patterns are likely to vary with age, but also across seasons and between years, and across environmental contexts.

In recent decades, several programs on dietary restriction in non-human primates, mainly rhesus monkeys, have been initiated. Although obviously confounded by captivity, artificial environments, and group compositions, these do mimic similar experimental designs among rodents (Bowden and Jones 1979; Bowden and Jones 1979; Ingram *et al.* 1990; Bodkin *et al.* 1995; Kemnitz *et al.* 1993; Roth *et al.* 1995; Weindrunch *et al.* 1995). Such studies will need to run their course over several more decades before meaningful interpretations of the results can be completed. However, alterations observed in biomarkers of physiological function are consistent with changes observed when rodents are calorie restricted (reviewed in Lane *et al.* 1997). For example, Weindrunch and colleagues (1995) placed 24 male rhesus (*Macaca mulata*) aged 0.6–5 years and 25 male squirrel monkeys (*Saimini sp.*) on a special diet (plus 40% extra of vitamins and minerals) and divided them into equal control and experimental (30% less calories per day than controls) groups. For the experimental rhesus group, body weight and crown-rump length were reduced by 10–20% after 1 year and skin folds and body circumferences by 20–25%. However, over the first 3 years, no effects of 30% calorie reduction among squirrel monkeys were observed; after 3 years, morphological alterations were observed (Weindrunch *et al.* 1995). Results from the National Institute on Aging study (Roth *et al.* 1995) of 200 rhesus and squirrel monkeys, one-half on a 30% reduced calorie diet, are also consistent with rodent studies. Restricted animals are smaller, show lower body temperature, glucose and insulin levels, mature slower and have slower age-related functional declines as measured by glucose utilization, reduced glycation, decreased damage from ROS, altered activities of stress hormones and proteins, and minimized growth and development (Roth *et al.* 1995). Roth *et al.* (1995) suggest that these multiple effects stem from increased efficiency in energy utilization leading to slower progress toward entropy as energy is processed more slowly, thereby reducing somatic dysfunction (p. 414). This suggests that the ability to reduce mortality and increase life span may be widespread among wild mammal populations. Non-human primate and rodent phyla are about as evolutionary diverse as mammals come (Austad 1997); thus, any senescence-related traits they share (plesiomorphs) may be universal among mammals. Slowed senescence, growth, and reproduction in settings of reduced calories may be part of a general mammalian adaptation to variable environments. These data suggest that similar regimens of dietary restriction may beneficially affect life span and enhance vitality in humans.

Elsewhere, studies of non-human primates have shown that many aspects of human senescence and CDCs reflect a general pattern of primate aging. For example, wild-caught chimpanzees and other non-human primates show signs of arteriosclerosis/atherosclerosis, as do humans, as early as their third decade of life (DeRousseau 1990). Also, arthritis, cataracts, and bone degeneration are common among free-ranging and caged non-human primates including macaques, chimpanzees, and gorillas (DeRousseau 1994). Furthermore, neurological degeneration and declining immune function apparently characterize old age in primates. For example, non-human primate models of Parkinson's disease were used to test the hypothesis that the implantation of dopaminergic cells and tissues containing growth factor into affected brain areas reduces symptoms and slows advance (King and Yarborough 1994). Non-human primate models are being used in ongoing research on multiple neurodegenerative, autoimmune, immunodeficient, and neuroendocrine conditions (Nimchinsky *et al.* 1999).

Applications to humans

Calorie restriction

Successful extension of life span among nematodes, insects, rodents, and, apparently, non-human primates suggests that human longevity may also respond to calorie restriction. Unfortunately, few natural experiments are available to examine whether calorie restriction, with adequate nutrition, prolongs human life. In natural settings, when calories are restricted to 70% of *ad libitum*, humans nutrition is usually also compromised for many essential vitamins and minerals. Human populations with inadequate calorie intake commonly experience one or more additional stressors (minority status, poverty, poor health care and sanitation, and/or more frequent exposures to infectious and parasitic disease), none of which bothers "laboratory rats" in their hermetically sealed environments with just the right combinations of nutrients, light, water, sanitation, and daily care. Compared with "wild-type" control individuals in low-calorie settings today, and likely including the majority of humans who have ever lived, there is one experimental group with more than adequate nutrition, high calorie intakes, *ad libitum* feeding, and rampant obesity. Residents of the world's cosmopolitan societies are the *ad libitum* group. The few low-calorie/adequate nutrition groups for comparison in these settings are confounded by additional religious or secular factors. Observed risk factors, disease patterns, and mortality in more cosmopolitan populations (e.g., obesity, over-nutrition, atherosclerosis, hypertension, hyperlipidemia, high ROS, and low antioxidant activity, leading to cardiovascular disease and circulatory disease, cancer, diabetes, osteoporosis, degenerative joint disease, and dementia) suggest that *ad libitum* fed humans are neither as healthy nor as long lived, on average, as they might be, given a different dietary regimen.

Several religious groups in the U.S.A. that promote more austere lifestyles may provide a comparison group. Both Seventh-Day Adventists and members of the Latter Day Saints (Mormons) restrict their use of meat, alcohol, and other drugs (caffeine and nicotine). In addition, members of both groups generally are expected to fast 1 day per week, thereby reducing their calorie intake by about 15%. Members of both groups show lower age-specific mortality rates at almost all adult ages, greater life expectancy, and a higher average age at death than the general U.S. population. In addition, Mormon women tend to have higher completed fertility than the general U.S. population of women. Neither religious group has produced, as yet, an individual who survived to or beyond the current maximum human age of 122 years. Thus, although calorie restriction may promote longer and healthier lives on average, as yet they have not extended the maximum life span as has been observed for rodents and insects. Factors that produce higher average life expectancy in populations of humans apparently are unrelated to our current maximum observed life span. Given that some members of all human populations show late-life survival and achieve over 80 years of age, it seems likely that biological propensities for late-life survival are widespread. Calorie restriction may allow more individuals to express their innate propensities, but not extend an already long maximum life span.

Another group who may show some effects of calorie restriction is vegetarians; however, they represent a wide variety of types (e.g., Vegan, ovolacto, fish eating, and sporadic). Additionally, since this is a life-style choice, data are multiply confounded by age at commencement of diet, periods of adherence, other lifestyle choices, and type of diet followed. However, at least among a sample of German vegetarians followed for 11 years, total mortality was one-half that observed in the general German population (Chang-Claude *et al.* 1992). Additional data comparing life spans of various categories of vegetarians with other dietary groups would help to clarify these associations, but a healthy low-calorie diet will probably reduce disease progression and slow senescent processes. Available evidence suggests that calorie restriction is likely to extend the average life span of various age groups of humans, particularly those who are currently overnourished and obese. It is not likely that most individuals in most areas of the world can attain both a calorifically restricted and nutritionally complete diet. Rather, adequate nutrition and calories, as opposed to reduced calories, will help the majority of the world's population to attain not only greater longevity, but also to cope with inequities and marginal life styles.

According to all available evidence, calorie restriction instituted during childhood should retard CDCs/senescence and increase average life span for humans. Calorie restriction with a nutritionally adequate diet may represent an untapped source for human life extension. Interestingly, calorie restriction only appears to lead to increased life spans among rodents (and perhaps non-human primates) when either fats or carbohydrates are restricted, but not protein. To improve their health and life expectancy as adults, it would seem prudent to decrease calorie requirements for children over 2 years of age, while increasing protein requirements. In many ways, a 30–40% calorie-restricted diet would be more like the diets experienced by the majority of our hominid ancestors compared with our current *ad libitum* diets. Our progenitors likely evolved in a more calorically restricted environment than that of cosmopolitan societies today. The advent of agriculture and herding (cultural developments) provided the high-calorie, carbohydrate-based diets, and, eventually, high-fat diets (similar to *ad libitum* fed rodents) and led to our current spectrum of risk factors and CDCs that frequently curtail longevity. It is unlikely that a diet of reduced calories but adequate nutrition will be promoted by any private or governmental agency as a panacea for age-related disease, premature senescence, and the CDCs observed so widely in cosmopolitan societies. Rather, any children eating a 30% calorie-restricted diet will be identified as stunted when they fail to achieve standards on growth charts. Similarly, no experimental protocols restricting infants and children to a low-calorie diet during their developmental years are likely to be approved.

Currently, nutritional restriction appears to be humankind's best hope for a fountain of youth. While more than one-half of the world's population struggles to maintain sufficient calorie and nutrient intakes, a smaller proportion ingests an overabundance of calories and often neglects to maintain adequate nutrition. Reductions in calorie intakes by 30% among most individuals in more cosmopolitan settings (Europe, North America, the Middle East, and the socially elite in most other nations of the world) would have immediate benefits in terms of life expectancy. Currently, subpopulations at 60–70% of *ad libitum* calories (comparable to the experimental group in rodent models) also suffer from poverty, poor nutrition, low-protein and -fat diets, and frequent parasitism, preventing any reasonable comparison with subpopulations in settings with a surfeit of calories (*ad libitum* fed group in rodents). Perhaps the best two comparative (experimental) groups are Seventh-Day Adventists and Mormons – both of whom show higher life expectancy and reduced mortality rates for all CDCs than does the general U.S. population of which they are a part. Their lower age-specific mortality at all ages results from lower mortality from all the major killer diseases (stroke, heart attack, cancer, and diabetes). In addition, prospective studies of dietary reductions in calories and fat in humans and non-human primates have shown associations with beneficial declines in multiple risk factors (atherosclerosis/arteriosclerosis, hyperlipidemia, obesity/fatness, blood pressure, and hyperglycemia), although there seem to be limits to the effectiveness of these modalities (Gerber and Crews 1999). Retrospective studies also support findings that those who maintain healthier life styles with respect to diet and calorie intake show less morbidity from CDCs and survive longer.

Selective breeding/assortative mating

Selective breeding and assortative mating for later reproduction, somatic maintenance, and survival have characterized human evolution since slowed growth and development, altricial infants, and pedomorphism became a part of human life history strategies. Increased reliance on pedomorphism and extension of human developmental phases through the second decade of life make late-life reproduction (after ages 20 and 30) a necessity. Delaying attainment of maximum reproductive potential (MRP) to the third decade of life (20–29) requires longer adult survival because child-rearing must follow child-birthing. Early in human evolution, those who could not reproduce at ages (twenties and thirties) that were beyond those of first reproduction among hominoids and early hominids (the pre-teens and teens) likely failed to fledge many offspring. Those who reproduced in their twenties or thirties would on average have a later age of attainment of MRP. Assortative mating for later-life reproduction would follow directly from selection against those who could not reproduce at later ages, given the extension of growth and developmental processes and reproductive maturity into the late teens that characterized human evolution. Members of any generation who could not reproduce at three times the age of first reproduction among existing apes (7 years) and much later would leave no descendants at all. As developmental processes occurred later and later in the human life history, the population would include only late developers and late reproducers, similar to rats being selectively bred for long life and late-life reproduction in today's laboratories.

Humans have been conducting a natural selective breeding experiment, similar to those on laboratory rodents, since they evolved bipedalism and culture and set themselves on an evolutionary trend toward life history extension (perhaps as long as 200 000 – 2 000 000 years) (see Smith 1986, 1991). Still, in addition to such species-wide longevity mechanisms, it is highly probable that "selective" matings among living humans could produce numerous inbred strains with significantly higher mean and maximum life spans than currently are achieved by any human population. Mechanisms that might lead to such increases could be similar to those in rodents: increased antioxidant defenses, reduced generation of ROS, decreased metabolic rate, and/or slower maturation and reproduction. However, since humans already use a majority of these, new mechanisms could be unlike anything currently observed in rodents. It is improbable that any experimental program of selective breeding for longevity will ever occur in humans. Human generation times and the length of time needed to identify long-lived, late-reproducing types to select for reproduction of the next generation preclude any such artificial selection, regardless of ethical considerations. Undoubtedly, assortative matings between long-lived, late-reproducing individuals have occurred, and continue to occur, in most populations. Also, some kindreds probably already produce offspring predisposed to be longer-lived individuals than those of other kindreds. However, in general, environmental influences of survival are likely to be so pervasive that most genetic heterogeneity in longevity is swamped (Service 2000a, b). Still, siblings of centenarians do survive longer on average than do those of non-centenarians.

Supporting the suggestion that multiple alleles for enhanced longevity are segregating in the human gene pool, are the many heritable conditions characterized by early manifestations of some of the physiological alterations common during senescence. Progeroid syndromes, Down's syndrome, pheny/ketinuria Tay–Sachs, and Huntington's disease are all characterized by reduced longevity, evidence that multiple genetic factors that shorten human life span are segregating within human populations. Conversely, other alleles at these loci must be neutral, while still others must enhance survival compared with the general background level. One hypothesis that has guided genetic theories of senescence for decades is that genetic factors for shortening life span will be far more common than those for lengthening it. Numerous paths for disturbing the complex and integrated human soma must exist; however, there must be far fewer ways available to improve it. In general, this leads to the notion that we most often inherit frailty from our parents, rather than longevity (Vaupel *et al.* 1979; Vaupel 1988). Stated differently, any human genetic predispositions for longevity may simply be the absence of genes that predispose to early death (Turner and Weiss 1994). Still, some variants at some loci must provide improved vitality and late-life survival, even if these are much more rare, say 1% of mutants, than those that hasten death. Over numerous generations, it is likely that alleles predisposing to slower senescence, by not enhancing or even countering changes that predispose to early death, have become more frequent in particular kindreds and populations, while predispositions to early death have increased in others. Since all mating boundaries are permeable and ultimately break down, all types must eventually remix across groups and compete with one another. This suggests that to the degree that "longevity-assurance" genes are more fit, in a Darwinian sense, they continue to spread through the human population.

If insect and rodent breeding studies provide any clues to human longevity and senescence, the most important one may be that there exists a large reservoir of untapped genetic variation associated with longevity and delayed senescence in wild populations. A second is that there are no indications as yet that the full extent of these genetic predispositions has been tapped; animal models continue to show improved longevity after decades of study. A final consideration is inheritance of longevity in humans. Previously, in normal, more or less randomly mating, human populations, only low correlations (about 30%) between the longevity of parents and their offspring were observed and heritability of life span was thought to be similarly low (about 30–40%) (Vaupel 1988, p. 277). Given even these levels of heritability, selective mating could still enhance human longevity. However, recent data suggest that heritability of life span may be as high as 60% or more among relatives who survive to adulthood. This suggests the potential to improve longevity among humankind is available within our genome.

As yet, no non-human primate models have been selectively bred for long life. In most primate centers, older animals are not maintained and late-life breeding does not occur. Among humans, one must look for natural experimental settings with postponed marriage and continued natural birth spacing that may provide partial tests of this model (Irish villagers, Anabaptist religious sects, and other religious semi-isolates, e.g., Mormons and Seventh-Day Adventists). Unfortunately, environmental, cultural, and lifestyle differences confound these comparisons. Given equal access to nutrition, health care, and other cultural constructs along with similar life styles with respect to health risks (e.g., smoking, sexually transmitted diseases, radiation, etc.), it is likely that, as with a reduction in calories to 60–70% of average, selective breeding of older mates will improve life expectancy in human samples. However, this is not likely to increase maximum life span greatly.

Why will selective breeding only slowly increase human maximum life span? Genetic variation for human life history is likely already stretched close to its limits to compensate for humankind's pedomorphism, immature neonates, and late reproduction. One axiom of evolutionary biology is that traits under high selective pressure often show less variation than do similar traits not influenced as strongly by natural selection. Maximum life span varies little across humanity's many populations, while life expectancy ranges widely and is easily altered by cultural, environmental, and lifestyle factors. Furthermore, if there was a specific set of alleles that when inherited together would alter greatly the maximum life span of an individual, it is likely that this phenotype would already have been recognized as the most long lived of humans. Last, given human mating structures (Lasker and Crews 1996), such a set of alleles would have already spread, or, if just arrived on the evolutionary scene, should spread rapidly throughout the human population.

Candidate alleles

Among nematodes, a total of about seven loci may determine senescence, making this and similar species unlikely to yield direct insights into the genetic basis of senescence in humans – in whom it has been estimated that up to 7% of the genome mediates some aspect of senescence (Martin 1978; Arking and Dudas 1989; Arking 1991; Rose 1991). (This number may be modified since the human genome project is reporting finding only about 30 000, as opposed to the expected 70 000–100 000, coding loci on the human chromosomes; this still suggests that 2100 loci mediate aspects of senescence.) At present, unlike nematodes and *Drosophila*, there are few convincing molecular correlates of longevity in humans. Among these are the apolipoprotein E (*ApoE*) locus where the three major alleles vary in frequencies between age groups and the ApoE2 protein type is more prevalent among the oldest-old in U.S. samples (Schächter *et al.* 1994). Additionally, the ApoE4 protein is present at significantly lower frequencies among older than among younger adults in several populations including Samoans and U.S. citizens (Cauley *et al.* 1993; Crews and Harper 1998). In the general population, the E4 protein type, the ε^*4 allele, is associated with higher serum cholesterol levels, more advanced atherosclerosis, and increased risk for late-onset Alzheimer's disease compared with other *ApoE* variants (Davignon *et al.* 1988; Corder *et al.* 1996; Strittmatter and Roses 1995). The association of specific *ApoE* alleles with both longevity and the incidence of chronic disease is another example of our inability to divide pathology from senescent processes. The angiotensin converting enzyme (ACE) locus also exhibits alleles that show differential associations with age. In several populations, including Samoans, the insertion allele is at a higher frequency in older compared with younger members of the same population (Crews and Harper 1998), suggesting that the deletion allele is associated with early mortality. However, in other samples, the D allele is at an elevated frequency among older samples, although associated with myocardial infarction during middle age (Schächter *et al.* 1994).

Other interesting candidates for life-lengthening/shortening in humans may occur in the U.S. Amish community. One kindred of men is noted for generally surviving longer than their spouses (see Holden 1987). Apparently, men of this kindred possess a deletion on the long arm of their Y-chromosome. This suggests that common alleles at one or more loci on the Y-chromosome increase frailty and reduce the life span of men or, conversely, that deletion of Y loci increases vitality and life span. Most alleles associated with life span appear to increase frailty/risk for some disease processes thereby promoting senescence. Most loci are likely to harbor some alleles that promote one or another CDC process and shorten life. However, most loci are likely not to be disease promoting;

rather, most are likely not associated with significant differences in life span, while a few may retard senescent processes. For example, a variety of mitochondrial DNA (mtDNA) mutants are associated with diseases that lead to shorter life spans (blindness, deafness, cardiac failure, diabetes, and renal dysfunction, along with Parkinson's and Alzheimer's disease; reviewed in Holt *et al.* 1988; Wallace *et al.* 1988; Wallace, 1992a, b). This wide variety of degenerative processes attributed to mtDNA is in part the basis for the mtDNA theory of senescence. Still, although multiple alleles may increase frailty, a lack of such frailty-increasing alleles does not appear to promote above-average longevity; rather, most alleles promoting vitality and senescence resistance are likely to approach fixation across the human species. However, a few such vitality-enhancing alleles may occur only in specific groups such as Amish kindreds or other reproductively isolated populations (e.g., *ApoE*-Milano in Italy). Conversely, a number of disease-/senescence-promoting alleles are found in almost all populations. Most of those observed appear to increase the risks for CDCs well after the ages of maximum reproductive output (20–40 years) and before the ages commonly thought of as late life (75+ or 85+). Those who survive to late life appear to be a highly select group with respect to the HLA, apolipoprotein, ACE, and mtDNA alleles they possess compared with those dying during middle age (40–60) or earlier in late life (60–75).

Reducing chronic degenerative conditions

CDCs (e.g., coronary heart disease, cancer, and diabetes) represent complex interactions of genes, environment, personal life history, and multiple cultural variables (e.g., medical technology, ritual activities, and mating systems). The exact etiology of CDCs is difficult to determine in any individual. Current research often focuses on the molecular origins of risk factors (e.g., hypercholesterolemia, high blood pressure, and obesity) for these conditions rather than the disease itself. Unfortunately, risk factors are as complex in origin as are their associated CDC. Multiple genotypes and environments predispose individuals to similar physiological alterations (e.g., increased salt retention, increased mobilization of fatty acids, and rapid blood pressure responses) that may alter a variety of risk factors. At present, clinical and research understanding does not allow the detection of subtle differences in CDC presentation and course secondary to different gene, environment, and gene–environment processes. For example, blood pressure is regulated by many loci with innumerable alleles acting on multiple physiological systems. Each such allele may show different gene–environment, gene–gene, and gene–culture interactions. This variation often is collapsed into a single "hypertensive phenotype" in clinical and epidemiological research (James and Baker 1990; Crews and Williams 1999). One of the best avenues for understanding and extending life expectancy in humans may be the identification of such variable phenotypes for detailed research. During the last half of the twentieth century, with only limited understanding of these complex etiologies, CDCs were slowed and life span extended. With new information on genotypes and phenotypes, trends postponing the onset and progression of CDCs will continue. Most recent reductions in mortality are likely to have resulted from reductions in behavioral risks (smoking, fat and salt ingestion, low physical activity) and medication. Further reductions may require interventions aimed at actual physiological processes.

Better understanding of complex etiologies and improved medical technology is not likely to eliminate any major CDC as a cause of death in the foreseeable future. However, understanding molecular influences and subphenotypes will provide more accurate diagnoses and treatments. Demographic modeling of life span with one or more causes of death either totally eliminated or reduced in frequency suggests that neither elimination nor reduction of any single CDC will lead to large gains in human life expectancy (Manton and Stallard 1984). Competing causes of death, particularly at older ages, preclude attainment of longer life spans. However, the earlier the eliminated cause of death acts, the greater is the gain in overall life expectancy. At later ages, multiple chronic conditions affect individuals and one of these "competing conditions" will still cause death within a short time period when others are eliminated. Remaining life expectancy at old age (70 years) is much smaller than at younger age (20 years); anyone saved late in life has little influence on overall population life expectancy. In addition to such considerations, it is obvious that much of humankind's mortality is already pushed to the later years of life, when competing chronic conditions abound. Furthermore, today's 60-, 70-, and 80-year-old cohorts in cosmopolitan settings have already benefited from large proportional reductions in total mortality and cause-specific mortality from infections and parasitic diseases, cardiovascular complications, neoplasms, dementia, and diabetes. Thus, this avenue for life extension in humans may be approaching some innate somatic limits. However, these limits may only reflect current sociocultural/environmental settings, not an unbreakable barrier.

Enhancing physiological defenses

Another possible avenue for human life extension may be altering the activity or representation of specific senescence-resistant or -enhancing loci, proteins, or enzymes. In non-human models, activity levels of the enzymes SOD and catalase are higher in long-lived than short-lived strains. Enhancing the activities of both loci by gene duplication in transgenic animals is associated with extended life spans. Enhancement of circulating SOD/catalase levels in humans may be possible through manipulations leading to the increased transcription of loci, decreased destruction of enzymes, or increased activity of enzymes. At present, it is unclear how to induce such alterations. As with other pharmacological interventions (e.g., ACE inhibitors and dopamine stabilizers), there is the possibility of developing an agent, or agents, that enhances activity or transcription of specific enzyme(s) (e.g., SOD and catalase). However, based on rodent models, a multitude of physiological alterations may be necessary to halt even a minority of, let alone all, senescent processes in all cell types and organ systems.

The current model linking CDCs and senescence suggests that further advances in human life extension may arise by slowing or halting senescence-promoting alterations in risk factors. Recent success include ACE inhibitors, beta-blockers, and diuretics to reduce blood pressure, the progression of arteriosclerosis, and cardiovascular complications, and pharmacologicals to control hyperglycemia and hyperlipidemia. On the horizon is a vaccine to protect susceptible genotypes using a monoclonal antibody that targets the neurofibrillary tangles of Alzheimer's disease. These therapies do not enhance function; rather, they target particular aspects of risk factors that promote specific diseases that are themselves senescent processes or senescence promoting.

Many believe that, as the elderly population increases, there will be a concomitant increase in frailty (Rosenwaike 1985; Stini 1990; Rakowski and Pearlman 1995). However, the oldest-old appear to show greater vitality than previously predicted (Alterm and Riley 1989; Perls 1995; Service 2000b). Even though increasing numbers of frail elders are today a public health concern, data do not support the hypothesis that they will be more frail. Those aging today are expressing the outcomes of their own life styles, diet, environment, culture, and social circumstances, such factors changed rapidly in the twentieth century. Senescence and related fraility should occur more slowly when the external environment is less hazardous (culturally controlled) and organisms are less likely to die from non-senescent processes (see Austad 1997, p. 31). In such settings, the force of natural selection declines less rapidly and maintenance of the soma for reproductive purposes should lead to greater senescence resistance. Through biocultural mechanisms, humans have over their evolutionary history reduced their susceptibility to environmental hazards, extended their premature period of life, and provided opportunities for late-life survival.

Senescence is a multifactorial process that arose along with the evolution of sexually reproducing organisms. Production of a "magic bullet" that halts all root causes of organismal senescence is extremely unlikely. Among humans, multiple counterbalancing, confounding, and interacting processes have come together to produce reproductive success with slowed growth, development, and maturation, dependent young, and very long, but finite life spans. Some of these processes have also led to late-life survival such that a woman may live twice as long post-reproductively (ages 50–120, 70 years) as she did reproductive (ages 20–50, 30 years). Similarly, some men continue to produce sperm and may father offspring into their ninth decade of life, well after the majority are long dead. Altering activities of a single or even several enzymes may have the effect of slowing progression of some risk factors and CDCs, allowing some to survive longer. However, this may have little influence on progressive senescence in other systems and do little to alter the overall pattern of human senescence.

Still, humankind's second best hope to increase life expectancy and life span may be enhancing physiological defense systems (Walford 1984). Proponents of antioxidants have for decades ascribed to the hypothesis that ingestion of a variety of vitamins and minerals (e.g., Vitamins E, C, and A, and the minerals selenium, zinc, and chromium) will improve defenses and extend life. As yet, research results have neither confirmed nor refuted this hypothesis. One fact that has been observed is that ingestion of large amounts of antioxidants may have the untoward effect of reducing the body's natural antioxidant defenses. As research continues, it seems likely that new ways to enhance humankind's antioxidant defenses and detoxifying abilities *in vivo* will be discovered and that these may constitute an additional mechanism for extending average and maximum life spans. However, the best-established way to increase defensive capabilities remains to maintain a healthy lifestyle. Exercise, low-fat/-calorie diets, mental activity, and maintaining moderate weight remain the best-known and proven methods for enhancing human physiological defenses at older and all ages.

Limits to human longevity?

"What is the maximum achievable human life span?" is a question that has intrigued generations of humankind. One widely read book set this at about 120 (*Genesis* 6:3) years, an age that 122 years is only slightly beyond. There are two possible views on human longevity: *there is a limit* to human life span or *there is no limit*. In the first case, human life span, and thus the life spans of *all* sexually reproducing species, is a genetically determined species-specific characteristic. If true, human longevity cannot increase beyond a specific limit. The proposed upper limit is currently set at 120 years (see Fries 1980, 1983, 1988; Olshansky *et al.* 1990; Olshansky and Carnes 1994; Clark 1999). In the second view, life span is responsive to environmental stimuli, is genetically labile, and no species-specific limit exists (Rose 1991). Since no limit exists, no species can achieve its limit. Rather, in this case, life span is the product of a species' available genetic variation interacting with a variety of specific environments (which, in humans, includes cultures). Thus, maximum life span is environmentally labile and varies as the selective pressures on populations change, just as do most phenotypic traits.

Proponents of species-specific, genetically programmed life spans predict a maximum of about 85 years for human life expectancy and about 120 years for maximum life span (Olshansky *et al.* 1990; Barinaga 1991; Olshansky and Carnes 1994; Clark 1999). Since the eighteenth century, an approximately 95% reduction in mortality from infectious diseases has occurred in the U.S.A. and other cosmopolitan societies. This has been associated with an increase in life expectancy from about 40 to 75 years (Olshansky *et al.* 1990) and decreased infant and child mortality (Fries 1983; Guralnick *et al.* 1988, Brock *et al.* 1990; Olshansky *et al.* 1990). In Japan, life expectancy today already approaches the upper limit set by some researchers – about 83 years among women. Less dramatic changes in older adult life expectancy have also been observed. Any additional increase in longevity at older ages will necessitate decreased mortality from CDCs during both middle (45–64) and older years (65+), where only small decreases in chronic disease mortality are thought to be possible (Fries 1983; Olshansky *et al.* 1990). In general, the model of limits to human life span represents an extension of the Hayflick limit and genetically determined senescence to humans from serially cultured cells (Barinaga 1991). Proponents of the limits to human life span model interpret Hayflick's replicative limit for fibroblasts *in vitro* as determining finite life spans for organs, and, thus, the organisms they make up. This would necessitate a biological limit at which the soma could no longer function due to a combination of continuing environmental insults and increasing physiological failure. According to Fries, the species-specific average life span (life expectancy) is the length of life that humans might expect to live (when Fries' theory was originally published, the average life span was 73 years – 77 for women and 70 for men; today, for women, it is 78 in the U.S.A. and over 83 years in Japan). Recently, proponents of these models have interpreted telomeric shortening as the biological mechanism determining species life spans – the ultimate molecular clock. The Gompertz plot of linearly increasing mortality with age has also been used to support the notion of an inherent limit to life span (see Figure 1.1). However, among nematodes, the Gompertz function slows by about 50% in *age-1* mutants and maximum life span is extended by 110% (Johnson 1990), showing that mortality rate-doubling times are alterable within species. In addition, in some species, mortality rate may fall at later ages, contrary to Gompertz predictions.

Today, many students of longevity subscribe to the idea that neither life span nor life expectancy is constrained to prescribed limits; rather, these respond to the environment. One group predicts that life expectancy may increase to 100 years during the twenty-first century (Manton *et al.* 1991, 1995), while a recent (2000 BC) bet between two gerontological theorists predicts a maximum life span between 130 (J.R. Carey) and 150 years (S.N. Austad) within 150 years. Investigators base these predictions on several observations. For one, maximum life span has been modified using several environmental manipulations and artificial selection in multiple laboratory models – rodents, insects, and worms (Curtsinger *et al.* 1992; Carey 1997). Even among humans, increased life expectancy after age 85 has been documented for Scandinavian and Austrian samples (Wilmoth *et al.* 2000; Doblhammer and Kytir 2001). Furthermore, mathematical estimates of human mortality hazard functions appear to level off or decline slightly after about age 85 (Weiss 1989a). Such data lead to the proposition that there is no limit to human life span and suggest that a series of stochastic risks limits human and other species' life spans in a probabilistic, not a deterministic, fashion (Weiss 1989a; Rose 1991; Wood *et al.* 1994).

Data from long-lived humans indicate decreased mortality rates at very old ages (80 years and over) (Perls 1995; Wilmoth *et al.* 2000; Doblhammen and Kytir 2001). One way to appreciate these rates is to understand that the large fraction of individuals who are today saved from deaths during their infant, child, and middle years may now show their inherent (genetic) frailty by succumbing to late-life CDCs at earlier ages than those with greater inherent vitality. Senescence may be inevitable and progressive (Olshansky and Carnes 1994); however, it is neither precipitous nor invariant as suggested by the concept of a maximum life span or an absolute limit to life expectancy. Rather, as progress is made toward eliminating and postponing the currently recognized CDCs of old age, new ones may emerge to become the leading causes of death among elders. This process allows age at death to increase in variance (rather than decrease), thereby preventing any rectangularization of survivorship among the elderly (see Myers and Manton 1984; Weiss 1990; Olshansky and Carnes 1994). Swedish mortality data over the last 50 years indicate that death rates at ages 85 years and older have decreased, while, in the U.S.A. a significant decrease in mortality at later ages, due to decreasing cardiovascular disease from 1960 to 1989, has also been reported (Verbrugge 1989). This decrease is attributed to control of hypertension, changes in nutrition, increased fitness, and personal health practices that have more to do with cultural developments than biology. Similarly, between 1861 and 1999, the maximum life span of the population of Sweden increased by 0.44 years per decade before 1969 and at the pace of 1.11 years per decade thereafter, with 70% of this increase due to reduced death rates after age 70 (Wilmoth *et al.* 2000). Increased maximum life span was 16% attributable to mortality reductions prior to age 70 and 12% to demographic factors; among cohorts dying out after 1969, 95% of the increase was due to mortality reductions after age 70. These data clearly question any fixed maximum for human average or maximum life span (Wilmoth *et al.* 2000). Not only is life extension possible, it has been happening at least over the past century as the age distribution of mortality has steadily risen and variance in age at death has increased.

Neo-Darwinian evolutionary theory does not support the hypothesis that specific loci and alleles controlling or limiting life span to some arbitrary maximum will evolve under the influence of natural selection. However, Williams' (1957) articulation of senescence and pleiotropy clearly predicts both the existence of alleles with late-acting detrimental effects and an accumulation of such ill effects at later periods in the life span. Such phenomena may be easily misinterpreted as genetic longevity characteristics evolved to limit life spans (Clark 1999). Under different environmental conditions, different samples of the same species may have very different average and maximum life spans. This is true of roundworms, fruitflies, mosquitoes, rodents, canines, primates, and humans (Resnick 1985; Arking 1991). Such observations tell us little about the biology of senescence, except that species vary as much in longevity as they do in age at reproduction, growth and development, adult size, and other life history characteristics. A single/simple genetic switch controlling life span does not exist.

Chapter synopsis

Physiological measures alter over the life span. During growth, development, and maturation, changes are predictable. After the attainment of reproductive adulthood (about 20 years) they are neither consistent nor constant across individuals. Some physiological functions improve with age while others decline Functional capacity may even improve in response to alterations in activity and dietary patterns during late life. During early years, death may be attributed to a single cause; after age 85, multiple insults contribute to mortality (Susser *et al.* 1985; Manton 1986a, b; Crews 1990b). Unfortunately, for the model developed earlier, it is not clear that loss of organ reserve is responsible for any of these insults.

Several simple environmental manipulations (e.g., calorie restriction, low temperatures) appear to rectangularize survivorship curves and shift them to the right. Few data counter suggestions that human survivorship curves can be rectangularized and shifted to the right as demonstrated for fruitflies, roundworms, mosquitoes, and rodents (Curtsinger *et al.* 1992). However, it may not be as simple to extend life in humans. Humankind has a peculiar biocultural adaptability unlike other organisms and differs in all life history traits from rodents (Lasker and Crews 1996). Using culture, humans already experience many of the same benefits as long-lived laboratory rodents. For many generations, they have been protected from predation, able to survive severe trauma, and have lived generally healthy long lives. Cultural processes long ago extended average and maximum life spans well beyond those of earlier generations. This process takes but a few generations among rodents. A reasonable suggestion is that current life spans enjoyed by humans developed over evolutionary time via mechanisms similar to those used to extend life span among laboratory rodents (e.g., controlled environments, nutritionally adequate diets, and selective breeding). Humans are already very long-lived, K-selected species. Any remaining untapped senescence-resistant biological traits are unlikely to be those associated with the extension in a short-lived, small-bodied, litter-bearing, r-selected species.

Among humans, non-random breeding for late-life reproduction likely began as soon as culture provided opportunities for direct investment in grandchildren and later-born children. Once survival to late life (arbitrarily set at 65+) increased the odds of siring children, or of grandchildren and other kin reproducing, any biological factors promoting senescence resistance would have had positive influences on fitness. As such kindreds outcompeted others in fledging offspring, senescence resistance would spread through populations.

Slowed, absent, or developmentally delayed maturation and senescence already characterize human somas. A new evolutionarily stable balance between delayed senescence and late reproduction may have been achieved in humankind's recent past – one suggestion is about 250 000–500 000 years before present, coinciding with *H. sapiens'* spread across the globe. As with any ESS, however, environment and culture change alter these temporary balances, allowing new strategies to emerge. The emergence of large cosmopolitan societies with average life spans approaching 80 years may reflect such an environmental shift.

Calorie restriction is a sufficient means to increase both average and maximum life span in laboratory rodents. Rather than calorie-restricted laboratory diets, human populations have over recent evolutionary times undergone multiple periods of starvation, calorie inadequacy, and malnutrition due to environmental fluctuations. Human reproduction and senescence processes may have already responded to "calorie-restricted" environments, such that today, in "calorie-unrestricted" settings, only a proportion of individuals show greatly reduced longevity and/or rapid senescence. Calorie restriction may add a few years to the average human life span, but have little, if any, effect on increasing the maximum life span of a large-bodied, late-reproducing, culture-bearing, k-selected hominid, compared with small-bodied, early-reproducing, r-selected rodents.

Another proven method to raise the life expectancy of rodents and insects is temperature modulation – survival improves at lower temperatures and decreases at high temperatures. Among the world's current frontrunners in human life expectancy are Norway and Sweden, both at higher latitudes, and Japan, a mid-latitude nation. However, as yet, there are no indications that life span has exceeded 122 years in these groups. In general, it does not seem likely that most life extension experiments in rodents or insects will have wide applicability to humans. This is largely because human cultural evolution has already modified our environment to such a large extent that similar modifications have probably occurred frequently enough that many such contributions already are expressed in human populations. How these vitality-enhancing or senescence-postponing loci currently are distributed across humankind is probably, in large degree, determined by population structure, which is determined by biocultural interactions that produce population variation in senescence, longevity, and CDCs. Simply put, although defining population subgroups is so difficult and value laden, this does not alleviate the scientific necessity to do so and to ask the question of whether there are different patterns of senescence and longevity across racial, ethnic, national, or local groupings of humankind. The answer appears to be an equivocal "no". Although life expectancy varies widely across populations around the globe, most such differences appear to be determined more by sociopolitical and socioeconomic factors rather than by innate differences in human biology. In no population has a reliable recorded age over 122 years been reported, although the claims are many. Understanding differences in survival among human populations today amounts to determining how political economy and individual opportunity structures (processes that likely have little to do with the biology of senescence) prevent many people from achieving their biological potential for life span, just as they prevent achievement of physical growth or intellectual potential.

In general, evolution is highly conservative, using the same basic regulatory, transcription, and control factors (e.g., homeobox genes, hormones, and growth initiators) to produce the basic animal design, but allowing ample opportunity for variable responses in the final product due to local factors and receptors (e.g., hormones, peptide growth factors, and cytokines). The latter may be altered quickly in response to environmental inputs. It is these more proximate factors that determine MRP, RS, senescence, and individual life span. Such species-specific factors overlay a well-developed set of evolutionary conserved biological processes and tools (pleisomorphies) that maintain the basic homology of animals and their systems of organ development (Charnov 1993;

Finch and Rose 1995). Apparently, it is local/proximate factors that are altered during environmental manipulations that extend both average and maximum life spans in model organisms. Population differences in the biology of human senescence also undoubtedly exist, as they likely do for risk factors for CDCs and obviously do for responses to pharmaceuticals. However, as with CDCs in general, biological risks for "early" or "postponed" senescence are likely to be multiple, variable in their expression, pleiotropic, and in epistatic relationships with other loci. The joint effects of SOD and catalase on longevity in transgenic insects and rodents provide a simple model of expected relationships. When either is increased individually, there appears to be little influence on longevity or life span. Conversely, when both are overexpressed, large increases in measures of "postponed senescence" are observed. In addition, rodents with either SOD or catalase knocked out do much more poorly than "normal" controls. Given the wealth of polymorphism at humankind's 30 000 or so "coding loci", not to mention the even larger amount of variable DNA elsewhere, the likelihood is that over- or underexpression at numerous loci may influence longevity, life span, and rate of senescence. Rodent, insect, and invertebrate studies provide human homologs as candidate loci for senescence-postponing alleles. These may be distributed differentially across populations, as are numerous other polymorphisms. Thus, although SOD and catalase should head any list of candidate loci for "postponed senescence" in humans, followed quickly by glutathione peroxidase, these may show reduced influences in large-bodied, K-selected humans. The first questions to answer are whether a variety of protein forms exist, and, if so, which are more frequent among the long lived of various human populations compared with younger cohorts and those who failed to survive past age 55 but survived to over age 30 or so.

It has been asserted that, to increase life expectancy further, it is first necessary to eliminate CDCs, a prospect that many believe is unlikely (Olshansky *et al.* 1990, 1998). If all circulatory diseases, diabetes, and cancer were eliminated, life expectancy at birth in the U.S.A. would only increase by 15.82 years for females and 15.27 years for males (Olshansky *et al.* 1990, 1998). If these estimates are accurate then life expectancy for women would increase to 93 years and to 85 years for men. Complete elimination of any known CDC is unlikely. Still, morbidity continues to be pushed to later years of the life span. Decreased age-specific morbidity and mortality allow a healthier cohort to survive into late life today than did even two generations ago. Some argue that, even if mortality from certain diseases was to decrease, other diseases would take their place and life expectancy would increase little at older ages (Sacher 1977; Brody 1983); others suggest that, even if death rates at age 85+ are reduced, currently too few people survive to over 85 to change life expectancy statistics significantly (Barinaga 1991). If true, this may represent only a current practical limit based on current culture and disease patterns, and not any biological limits (Olshansky *et al.* 1990; Barinaga 1991; Olshansky *et al.* 1998). However, as yet, there are no indicatons of "other diseases" destined to take the place of our current set of CDCs at later ages. No new CDCs are arising among today's "oldest-old".

6 *Discussion and perspectives*

Human senescence

Humankind's existence depends on evolutionary balances in the pace of conception, gestation, growth, development, maturation, and reproductive effort. The confluence of biology, environment, culture, and contingency that sculpted human life history did not fashion senescence. Instead, these interactions set minimum limits on the life span required to progress from conception to the fledging of human offspring (the minimum necessary life span (MNLS)). Somatic systems, from cells to organs, are set to the pace at which the necessary processes of life unfold (Weismann 1889; Finch and Rose 1995; Austad 1997). Successful alleles and genomes have had to predispose for phenotypes (somas) capable of outlasting their MNLS. The existence of a MNLS sets the stage for the cumulative, progressive, irreversible, and degenerative changes that we label senescence in somas designed to complete reproductive effort within a limited span.

Species' survival and reproductive strategies are mixed together in a complex web of evolutionary tradeoffs. Multiple synergistic, counterbalancing, and random alterations in molecular organization and physiological function have occurred over evolutionary time. Natural selection has molded these such that the average individual achieves about average relative fitness in competition with its conspecifics. Natural selection did not, however, directly produce senescence. Senescence results because natural selection lacks the ability to affect changes in allele frequencies once the period of reproductive effort is complete, declining in strength as the maximum reproductive potential (MRP) of organisms falls. Natural selection is based on differential survival and fitness. As age past MRP increases, investment switches from developing and maintaining the reproductive organ, a soma, to producing and fledging offspring, the germ line. Thus, alleles enhancing reproductive effort (including growth, maturation, mate acquisition, conception, gestation, and parental care) and relative fitness (including reproductive success (RS) and inclusive fitness (IF)) outcompete any that enhance survival at the expense of the former. As MRP declines toward zero, natural selection cannot eliminate alleles detrimental to the soma or increase the frequencies of those enhancing its maintenance, repair, and defensive systems.

Nor should selection act against any alleles that enhance survival and somatic maintenance once reproduction of the germ line is complete. Late-life reproductive effort by men benefits such alleles.

Senescence is not an adaptation developed through natural selection, because it does not produce organisms capable of achieving greater fitness in their environments. Simply stated, all organic physiological systems decline from their optimum, achieved around the point of MRP, with increasing age due to functional losses inherent in their molecular structures. The easiest way for natural selection to insure a soma outlasts its MNLS is to program all needed systems to overshoot their required capacity. Redundancy and quality are two ways to ensure individual cells, tissues, and organs outlive their usefulness. Development of better defensive, maintenance, housekeeping, and repair systems is another. Providing organisms with wide latitude in physiological function such that survival and reproduction can be achieved in a variety of environments is yet another way. In general, all processes contributing to phenotypic development and attainment of competencies promote somatic survival and reproduction. Along with multiple random and environmental factors, these also determine predispositions to senescence and life span.

All members of a species are constrained by previous life history adaptations, but the pace of senescence and life span differs widely across individuals. As mortality hazards, genetic predispositions, and environment change, new combinations and possibilities must arise that further slow senescence and extend life span. Among humans, culture and life style have been added to this mix of predisposing factors. Evolutionarily determined survival and reproductive strategies interact with culture and multiple random factors to produce human senescence. Humanity's 30 000 coding loci, each with a variety of possible alleles, and multiple ecological settings produce a broad array of possible interactions. This variation has prevented any single senescence-inducing gene, molecular clock, protein, or error to control senescence. The same is true for all sexually reproducing species. In species for which this does not appear to be true (e.g., salmon), tradeoffs between reproductive effort and somatic maintenance still structure senescence and life span. Such tradeoffs may produce such strong coordination between reproductive effort and death that a switch seems plausible. However, for numerous species of insects and ocean-dwelling life forms, including salmon, reproductive effort is a one-time event. Salmon somas are reminiscent of the metamorphoric forms of moths and butterflies, where adults are necessary only for reproduction. Having no mouths with which to sustain them selves, adults are destined to die following a single bout of reproductive effort. The neuroendocrine cascade that prompts salmon to migrate and spawn also causes them to cease feeding and to put all remaining effort into spawning. Salmon somas are not maintained following reproductive effort any more than are those of moths or butterflies because the cost of such somatic investment will not improve RS and relative fitness, necessitating that all effort goes into a single bout of reproduction.

Among humans and all sexually reproducing species, over time, external and internal processes increasingly damage the soma, necessitating its mortality. To the degree that such degenerative processes reduce relative fitness, natural selection prevents their increase. When somatic damage fails to reduce relative fitness, natural selection cannot eliminate the degenerative processes. Consequently, degeneration accumulates with age. This progressive somatic damage is constrained only by the intrinsic vitality and frailty of individual organisms, their repair and maintenance systems, and the prevailing environment and culture. As with most other quantitative aspects of the human phenotype, wide latitude in somatic and organ reserve capacity (RC), repair, defense, maintenance, housekeeping, antioxidant, and communication systems are tolerated. This underlying variation is expressed as differential responses to stressors throughout life and differences in function and adaptability during later decades. Latitude in somatic structures compatible with survival and reproduction through the MNLS ultimately manifests as differential responses to stressors and alterations in physiological function (senescence). Differential inheritance and variable expressions of senescence-enhancing and senescence-slowing alleles in individuals are dependent on biology (e.g., pleiotropy and epistasis), random events (e.g., historical contingencies, population structure, and exposure to infectious disease), cultural differences (e.g., mating systems and religion), and environmental circumstances (e.g., diet and stressors). These circumstances produce variable rates of senescence in different cells, tissue, organs, and individuals. They also allow senescent alterations to accumulate in a cumulative, progressive, sequential, and apparently age-determined fashion. Senescence is not age determined; individuals of the same age show very different degrees of senescent change. It is age related, however, because it requires at least some time for even the most rapid senescent-enhancing alleles to show their effects.

Over evolutionary time, the phylogenetic line leading to modern humans must have produced many alleles predisposing to slower or accelerated progression of somatic damage and senescence. To the degree that they either reduced or enhanced relative fitness, selection pressures would eliminate or increase their frequencies. So long as human life remained short, brutal, and dirty, most such alleles probably did not influence relative fitness. Frequencies of these alleles have drifted along, attaining random and variable levels in different groups. Through this process, diverse sets of senescence-enhancing and senescence-slowing alleles have accumulated over evolutionary time in all species, and subpopulations thereof. So long as no outside forces alter survival probabilities and mortality hazards, these senescence-slowing or -enhancing predispositions are not observed. Such alleles provide all species' genomes with a reservoir of innate propensities toward variation in degenerative processes, pace of senescence, and variable life spans. Selective breeding for long- or short-lived strains of fruitflies, nematodes, and rodents is successful because it taps these genetic reservoirs. In wild (natural) populations, many factors prevent those endowed with allelic propensities to rapid or slowed senescence from supplanting those predisposed to a more moderate pace. In free-living, naturally reproducing environments without selective breeding, propensities to rapid and slow senescence also are distributed randomly, tending to cancel one another out.

Alleles promoting slower senescence generally also promote less early-life reproductive effort. Alleles predisposing to early maturation and reproductive effort are commonly associated with greater relative fitness. Propensities to both early and late reproduction are easily selected for in the artificial settings of modern experimental laboratories; the former are associated with shorter, and the latter with longer, life spans. In natural living populations with high mortality, predispositions to early and rapid attainment of MRP, birthing, and fledging of offspring outcompete those for later and slower reproduction and senescence. Often, alleles predisposing to slow or rapid senescence are linked to others or themselves carry pleiotropic effects necessary for, or incompatible with, survival to maturity and reproduction; those linked to or themselves predisposing to early-life benefits are usually at the highest frequency in natural populations. By selecting only those who can reproduce at ages above those associated with the greatest fertility in the wild to reproduce in the laboratory, the benefits of early-life survival and fecundity are eliminated. This allows alleles associated with slower maturation, later fertility, and slower degeneration of physiological function to increase in frequency.

Not everyone sees a major role for antagonistic pleiotropy in human senescence (Masoro 1996; Harman 1999). Still, the concept is theoretically persuasive and empirically supported (Williams 1957; Luckinbill *et al.* 1984; Rose 1984; Albin 1988, 1994; Partridge and Fowler 1992; Zwaan *et al.* 1995; Austad 1997; Westendorp and Kirkwood 1998; Harper and Crews 2000). Multiple aspects of human biology and physiology suggest antagonistic pleiotropy. At the very least, sex hormones (Grossman 1985; Adams *et al.* 1995; Wilding 1995), metabolism of glucose (Cerami 1985; Monnier *et al.* 1991), and angiotensin converting enzyme and apoliopoprotein E alleles (Schächter *et al.* 1994; Crews and Harper 1998) are good candidates. Early-life benefits of these are obvious, and all predispose to late-life detriments. The low number of alleles hypothesized to show antagonistic pleiotropy does not imply their rarity. After all, allelic variation is well studied for only a few hundred loci and only in a few populations. Those identified as candidates thus far are all associated with major differences in clinical risk factors. In most cases, antagonistic pleiotropy is likely to be more subtle. Numerous genetic influences on life span have been identified. Housekeeping genes (e.g., superoxide dismutase (SOD), catalase), cell cycle controllers (e.g., p. 53), cellular receptors (e.g., IRS-1 and ILGF), signal transducers (e.g., cytokines), DNA repair systems (e.g., polymerases and excision factors), and proteins active in the neuroendocrine (e.g., protein hormones and neurotransmitters), immunological (e.g., human lencocyte antigens (HLA)), and metabolic (e.g., apolipoproteins and glucokinase) systems all modulate senescent degeneration and life span. Numerous chronic degenerative conditions (CDCs), including MODY, type II diabetes, Huntington's disease, cancer, cardiovascular disease, and dementias, are secondary to genetic predispositions that increase frailty and risk. Humankind's 3 billion or so base pairs of DNA and 30 000 coding loci are likely to include innumerable alleles predisposing to rapid (frailty) or slow senescence (vitality).

Obviously, immortality has not been a viable evolutionary strategy for any sexually reproducing species on Earth. All, including humans, create disposable somas (Kirkwood 1990). Although immortality has not been a viable strategy, extended somatic development, late attainment of MRP, low reproductive rates, high parental investment in offspring, long life, and grandparenthood have developed in multiple mammalian species (e.g., elephants, pilot whales, chimpanzees, gorillas, and humans). This set of characters appears to represent an interdependent, interconnected adaptive suite of longevity characteristics. It is not dependent on language or culture; rather, this suite typifies a variety of non-verbal, non-cultural species.

As a basic animal adaptation, the disposable soma provides a blueprint for finding mechanisms, processes, proteins, loci, and alleles important in modulating senescence and life span at all levels of somatic integration (Kirkwood 1995, 2000; Kirkwood and Austad 2000). Identified mechanisms include oxidation, mutation, wear-and-tear, stress and stressors, infections, parasites, and other environmental stressors. Processes include the accumulation of metabolic byproducts, damage to DNA, proteins, cells, and organs, loss of DNA, cells, RC, and competencies attained earlier in life, and wear-and-tear/allostatic load on a mortal soma. The biological bases for these multiple alterations are losses, alterations of expression, or inherited variations in loci responsible for maintenance, energy, production, protection, reproduction, and repair of somatic cells and tissues during growth, development, maturation, and reproductive adulthood. These systems harbor numerous alleles that influence the phenotypic expressions of senescence. Following the attainment of MRP and reproductive adulthood (in humans, commonly ages 20–35), these systems show highly variable but progressive dysregulation, culminating in somatic death. Among free-living species, alleles predisposing to greater reproductive effort swamp those predisposing to somatic maintenance, which may never be expressed. Only when extended survival also predisposes to greater IF are senescence-delaying alleles advantaged. Numerous alleles enhancing both relative fitness and vitality have arisen over evolutionary time. These have produced the variety of life history schedules and life spans (e.g., tortoise, 150 years; humans, 122 years; fruitflies, 4 weeks; mice, 4 years) observed among extant species. For a mammalian primate species, modern humans live a long time. This is in a large degree because humans have added biocultural influences to the suite of interdependent, interconnected adaptive characters exhibited by other long-lived mammals. Socioculturally, humans have for many generations provided opportunities for individuals to express their senescence-delaying allelic pre-dispositions and thereby enhance their relative fitness (both Darwinian and IF).

Understanding human senescence and longevity requires more than understanding the biology of senescence. Explorations of humankind's unique biocultural adaptations and how these have altered life history are required. This is where humans are unique in studies of senescence. By continually elaborating culture, humankind produced multiple "biocultural selective pressures". These have been sufficient to slow intrauterine and post-natal growth and development, rates of maturation, attainment of MRP, reproduction and fledging of offspring, and senescent dysfunction. During the evolution of humans, alleles predisposing to somatic maintenance have outcompeted alternatives. Biocultural interactions during hominid/human evolution placed sufficient pressures on human life history to predispose most not only to attain the species' MNLS but to survive into their seventh decade of life. These pressures allowed elders (today, those over 65; several millennia ago, those over 50; 100 000+ years before present, those over 40 years) to invest in their descendants (the Grandparent Hypothesis), and/or to survive sufficiently long to increase their own relative fitness. Any reproductive success by elder males places strong selective pressure on survival. In all settings, men show greater variance (range zero to hundreds or even thousands, with a highly skewed distribution) in fitness than do women (range approximately 0–50, less skewed). Mammalian males have virtually un-limited reproductive possibilities compared with females. Physiological limits on women's reproduction suggest they have less to gain compared with men from slow senescence and an extended life span. Eventually, cultural develop-ments allowed older men with greater access to reproductive-age females than other age groups of men, as observed among many more traditional-living peo-ples today. This produced something uniquely human – bioculturally driven reinforcing selection. As culture provided opportunities for longer survival, greater reproductive success by older men increased the fitness of alleles as-sociated with late-life vitality. In turn, late-life vitality allowed older men with more opportunities to enhance their relative fitness, thereby promoting even greater representation of their offspring in the next generation.

Additional biocultural pressures followed when culture began tipping the prevailing evolutionary balance such that natural selection acted more through differential fitness post-maturity than through survival to the age of MRP and differential reproduction. Today, 95% of persons born live to reproductive age in more cosmopolitan populations. Among modern "hunter–gatherers", only about one-half of children born survive to reproductive age (Lancaster and King 1985). This proportion was likely to have been lower among hominid and early human populations, as it is among extant wild-living chimpanzees. Culture has not eliminated selective pressures – up to 15% of couples are infertile in some cosmopolitan settings indicating strong selective pressures. Similarly strong selective pressures are likely to act to maintain the soma through its MNLS to allow rearing altricial infants, long-term child dependence, extended birth intervals, and parental investment. Among humans, those who survive and reproduce longer are likely to contribute more alleles to the next generation. The relative fitness (RS + IF) of long-lived (over 50) men may vary more from the mean than that of women (IF only). Men reap larger gains from slowed senescence and longer life spans than do women.

Mechanisms/processes

Gerontological research has three major goals: explaining why, determining how, and halting, slowing, or otherwise intervening in human senescence. Evolutionary theory explains why, but not how, senescence arises in sexually reproducing species. To understand how senescence occurs, mechanisms and processes contributing to senescence must be detailed. These mechanisms include the basic molecules of life, variable DNA sequences, and transcription, translation, error, and repair rates, along with variations in antioxidant levels, hormone action, cell structure, redundancy, and RC of cells and organs. The specific mechanisms important in senescence may vary widely from species to species (Rose 1991; Finch and Rose 1995). They are likely to be simpler in less complex (e.g., yeast and invertebrates) than in more complex species (e.g., rodents and mammals). The mechanism and interactions involved in creating somas of several hundred or thousand cells are of a different order than creating those of several billion. Multiple co-adapted and overlapping systems provide a set of molecular checks and balances regulating homeostasis in complex systems, some of which are not found in simpler systems (see James and Baker 1990; Crews and Williams 1999 for a blood pressure model). Cumulative degenerative alterations in these lead to observable somatic alterations that accumulate in an age-related fashion. Among these are oxidation of DNA, proteins, lipids, blood proteins, and supporting structures, alterations in neurological, endocrine, immunological, cardiovascular, and organ function, loss of bone, muscle, body mass, hypothalamic–pituitory–adrenal (HPA) regulation, and responsiveness to a variety of stressors, and increased mortality. These processes are observed across multiple species and produce similarities in senescent phenotypes.

Molecular, cellular, and non-human models of senescence are useful for determining mechanisms that contribute to senescence in species. However, mechanisms observed in disarticulated cells or invertebrates and rodents may not apply directly to whole organisms or humans. Recall the example of enhanced expression of SOD. In fruitflies, it failed to improve life span and in mice it actually decreased life span, but, when catalase was jointly overexpressed, *Drosophila melanogaster* life span was extended as expected (Orr and Sohal 1994). Among humans, dietary supplements of antioxidants may reduce their endogenous production. Multiple and complex relationships among biology, environment, culture, and life style contribute to human senescence. Simple mechanisms such as more cysteine, growth hormone, telomerase, SOD, or catalase are not likely to enhance longevity in an already long-lived species with wide phenotypic variation.

Extrapolation of associations between specific alleles and life span observed among invertebrates and rodents to human homologs is not appropriate. Alleles at some loci in nematodes and fruitflies appear to be simple switches that increase life span by 50%, 100%, or more. Their homologs in humans may have very different functions, along with new patterns of epistasis, pleiotropy, regulation, and expression. What enhances longevity in short-lived animals may not do so in a long-lived, K-selected species with a greater investment in somatic maintenance and repair. Homologous loci in long-lived, large-bodied species have probably adapted new functions quite different from those in short-lived ones. They also may respond differently to neuroendocrine inputs and local metabolic environments. An integrated systems approach to senescence is needed to study long-lived humans. Long-lived, large-bodied species invest more in somatic development and maintenance than do short-lived ones. Because patterns of senescence differ across cell types, organs, environments, and species, only integrated studies of senescence that include DNA, cell, organ, and organ systems within individuals and across population levels will provide a clear understanding of human senescence. Comparative studies with other large-bodied mammals also are likely to contribute to understanding human senescence in ways that invertebrates and rodents cannot.

Human senescence and long life spans seem dependent on alterations in regulatory systems and mechanisms organs have developed to sustain themselves in a disposable soma (Kirkwood 2000). How adaptations to the constraints of time (e.g., time to gestate, grow, develop, mature, mate, reproduce, and fledge offspring) are expressed determines intrauterine and post-natal development, life history, and, ultimately, life spans in all species. Evolutionarily, the complex bodies our genes build and the cultures we elaborate with our somas are no different from elaborate tail feathers, large horns, or large body sizes in other species, mechanisms which are designed to pass DNA on to the next generation. Those who complete reproductive tasks more efficiently, quicker, or more frequently generally have greater relative fitness and more representation of their alleles in future generations than those who are slower. Most individuals carry a random configuration of alleles. These predispose the majority to average survival or an average mortality hazard in the environment their forbearers inhabited. Human genetic predispositions to somatic survival are not static; they are molded by the particular cultures and environments in which somas live. Genotypes with identical mortality hazards from infancy through to old age may have very different morbidity patterns and ages at death. This sometimes gives the impression that mortality differentials and life span are unrelated to genetic heterogeneity, when in fact they may be very closely related (Service 2000a, b).

Sex differences in life span

Females do not necessarily live longer than the males of a species (Gavrilov and Gavrilova 1991, 2001). Sex differences in life span are species and environment specific and, in humans, culturally modulated. In non-cosmopolitan settings, where survivorship curves for men and women are about equal, the latter die more frequently from reproductive complications compared with places where women outlive men. Over most of human evolution, upon attaining MRP through the reproductive period, childbearing was a major stress facing women. In more traditional settings, maternal mortality is still high. During earlier phases of human evolution and in today's non-cosmopolitan settings, women tend to expend more of their total somatic capacity on reproductive effort, retaining little RC with which to maintain their somas into late life. In more cosmopolitan settings, maternal mortality is ameliorated by culture, while bearing and fledging offspring are less draining on somatic resources. Investing less in offspring, women are able to retain more RC throughout their reproductive years. This may then be diverted to somatic maintenance, defense, and repair. In such cosmopolitan settings, men tend to die more rapidly than women except at the most advanced ages (Waldron 1983; Verbrugge 1984, 1989).

When both sexes are protected from early mortality and women from maternal mortality, men die more quickly, except at extreme ages; without such protection, women die more quickly. This difference has generated many speculations and theories. Over most of their life spans, men generally exhibit greater phenotypic variability than do women (Geodakian 1982; Stinson 1985). Even among children, girls appear to be more buffered from the environment than are boys. For example, among Guatemalan children, girls vary less in average height and weight across SES levels than do boys (Sullivan 2002). Men appear to be more susceptible than women to a variety of environmental factors during not only growth and development, but throughout life (Stinson 1985). Men also tend to engage in more risk-taking behaviors, including male–male competition, intergroup conflicts, exposures to predation, accidents, trauma, and during mating effort. These activities increase investment in reproductive effort, while decreasing RC and increasing mortality hazards (Williams 1957; Rose 1991, p. 95). Sex differences in life span in cosmopolitan settings also may reflect allometric associations with other aspects of physiology. Once body weight is controlled, no significant differences in longevity of males and females are observed across a variety of mammalian species (Prothero and Jürgens 1987). However, in these settings, all senescent processes appear to be more rapid in men, who show the full spectrum of CDCs and risk factors at earlier ages than women. Over evolutionary time, sex differences in reproductive strategies likely contributed significantly to variability in senescence. Natural selection may have favored adult vigor and high adult reproductive effort over somatic maintenance in large-bodied mammalian males. As men began enhancing their relative fitness at older ages, this pattern may have altered leading to enhanced somatic survival.

Women's seeming innate somatic advantage over men is unique to cultural settings of low fertility, low maternal mortality, late reproduction, low infant mortality, low investments in lactation, high plasma estrogen over most of a female's life, and retention of RC into late life. When childrearing was more stressful, women may have developed somatic systems more efficient at converting their environment into useful and stored resources and offspring than did men. Once culture provided opportunities for late-life survival, the benefits of late-life reproduction would be greater for men than women. Even slight improvements in relative fitness place strong selective pressures on the retention of reproductive function and late-life survival in men. Recent rapid changes in culture and alterations in pre-existing evolutionary and biocultural balances as culture became humankind's major adaptation likely have contributed to currently observed sex differences in life span, a pattern that did not characterize most earlier human populations.

Human senescence, health, and disease

As senescence itself, CDCs are multifactorial. Genetic, behavioral, sociocultural, and environmental risks contribute to their highly variable etiologies. As cells and organs complete their primary functions, they also undergo incremental degenerative alterations in their components. Such alterations are inherent aspects of the biochemistry and life history of sexually reproducing organisms. Molecules and systemic processes promoting the development and maintenance of reproductive somas are liable to dysfunction, and metabolism itself produces harmful byproducts. Somatic dysfunction accumulates with survival, but is not time independent – different structures and individuals accumulate dysfunction at different rates. With both random and inherent variation in dysfunction combining in an already highly variable species, multiple pathways and rates of functional loss occur. These present as risk factors and CDCs.

CDCs and senescence are so inextricably linked that, to some, CDCs are the visible manifestations of senescence. All current mechanistic theories of senescence are based on alterations, decrements, and losses of function that increase risks for CDCs. Synthesis of theories and methods across disciplines examining human variation, CDCs, aging, and senescence is contributing to the specification of specific senescent human phenotypes. Several categorizations of the elderly are useful in gerontological research. Definition of elders as those aged 65+ years was a late nineteeth, early twentieth century criteria that became associated with social security. Later, a three-way classification into the young–old (ages 65–74), old–old (75–84), and oldest-old (85+) was developed to emphasize that elders are a highly variable population. Healthy/diseased, dead/alive, and frail/vital are more related to actual function and losses associated with CDCs. Senescent phenotypes may be defined in a number of ways. For example, frailty, a state of vulnerability and increased probability of death, has been defined using a suite of traits (e.g., unintentional weight loss, self-reported exhaustion, weak grip strength, slow walking pace, and low physical activity) (Fried *et al.* 2001). Incontinence, diabetes, and lack of ability to complete activities of daily living (ADLs) are additional candidates for assessing frailty. Frail phenotypes might then be compared with long-lived or high-vitality phenotypes. Long-lived phenotypes could include centenarians and their families, such as those identified as part of the New England Centenarian Study or in genealogies from Utah, U.S.A. (Perls 1995; Puca *et al.* 2000; Kerber *et al.* 2001; Perls 2001).

Masoro (1996) along with McEwen (1998) and colleagues (McEwen and Steller 1993; Seeman *et al.* 1997; McEwen and Seeman 1999) suggest that the body undergoes a cumulative, multisystem physiological toll as it responds to everyday stressors over the life span. This predisposes individuals to homeostatic failure. A proposed measure for this toll is "allostatic load". Determined on the basis of ten parameters reflective of systemic regulation of the soma (i.e., systolic and diastolic blood pressure, waist hip ratio, high-density lipoprotein (HDL)-c, total-c, glycated hemoglobin, serum DHEA-S, and 12-hour urinary cortisol, noradrenaline, and adrenaline excretion levels), allostatic load is thought to measure to some degree "the price that the body may ultimately pay for its adaptational efforts" (Seeman *et al.* 1997, p. 2263).

Allostatic load is reminiscent of wear-and-tear theories of senescence. Allostatic load reflects the soma's innate inabilities to withstand stress continually, maintain RC, and sustain continued survival. Allostatic load measures loss of homeostasis by the disposable soma due to continued response to life's stressors, is lower in high-functioning elders, and is correlated with cognitive performance, declines in physical performance, and the incidence of cardiovascular disease (Seeman *et al.* 1997, p. 2264). Allostatic load is closely tied to cardiovascular risk factor profiles (six of ten component variables), but, it correlates with non-cardiovascular events and measures and may reflect general frailty and vitality. At the very least, allostatic load provides a first approximation to measuring human senescence that may be tested with prospective data for associations with dysfunction, disease, and life span. Baseline data to test such associations are probably already available for many samples. Addition of genetic (e.g., *ApoE*4*, ACE D), biocultural/sociobehavioral (e.g., social status, fertility, age, sex, and birth weight/length), and frailty (e.g., grip strength, walking pace, and ADLs) measures to the components of allostatic load should increase its predictive power. As a multifactorial construct, allostatic load may contribute to geriatrics, gerontology, human biology, human variation, epidemiology, anthropology, and public health.

Aging and senescence

Senescence and aging often are used as synonyms. Unfortunately, this implies that biological senescence is a chronological process. Morbidity and mortality do increase with age within cohorts and loss of individual physiological function does progress in an age-related fashion. Still, for the most part, the underlying biological processes are independent of time and only secondarily correlated with chronological age. Senescence results from dysfunction of multiple intracellular mechanisms that cause progressive declines in organ function and capacity. Collectively, loss of biological integrity across multiple organs produces systemic declines in somatic homeostasis, leading to CDCs and an increased risk of death. This loss of homeostasis is measurable as allostatic load, frailty, and increases in risk factors for CDCs. Individuals vary in functional capacity, RC, training, exposures, fetal development, post-natal growth and development, familial environment, culture, lifetime diet and nutrition, genes, and sociocultural expectations, among other things, throughout life. Multiple interactions across these systems ensure that physiological loss, mechanical and functional impairments, and the pace of senescence differ between individuals. Survivors to age 85 generally show high physiological function after most in their cohort have already senesced and died (Perls 1995; Rakowski and Pearlman 1995; Perls 2001). Chronological age poorly measures the progressive loss in sensory, neural, immune, metabolic, and hormonal functions that accompany senescence. Environmental and sociocultural factors pattern human life history, while genes promote plasticity and flexibility in response to variable environments. Although age at death may seem a simple, precise, and quantitative endpoint, it poorly reflects the systemic interactions of genes, memes (units of cultural inheritance; Dawkins 1976), environmental, and sociocultural factors that produce senescence, death, and life span. Given such caveats, using chronological age and age at death as independent and dependent variables in models of CDCs and senescence has revealed the basic evolutionary biology and physiology of human senescence.

Transitions through life history stages (from gamete to centenarian) and attainments of required competencies are modulated at all levels of biology. Among humans, biocultural factors modulate life history transitions. Some transitions require relatively standard lengths of time to complete (e.g., gestation, weaning, and reproductive maturation). Early-life transitions (e.g., gastrulization and gestation) generally occur at less variant ages than later-life ones (e.g., cessation of growth, menopause, and male reproductive decline). Although such life history events often require a minimal time to attain, this does not mean they are age determined. Following birth, and more so following the attainment of MRP, the time needed to attain the next life history stage becomes increasingly variable, as do multiple physiological measures. Throughout life, individual differences in frailty and vitality lead some to succumb to life's stressors before others, even as embryos and fetuses. Loss of phenotypes declines rapidly after the neonatal period, but increases again following the mid-childhood growth spurt and attainment of MRP. Within cohorts, increases in physiological, phenotypic, and behavioral heterogeneity after maturity through to about age 65–70 are poorly related to chronological age. After age 70, phenotypic heterogeneity tends to decline and allele frequencies at a number of specific loci become skewed compared with those observed at younger ages. This suggests that the cohort is/was composed of a variety of rapid-, moderate-, and slow-senescing phenotypes. Those who survive to their mid-eighth decade are a select group, composed of slower-senescing, more moderate phenotypes.

There is wide variation in physiological target values (homeostatic ranges) compatible with survival through the periods of MRP, fledging of offspring, and reproductive adulthood (ages 20–49). Innate abilities to maintain physiological function at moderate values become more important as survival to 50 years plus becomes more possible and is particularly necessary for late-life survival (beyond 75 years). After age 65, extreme phenotypes succumb more frequently, causing coefficients of variation for physiological measures to decline. Elders aged 70+ with high vitality, low frailty, and high RC continue to maintain systematic homeostasis and survive – they are the slow senescing. Others at age 60 with low vitality, high frailty, and low RC may lose homeostasis following a minor stress or exposure to an infectious organism and die – they represent the more rapid senescing.

Use of age and life span as both independent and dependent variables has been criticized (Cristofolo *et al.* 1999; Harper and Crews 2000). More precise descriptions of senescent processes and phenotypes are necessary. Proposed biomarkers and indices of biological and functional age are for the most part dependent on chronological age, thereby failing to differentiate between aging and senescence (Adelman *et al.* 1988; Borkan 1986). Indices of frailty and allostatic load provide yardstick proxy variables with which to assess the systemic effects of senescence as functional loss. Ultimately, it will require complex multivariate modeling to process the range of senescent alterations that produce variable senescent phenotypes. Only well-defined phenotypes will allow further specification of senescence. Fortunately, many physiological changes associated with short life span are documented. Data on who survives, along with patterns of physiological change in survivors and decedents, have been obtained in a variety of sociocultural/political and ecological settings. In settings with inadequate housing, healthcare, and nutrition or overcrowding and poverty, senescence may proceed more rapidly than in those with adequate socioeconomic supports (Sorlie *et al.* 1995). In cosmopolitan settings, grandparents and great-grandparents frequently contribute to the well-being and survival of their children and other kin. In either setting, to the degree that they improve relative fitness, alleles enhancing somatic survival will increase in the general population. In humanity's bioculturally constructed ecological settings, different genotypes may attain the same relative fitness, while producing somas varying widely in survivability. The disposability (frailty and pace of senescence) of the soma may often have little relationship to its reproductive success or relative fitness. In cosmopolitan settings, men appear to have relatively more disposable somas than women. Similarly, mice have relatively more disposable somas than cats and dogs, which have still more disposable somas than humans. Having a more disposable soma defines an "early- or rapid-senescing phenotype". Such individuals may be compared with "late- or slow-senescencing phenotypes" who survive into late life (ages 80+).

Finch (1994) proposed three senescent phenotypes – rapid, gradual, and negligible – to describe species differences in life span/life history. These may also be applied within species where they are comparable to "more", "moderately", and "less" disposable somas. This directs attention toward defining rapid-senescing phenotypes, those who die at earlier ages (50–64 years), to compare with slow-senescing (dying after 80+ years) and moderate-senescing phenotypes (who die around 65–79 years). Since these phenotypes are based on when people die, retrospective and longitudinal research is needed to identify them. In most study designs, classification ends with "young" and "old". Even when there are three (young, middle aged, and old) or four such groups (30–44, 45–59, 60–74, 75+), classification remains age based. Early-senescing individuals of any age are grouped with moderate- and slow-senescing individuals of the same age. Extremely early-senescing or decreased somatic survival types are already missing before cohorts attain middle age. As in other areas of epidemiological and genetic risk assessment, the senescent phenotype is not well defined (Crews and Williams 1999). Valid, pragmatic, and well-defined phenotypes are necessary for comparative studies. For some studies, controls may be those who survive to ages 65–74 (moderate-senescing) or 75+ years (slow-senescing) without major health problems or decrements; the cases are those who succumb to senescent/CDCs between 45–64 years. The latter group likely includes multiple rapidly-senescing phenotypes, who, due to genetic predispositions to CDCs, early senescence, and/or susceptibilities to environmental stressors, fail to complete their sixth or seventh decade of life. Additional phenotypes may be defined with the use of allostatic load (low, medium, and high percentiles), genotypes (*ApoE4*, ACE I/D, and presenilins), and functional assessments (ADLs/IADLs, spirometry, grip strength, and heat tolerance). Identifying precise phenotypes allows more precision when examining CDCs and senescence. It will also help to lay to rest the idea that senescence is measured by age. New, valid, and well-defined classifications of senescing humans based on functional and observable decrements will also leave common (e.g., old, aged, elder, aging, and senior citizen), but less precise terms to social science, the lay public, and politicians. Aging is a sociocultural–political construct, not an aspect of biological senescence. Aging is part of how we and others perceive the outward manifestations of the internal processes of senescence.

Old age: uniquely human?

Elsewhere, it has been suggested that making "old age a special period of human life" is a self-serving political ploy (Hendricks and Achenbaum 1999, p. 37). This may sometimes appear to be true in cosmopolitan sociopolitical settings where most gerontological studies have been conducted. Most elders, however, reside in more traditional sociocultural settings where extended families are the norm and family needs are placed before individual ones. In such settings, elders are also perceived as being different from other members of the population and are allotted specific sociopolitical, religious, and familial roles and functions, some even after death. All cultures note the physiological and functional changes associated with survival into the sixth and later decades (Cowgill 1986; Crews 1993b). Even among our non-human primate relatives, "old age" appears to be a time of changing social roles (Corr 2000). Old age is not an arbitrary or manufactured construct developed as "a special period of life" by gerontologists and the public in modern societies (Hendricks and Actenbaum 1999, p. 37). Old age and senescence are sociocultural and biological realities in all human societies and at least for some non-human primates, and perhaps even for elephants and pilot whales. Historically, anthropology has focused on the continuity of life stages (e.g., growth/development, maturation, reproduction, parental investment, and adult morbidity/mortality) and life history in a cross-cultural and evolutionary context. A focus on old age and senescence as the final life stages in humans and primates provides a broader perspective for comparisons with current cosmopolitan societies (Cowgill 1986; Holmes and Rhodes 1990; Crews and Garruto 1994; Harper and Crews 2000). Understanding how old age and senescence are constructed today requires examinations across extent sociocultural settings, among other primates, among our hominid forbearers, and among other mammals.

Development of more precise senescencing phenotypes will help to promote comparative research across human populations and other species. All sexually reproducing species adapt to the stress of survival through time in a hostile environment. Many must have retained senescence-retarding mechanisms developed by earlier forms (plesiomorphies), while also evolving new responses that are unique (apomorphies) or similar to other species (symphomorphies) (Austad 1997). In evolution and biology, many theoretical advances during the nineteenth and twentieth centuries grew out of basic comparative research across species (Austad 1997), including that of Darwin (1859). Comparative studies of senescing organisms across species have been little used by biological gerontologists in recent decades. Biological gerontology is mired in studies of cellular biology, molecular biology, and genetics in rodents, flies, and worms (Austad 1997). Others have suggested that a resuscitation of comparative research and whole organism biology may be fruitful in human biological gerontology (Austad 1997; Kirkwood 2000). Biological anthropologists and human biologists are strategically positioned to foster comparative gerontology. Although they have an intellectual history of comparative research on reproduction, adult morbidity and mortality, and menopause across human and primate populations, they have only in recent decades turned this focus to senescence and late life. A variety of mammalian species, such as large marsupials (e.g., the kangaroo and wallaby) or other human-sized large-bodied mammals (e.g., sheep, deer, and bear), have seldom been compared with humans (Austad 1994, 1997). Such studies will improve our understanding of what are the plesiomorphic, apomorphic, and symphomorphic aspects of human and mammalian senescence.

Among mammals, humans rank with the slowest in rates of senescence (their Gompertz model mortality rate doubling time (MRDT) is now about 10 years) and reproduction, have the largest post-maturation to pre-maturation life span ratio, and females live as reproductive adults for a smaller proportion of their total lives. Women also produce proportionally lower neonatal mass given maternal size (about 6%) than most mammals (mice, 50% and rats, 17%) and have relatively short periods of gestation (266 days compared with 18 months for elephants and some whales) that are almost identical to other apes (Austad 1997). Humans also live about four times longer than would be predicted for a standard mammal of similar body size (Austad 1997). This counters earlier suggestions that human life span is not remarkable for a similar-sized mammal (Cutler 1980; Weiss 1989b). Senescence in long-lived, K-selected, slow-reproducing, culture-bearing humans also must differ greatly from those of r-selected, rapid-reproducing laboratory animals (e.g., worms, flies, and rodents). Still, human life span and senescence may not be that remarkable for a large-bodied primate.

In general, all extant large-bodied primates fall above predictions for life span based on mammalian body size, brain size, and other morphological indices (symphomorphic trait). Humans show a greater deviation from prediction than others (apomorphic trait). Survival through to the age of female reproductive decline characterizes humans, chimpanzees, gorillas, bonobos, baboons and macaques (symphomorphy). All extant moderate- to large-bodied primates show declining female reproduction during the fourth decade of life, long life spans (35 to >50 years), and relatively modest reproduction/fertility. These likely represent plesiomorphic traits for this group. All human females who survive past age 50 show a "post-reproductive" phase in their life history. Some suppose that this is uniquely human, but it is not. Among primates and land mammals, humans are unique in that most men and women survive beyond the age of 50 (although pilot whales may survive as long), and they often survive into late life (75 and beyond). Both men and women enjoy late-life survival. This apomorphic trait sets humans apart from other primates. On average, women in cosmopolitan settings survive about 30 years, and as much as 72 years, beyond their last reproduction. Men have an almost equal period of late-life survival, with the oldest living over 119 years. Declining reproductive function after about 35 years of life is common to all medium- to large-bodied mammalian females who survive sufficiently long (plesiomorphic trait). In nature, survival of either sex beyond the age of complete reproductive loss in mammalian females of the species is unusual. This derived apomorphic trait has been observed with high frequency only in humans and pilot whales. It is late-life survival that needs explanation in humans, not the loss of reproductive capability, which characterizes all medium- to large-bodied mammalian females (plesiomorphy).

Humans do not share patterns of growth, development, maturation, and reproduction, or their perpetual use of culture and their free upper limbs to interpret and manipulate their material world, with large-bodied mammals or primates (apomorphic traits). Multiple aspects of human biology, life history, senescence, and morbidity flow from these unique biological and biocultural attributes. To a large extent, such uniquely human factors determine how much RC, vitality, and frailty any individual 50-year-old or 60-year-old will have. However, the underlying biological processes of dysfunction and repair that produce senescent alterations likely are common among large-bodied mammals (plesiomorphies). Thus, processes whereby humans became unique may be viewed as adaptations to underlying loss of function common to all organisms with disposable somas and immortal germ lines. Comparative studies across multiple species are useful for understanding commonalities in senescent processes across species and the uniqueness of adaptations within different species. Only longitudinal studies of senescing phenotypes will reveal uniquely human responses to the stress of survival through time. Neither senescence nor old age are life history phases unique only to humans. These are biological universals for any organism that survives sufficiently long, including laboratory-reared worms, insects, and rodents who never would survive so long in the wild. Among humans, the aged as a social category are a cultural universal on a par with sex, gender, motherhood, sickness, and death. Biologically and physiologically, senescence is as much an aspect of life history as are conception, gestation, growth and development, and menopause. Senescence is an inherent life history process, and old age/late life is the final stage in human life history, a period of waning competencies and loss of function. The very processes that sustain and protect individual life produce internal damage, are disrupted by external factors, and progressively fail to maintain a homeostatic balance. This leads to cumulative, progressive, irreversible, and degenerative alterations in physiological function and organ capacity.

Theory in gerontology

Bengtson and Schaie (1999) recently lamented the lack of theory formation and enumerated the problems currently hampering theory development in both general science and "modern gerontology". As in any scientific discipline, theory building in gerontology requires the construction of explicit explanations to account for systematically observed (empirical) results (Bengtson *et al.* 1997, p. 5). These are then used to develop new testable hypotheses. Theories explain the why of observation, while models illustrate observed relationships – they show how or what happens, but do not explain why (Bengtson *et al.* 1997, p. 6). These two concepts may be stated biologically as evolutionary (ultimate) and proximate causes (mechanistic models) (Austad 1992; Crews and Gerber 1994; Austad 1997; Gerber and Crews 1999; Harper and Crews 2000). In biological senescence, evolutionary theory explains why the force of natural selection declines with increasing age as the probability of reproduction and contributions to relative fitness decline. Declining reproductive potential in sexually reproducing species limits the ability of natural selection to prevent late-life somatic dysregulation. Empirical tests of associated hypotheses strongly support this evolutionary theory (Rose 1991; Finch and Rose 1995; Austad 1997; Service *et al.* 1998; Service 2000a).

Evolutionary theory is the foundation on which the biological sciences, medicine, biotechnology, and genetics are built. Evolutionary theory is equally important for understanding the biology of senescence, sociocultural aspects of aging, and the "aged" (Gerber and Crews 1999). There is an arbitrary divide within gerontology between the social and the biological. Much research in social gerontology is atheoretical, driven by the search for interventions, applications, and solutions; research on the biology of senescence is often basic science designed to test hypotheses and reformulate theory (see Bengston *et al.* 1997, p. 5). There are multiple reasons for this divergence. Humans maintain a variety of social settings and belief systems. Often, each claims its basic superiority to all others (e.g., Christians/Muslims, communists/capitalists, Freudians/Jungians, and evolutionists/creationists). Additionally, research in social sciences is seldom exactly replicable and is often undertaken in pursuit of a particular sociopolitical agenda. Conversely, biological and medical experiments are more replicable and often there is only one possible result. In biological and natural sciences, theory may also be advanced to law (similar to the Law of Gravity, the Laws of Thermodynamics, and the Speed of Light). Social behavior varies between individuals and anyone may observe, perform, meddle with, or report on social factors. Genetics, molecular biology, and biochemistry are governed by laws and constants and require unique knowledge and experience to study, meddle with, or report about. Everyone has an opinion of how social relations work and how to fix them, what is politically best, and whether evolution is or is not a fact (law). Few without training purport to understand carbon bonding, genetics, statistics, computer programming, biochemistry, or medicine. Among humans, sociocultural–political–economic factors structure and determine biology and health. During the evolution of humans, multiple biocultural interactions influenced their biology and produced the long lives enjoyed by so many.

Over evolutionary time, multiple, variable, and counterbalancing mechanisms associated with early life survival and reproduction have become entrenched in the genomes of all sexually reproducing species. Tradeoffs between and among these (e.g., DNA content, antioxidant and repair capabilities, cell numbers, sizes, and functions, organ size and redundancy, and survival and reproduction) in producing a soma prevent any one aspect from attaining perfect function, setting the stage for senescence. The soma is disposable because its parts were designed to function in an integrated manner only so long as necessary to achieve life's goals (Weismann 1889; Williams 1957; Kirkwood 1990; Rose 1991; Wood *et al.* 1994). After about the species' MNLS, somas become increasingly susceptible to and more variable in their responses to damage, loss, and dysregulation. For many phenotypic traits, related species tend to share similar characteristics (symplesiomorphies and homologies). This may not be so for the mechanisms of biological senescence (Rose 1991; Austad 1992; Finch and Rose 1995; Austad 1997). For instance, except in cases of antagonistic pleiotropy, alleles specifically promoting somatic dysregulation are not likely to be more fit than those that do not. There are no innate allelic, mechanisms promoting senescence. What we have are innate defenses to prevent senescence, halt dysfunction, and maintain the soma for as long as possible, given basic tradeoffs among organs, survival, and reproduction. Over most of evolutionary time, alleles that promoted somatic dysregulation or improved regulation, but did not influence relative fitness, were randomly distributed in species genomes and were likely to be highly variable. Alleles improving survival eventually incorporated into various species' genomes to modulate survival and extend life span would have been, and continue to be, a random sampling of those available. Variable life history patterns observed, even between phylogenetically close species (humans–chimpanzees, modern humans–Neanderthals), further suggest that life history traits respond rapidly (within the limits of phylogenetic inertia) to alterations in ecological circumstances. Rapid alterations in life history traits may result from differential timing of neuroendocrine events, variability in release, processing, or responses to hormones, through variable major histocompatibility complex (MHC) activation, or through any alleles that modify these properties (Finch and Rose 1995). Alleles favored by natural selection for developing derived life history traits (apomorphies) may set even closely related species apart in patterns of growth, reproduction, and senescence. Humans show multiple derived states, fetal-like and altricial newborns, slow maturation, rapid and complex neurological development, late maturation and attainment of MRP, pedomorphism, long birth intervals, childhood, adolescence, never-ending parental investment, menopause, and late-life survival. Alleles promoting this suite of longevity characteristics underlie the mechanisms that determine humankind's unique pattern of late-life survival and senescence (apomorphic traits). This suite of traits overlies other earlier mechanisms inherited from humankind's progenitors and shared in common with other primates (symphamorphies) and mammals (plesiomorphies). A major goal of theory formation must be to determine which mechanisms for delaying senescence are human apomorphies, versus plesiomorphies and symphamorphies. Determining specific alterations and dysfunctions in humans (senescent phenotypes) and determining their distributions across primate and mammalian species should help to identify the mechanisms of senescence specific to humans, and those general to primates and mammals.

Current research directions

Investigating age-related physiological and somatic alterations, gerontology naturally overlaps broadly with physical and biological anthropology interests in human variation. Until recently, research efforts in gerontology were heavily focused on "normative studies" and "normal aging" (Comfort 1979; Shock 1984, 1985; Finch 1994). This is being replaced by an emphasis on individual and population variation and the view that senescence is a system-wide series of, often random, alterations in physiological function (Crews 1990a; Crews and Garruto 1994; Austad 1997; Bengston et al. 1997; Arking 1998; Cristofalo et al. 1998). "Modern gerontology" is currently focusing on the "aged" as a highly variable group (replacing older emphases on functional problems), senescence as a developmental process of longitudinal change, and senescence as an aspect of species' structure and response to evolutionary pressures (paraphrased from Bengtson et al. 1997, p. 9). Gerontology's renewed focus on evolution, development, and variation recalls long-term interests of physical and biological anthropologists on variation in growth, development, maturation, reproduction, and adulthood across human and primate populations residing in variable cultural and ecological settings. For decades, biological anthropologists have applied the species- and primate-wide multidisciplinary studies now proposed for gerontology (Baker 1982; Little and Hass 1989; Bengtson et al. 1997). Comparative studies of senescence across human populations and between humans and primates are a hallmark of anthropological research. Extension of this research to additional mammals and marsupials of similar size and/or life history traits will aid in understanding differences between humans, primates, and other animals. These new and renewed emphases represent a potentially productive arena for bioanthropological gerontology to develop.

No linear, univariate, or simple multivariate models provide the complexity necessary to understand the multiple mechanisms and processes underlying senescence even in cultured cells or simple organisms. Mechanisms and processes are likely to vary across cell type, tissue, organ, species, and higher order phyla. Renewed interest in physiological systems, whole organism biology, and interacting biological systems represents opportunities for pursuing integrated models of senescence. With multiple overlapping control systems, unless controlled by alternative processes, loss or failure in any segment leads to altered function in others. Similarly, a variety of up- and/or downregulation in interrelated systems to compensate for loss in one function necessarily alters others. An example is regulation of blood pressure (BP) and fluid volume. Multiple alleles, proteins, minerals, hormones, organs, and systems influence BP. Over the majority of one's life span, these maintain a balance. However, for many, BP becomes dysregulated with increasing age. As various aspects of regulation fail, others compensate even while senescence of the entire system proceeds and BP is maintained at an adequate level (James and Baker 1990). Eventually, more regulators fail, BP increases, and the system senesces. Which component of a complex somatic system fails first is secondary to inherited propensities, environment, culture, and, to a very large degree, random factors. Senescence is similar to growth and development – there are multiple regulators. However, growth and development are target-seeking processes, senescence occurs after the target – reproductive maturation, MRP, reproductive adulthood, and fledging of offspring – has been achieved. Senescence is an unscripted phase of life history. Senescence repfesents the secondary and tertiary effects of biological processes developed to achieve survival, maturity, and reproduction; with continued use these, show their inherent fragility. The senescent phase of life history develops from transitions and attainments of competencies occurring during earlier phases of life history.

Conclusion

Senescence is as complex an aspect of the phenotype as are gestation, maturation, and reproduction. No simple molecular switch exists nor will any single discipline unravel the causes of senescence. Understanding senescence necessitates multidisciplinary and transdisciplinary approaches. Human senescence responds to multiple genetic, environmental, and cultural factors. Research on its causes, patterning, and consequences has traditionally been carried forward by a diverse group of researchers combining a variety of backgrounds, while sharing ideas, methods, and theory across disciplines. Only recently have graduate programs in gerontology become available. Biology, sociology, anthropology, physiology, medicine, genetics, and psychology are all well represented in the gerontological literature. This integration of disciplines, research designs, and methodologies has made gerontology the premier transdisciplinary science (those sciences bringing together a wide variety of areas to focus on a single problem; Baker 1982), along with biological anthropology and human biology (Baker 1982; Little and Haas 1989; Crews and Garruto 1994; Crews 1997).

Much progress has been made toward outlining the proximate mechanisms and processes of human senescence. Some suggest that this outline is nearly complete, but it is not (Masoro 1996). Many problems such as variable rates of senescence, proximate mechanisms, and sex differentials in longevity have yet to be solved. Specific mechanisms of senescence have not been detailed. As yet, there is not even an acceptable measure of senescence. Proximate explanations have not sufficed to answer questions about rates, timing, or differences in senescence and life span. These discrepancies highlight the need to continue research for proximate mechanisms of human senescence. Humans are the product of 5 billion years of earthly existence, more than 3 billion years of organismal evolution, over 600 million of sexual reproduction, 65 million years of primate evolution, over 5 million of bipedal evolution, and 500 000 or more years of *sapiens* evolution. From our bipedality and versatile upper limbs and our patterns of development, to our use of culture and eventual causes of death, this heritage helps to explain our current processes of senescence. Any proximate theory of senescence must be in accord with this evolutionary background to elucidate the mechanisms of human senescence. This multifactorial complexity need not overwhelm the search to understand senescence; rather, the entire process of human senescence and its attendant CDCs can only be understood within the context of evolutionary biology and historical contingency.

The life history pattern that characterizes humankind in modern cosmopolitan societies did not appear fully formed on the evolutionary stage in recent millennia. The biological basis for humanity's current life history and longevity evolved over time in a complex process of biocultural adaptation to changing environmental circumstances, ecological exigencies, and historical contingencies. Either sequentially or concurrently, humans developed bipedalism, free upper limbs, physically altricial infants, dependent children, pedomorphic adults, slow maturation, late attainment of MRP, greater encephalization, a highly integrated neocortex, language, and complex sociocultural–political structures. In conjunction, and as part of this complex set of alterations, human life history was rewritten. Opportunities for late-life survival (post-reproductive survival in women), through enhanced RC and redundancy, likely were patterned into human biological processes well before anyone actually survived so long. Ultimately, cultural developments proceeded to the point where some did survive into their fifth and later decades. This exposed multiple innate differences between individuals in the pattern, pace, and extent of somatic degeneration after the attainment of MRP, differences not obvious and perhaps never seen when few survived their fourth decade of life. In earlier generations, most individuals likely succumbed to other risks – trauma, predation, illness, and violence – well before having the opportunity to senesce. Only after cultural processes provided opportunities for some to survive sufficiently long to express their innate differences in somatic maintenance, did senescence begin to curtail the lives of individuals.

Humankind is already close to unraveling the DNA sequences of all 13 601 genetic loci carried by fruitflies and the 9 706 genes of the roundworm. The entire human genome, of about 30 000 coding loci, is also now mapped. Several loci in fruitflies include alleles that greatly extend life span. One, dubbed Methuselah, extends life by 35%, and it is only one member of a family of at least ten similar loci. Similarly, several loci (e.g., *daf-1* and *age-1*) extend life spans in roundworms. What applicability such results have to human senescence is unknown, although it may be minimal given the wide evolutionary gap between these species and humans. Fruitflies and roundworms are ideal animal models for examining less complex systems where minor alterations in physiology or care may have profound effects on longevity and life span. A similar argument may be made for mice and rats, although they are more complex. Still, it is doubtful that alterations of such minor factors will greatly influence human senescent processes. In humans, the number of possible interactions among 30 000 coding loci is orders of magnitude higher. Still, many examples of conserved loci exist across a variety of species (e.g., SOD, homeobox genes, and MHC/human lencocyte antigen loci). Although homologous sequences of homeobox loci are conserved in fruitflies, mice, and humans, their patterns of activity (on and off), and the phenotypes they produce are quite different and associated with very different life histories. Studies of rodents, fruitflies, and nematodes may provide useful insights into plesiomorphic aspects of senescence. Mechanisms enhancing or preventing senescence that have arisen in these species add to our understanding of organismal evolution on Earth. However, they may provide little insight into the more recently evolved apomorphic aspects of human senescence.

It is doubtful that humankind will ever become a species of Methuselahs. However, at the beginning of the twenty-first century, some populations average almost eight and one-half decades of life. These later years of human life are also more productive than past experiences would have predicted. As with monetary investments, past trends should be interpreted with caution, particularly given the imprecision that accompanied the projections of life span and survival underlying current social security systems. Today's centenarians are experiencing lower mortality and morbidity rates than predicted even a decade ago and there are quite a few more members of each cohort reaching 100 years of age. In the last decade of the twentieth century, the oldest known person, Jeanne Calment of France, died at the age of 122 years. Having been born in 1875, Ms. Calment was representative of an earlier phase of human cultural development, before overnutrition and low physical activity, and may represent only the extreme for survival, given then current cultural developments. Projecting the growth of centenarians in the population and their survival probabilities from current trends suggests that soon someone will exceed Ms. Calment's record (a recent wager between S. Austad and B. Carnes proposed a 150- or 130-year life span within 150 years). In cosmopolitan settings, population growth among persons aged greater than 80 years is already proportionally more rapid than in any other age group. These elders are also living longer.

How one achieves the age of 100 remains somewhat mysterious. Unfortunately, for most, longevity is only about 30–40% heritable at birth. Fortunately, for those in long-lived kindreds, life span among those who do reach age 45 is more highly heritable (as much as 70%) and centenarians have more long-lived relatives. Numerous genetic factors contributing to relative differences in frailty and vitality are segregating in the human gene pool. To live a long life, it probably helps to obtain the best alleles at conception and to be very lucky and live in the right environment thereafter. Following your family physician's and public health practitioner's advice (e.g., exercise regularly, eat fewer calories, ingest more fiber and less fat, maintain moderate weight, and not to smoke) is also likely to improve longevity for most people. For those of us who do most of these, additional methods may include increasing our intake of cystine along with other antioxidants and their precursors, reducing calories to 70–75% of what one might ingest *ad libitum*, and maintaining active physical, intellectual, and social life styles. For all organisms, biological predispositions are played out in a specific environment, but only among humans do the added dimensions of biocultural and sociocultural constraints and opportunities influence an individual's life span, and it is only humans who make life style choices that enhance their opportunity for survival or increase their chances of death.

References

Adams, M.R., Williams, J.K., and Kaplan, J.R. (1995). Effects of androgens on coronary artery atherosclerosis and atherosclerosis-related impairement of vascular responsiveness. *Arteriosclerosis Thrombosis and Vascular Biology* 15: 562–70.

Adelman, R., Saukm, R.L., and Ames, B.N. (1988). Oxidative damage to DNA: relation to species metabolic rate and life span. *Proceedings of the National Academy of Science USA* 85: 2706–8.

Albin, R.L. (1988). The pleiotropic gene theory of senescence: supportive evidence from human genetic disease. *Ethology and Sociobiology* 9: 371–82.

Albin, R.L. (1994). Antagonistic pleiotropy, mutation accumulation, and human genetic disease. In *Genetics and Evolution of Aging*, M.R. Rose and C.E. Finch (eds.). Amsterdam: Kluwer Academic Publishers.

Alexander, R.M. (1998). Mammalian life history. In *The Cambridge Encyclopedia of Human Growth and Development*, S.J. Ulijaszek, F.E. Johnston, and M.A. Preece, (eds.). Cambridge: Cambridge University Press, p. 98.

Alfie, J., Waisman, G.D., Galarza, C.R. *et al.* (1995). Relationships between systemic hemodynamics and ambulatory blood pressure level are sex dependent. *Hypertension* 26: 1195–99.

Allsopp, R.C. (1992). Models of initiation of replicative senescence by loss of telomeric DNA. *Experimental Gerontology* 31: 235–43.

Allsopp, R.C. (1996). Models of initiation of replicative senescence by loss of telomeric DNA. *Experimental Gerontology* 31: 235–43.

Aloia, J.F., Vaswani, A., Kuimei, M., and Flaster, E. (1996). Aging in women – the four-compartment model of body composition. *Metabolism* 45: 43–8.

Alterm, G. and Riley, J.C. (1989). Frailty, sickness, and death: models of morbidity and mortality in historic populations. *Population Studies* 43: 25–43.

Alvarez, H.P. (2000). Grandmother hypothesis and primate life histories. *American Journal of Physical Anthropology* 113: 435–50.

Alzheimer's Disease and Related Disorders Association, Inc. (ADRDA) (1996). The use of *apo E* testing in Alzheimer's disease. *Research and Practice*.

American Heart Association (AHA). (1988). The Joint National Committee on Detection, Evaluation, and Treatment of High Blood Pressure: the 1988 Report. *Archives of Internal Medicine* 148: 1020–38.

Ames, B.N., Shigenaga, M.K., and Hagen, T.M. (1993). Oxidants, antioxidants and the degenerative diseases of aging. *Proceedings of the National Academy of Sciences USA* 90: 7915–22.

Anderson, J.J. and Pollitzer, W.S. (1994). Ethnic and genetic differences in susceptibility to osteoporotic fractures. *Advances in Nutritional Research* 9: 129–49.

Anderson, K., Launer, L.J., Dewey, M.E. *et al.* (1999). Gender differences in the incidence of AD and vascular dementia. The EURODEM Studies. *Neurology* 53: 1992–7.

Anderson, R.N. (2001). United States Life Tables, 1998. *National Vital Statistics Report*, 48. Hyattsville, M.D.: NCHS.

Andres, R. (1971). Aging and diabetes. *Medical Clinics of North America* 55: 835–46.

Andres, R. (1980). Effects of obesity on total Mortality. *International Journal of Obesity* 4: 381.

Anemone, R.L., Moorey, M.P., and Siegel, M.I. (1996). Longitudinal study of dental development in chimpanzees of known chronological age: implications for understanding the age at death of plio-pleistocene hominids. *American Journal of Physical Anthropology* 99: 119–33.

Arking, R. (1991). *Biology of Aging: Observations and Principles*. Englewood Cliffs N.J.: Prentice-Hall Inc.

Arking, R. (1998). *Biology of Aging: Observations and Principles*, 2nd edition. Englewood Cliffs, N.J.: Prentice-Hall Inc.

Arking, R. and Dudas, S.P. (1989). A review of genetic investigations into aging processes of *Drosophila. Journal of the American Geriatrics Society* 37: 757–73.

Arking, R., Buck, S., Berrios, A., Dwyer, S., and Baker, G.T. (1991). Elevated paraquat resistance can be used as a bioassay for longevity in a genetically based long-lived strain of *Drosophila. Developmental Genetics* 12: 362–70.

Armelagos, G.J. (1991). Human evolution and the evolution of disease. *Ethnicity and Disease* 1: 21–5.

Armstrong, D. (1984). Free radicals involvement in the formation of lipopigments. In *Free Radicals in Molecular Biology, Aging, and Disease*, D. Armstrong, R.S. Sohal, R.G. Cutler, and T.F. Slater (eds.). New York: Raven Press, pp. 129–42.

Austad, S.N. (1992). On the nature of aging. *Natural History* 2: 25–57.

Austad, S.N. (1994). Menopause: an evolutionary perspective. *Experimental Gerontology* 29: 255–63.

Austad, S.N. (1997). Comparative aging and life histories in mammals. *Experimental Gerontology* 32: 23–38.

Baker, P.T. (1979). The use of human ecological models in biological anthropology: examples from the Andes. *Collegium Anthropologicum* (Zagreb, Yugoslavia) 3: 157–71.

Baker, P.T. (1982). Human population biology: a viable transdisciplinary science. *Human Biology* 54: 203–20.

Baker, P.T. (1984). The adaptive limits of human populations. *Man* 19: 1–14.

Baker, P.T. (1990). Human adaptation theory: successes, failure and prospects. *Journal of Human Ecology* Special Issue 1: 41–50.

Baker, P.T. (1991). Human adaptation theory: successes, failure and prospects. *Journal of Human Ecology* Special Issue 1: 41–50.

Baker, P.T. and Baker, T.S. (1977). Biological adaptations to urbanization and industrialization: some research strategy considerations. In *Colloquia in Anthropology, Volume I*, R.K. Wetherington (ed.), Fort Burgwin Research Center, pp. 107–18.

Baltes, M.M., Wahl, H.W., and Schmis-Furstoss, U. (1990). The daily life of elderly Germans: activity patterns, personal control, and functional health. *Journal of Gerontology* 45: P173–9.

Barker, D.J.P. (1989). The rise and fall of Western disease. *Nature* 338: 371–2.

Barker, D.J.P. (1998). *Mothers, Babies and Health in Later Life*, 2nd edition. New York: Churchill Livingston.

Barker, D.J.P., Bull, A.R., Osmond, C., and Simmonds, S.J. (1990). Fetal and placental size and risk of hypertension in adult life. *British Medical Journal* 301: 259–62.

Barker, D.J.P., Hales, C.N., Fall, C.H.D., Osmond, C., Phipps, K., and Clark, P.M.S. (1993). Type 2 (non-insulin dependent) diabetes mellitus, hypertension and hyperlipidemia (syndrome X): relation to reduced fetal growth. *Diabetologia* 36: 62–7.

Barker, D.J.P., Osmond, C., and Golding, J. (1989b). Height and mortality in the counties of England and Wales. *Annals of Human Biology* 17: 1–6.

Barker, D.J.P., Winter, P.D., Osmond, C., Margetts B., and Simmonds, S.J. (1989a). Weight in infancy and death from ischaemic heart disease. *Lancet* 577–80.

Barinaga, G.T. (1991). How long is the human life span? *Science* 254: 936–8.

Barley, J., Blackwood, A., Carter, N. *et al.* (1994). Angiotensin converting enzyme insertion/deletion polymorphism – association with ethnic origin. *Journal of Hypertension* 12: 955–7.

Barzilai, N. and Shuldiner, A.R. (2001). Searching for human longevity genes: the future history of gerontology in the post-genomic era. *Journals of Gerontology Series A, Biological Sciences and Medical* 56: M83–7.

Baxter, J.D. (1997). Introduction to endocrinology. In *Basic and Clinical Endocrinology*, 5th edition, F.S. Greenspan and G.J. Strewler (eds.). Appleton and Lang.

Beall, C.M. (1994). Aging and adaptation to the environment. In *Biological Anthropology and Aging: Perspectives on Human Variation Over the Life Span*, D.E. Crews and R.M. Garruto (eds.). New York: Oxford University Press, pp. 339–69.

Beall, C.M. and Steegman, A.T. (2000). Human adaptations to climate: temperature, ultraviolet radiation and altitude. In *Human Biology: An Evolutionary and Biocultural Approach*, S. Stinson, B. Bogin, R. Huss-Ashmore and D. O'Rourke (eds.). New York: Wiley-Liss, pp. 163–224.

Beall, C.M., Blangero, J., Williams-Blangero, S., and Goldstein, M.C. (1994). A major gene for percent of oxygen saturation of arterial hemoglobin in Tibetan highlanders. *American Journal of Physical Anthropology* 95: 271–6.

Bendich, A. (1995). Criteria for determining recommended daily allowances for healthy older adults. *Nutrition Reviews* 53: S105–10.

Bengtson, V.L. and Schaie, K.W. (eds.) (1999). *Handbook of Theories of Aging*. New York: Springer Publishing Company.

Bengston, V.L., Burgess, E.O., and Parrot, T.M. (1997). Theory, explanation, and a third generation of theoretical development in social gerontology. *Journal of Gerontology: Social Sciences* 522: 577–8.

Berkovitch, F.B. and Harding, R.S.O. (1993). Annual birth patterns of savanna baboons (*Papio cynocephalus anubis*) over a ten-year period at Gilgil, Kenya. *Folia Priatol*, 61: 115–22.

Bindon, J.R. and Crews, D.E. (1993). Changes in some health status characteristics of American Samoan men: a 12-year follow-up study. *American Journal of Human Biology* 5: 31–8.

Bindon, J.R., Crews, D.E., and Dressler, W. (1992). Lifestyle, modernization, and adaptation among Samoans. *Collegium Anthropologicum* 15: 101–10.

Bindon, J.R., Knight, A., Dressler, W.W., and Crews, D.E. (1997). Social context and psychosocial influences on blood pressure. *American Journal of Physical Anthropology* 103: 7–18.

Bockxmeer, F. (1994). ApoE and ACE genes: impact on human longevity. *Nature Genetics* 6: 4–5.

Bodkin, N.L., Ortmeyer, H.K., and Hansen, B.C. (1995). Long-term dietary restriction in old-aged rhesus monkeys: effects on insulin resistance. *Journal of Gerontology* 50A: B142–7.

Bogin, B. (1998). Patterns of human growth. In *The Cambridge Encyclopedia of Human Growth and Development*, S.J. Ulijaszek, F.E. Johnston, and M.A. Preece, (eds.). Cambridge University Press, pp. 91–95.

Bogin, B. (1999). *Patterns of Human Growth*. 2nd edition. Cambridge: Cambridge University Press.

Bogin, B. and Smith, H. (1996). Evolution of the human life cycle. *American Journal of Human Biology* 8: 703–16.

Bolzan, A.D., Brown, O.A., Goya, R.G., and Bianchi, M.S. (1995). Hormonal modulation of antioxidant enzyme activities in young and old rats. *Experimental Gerontolology* 30: 169–75.

Borkan, G.A. (1986). Biological age assessment in adulthood. In *The Biology of Human Aging*, A.H. Bittles and K.J. Collins (eds.). New York: Cambridge University Press, pp. 81–94.

Borkan, G.A. and Norris, A.H. (1980). Assessment of age using a profile of physical parameters. *Journal of Gerontology* 35: 177–84.

Borkan, G.A., Hults, D.E., and Mayer, P.J. (1982). Physical anthropological approaches to aging. *Yearbook of Physical Anthropology* 25: 181–202.

Boullion, R., Bex, M., Van, E. Herck *et al.* (1995). Influence of age, sex, and insulin on osteoblast function: osteoblast dysfunction in diabetes mellitus. *Journal of Clinical Endocrinology and Metabolism* 80: 1194–202.

Boult, C., Kane, R.L., Louis, T.A., Boult, L., and McCaggrey, D. (1994). Chronic conditions that lead to functional limitations in the elderly. *Journal of Gerontology* 49: M28–36.

Bowden, D.M. and Jones, M.L. (1979). Aging research in nonhuman primates. In *Aging in Nonhuman Primates*, D.M. Bowden, (ed.). New York: Van Nostrand Reinhold.

Bowling, A.C. and Beal, M.F. (1995). Bioenergetic and oxidative stress in neurodegenerative diseases. *Life Sciences* 56: 1151–71.

Brach, C. and Fraserirector, I. (2000). Can cultural competency reduce racial and ethnic health disparities? A review and conceptual model. *Medical Research and Review* 57: 181–217.

Breitner, J.C., Welsh, K.A., Gau, B.A. *et al.* (1995). Alzheimer's disease in the National Academy of Sciences-National Research Council Registry of aging twin veterans III. Detection of cases, longitudinal results, and observations on twin concordance. *Archives of Neurology* 52: 763–71.

Brennan, P.A., Grekin, E.R., and Mednick, S.A. (1999). Maternal smoking during pregnancy and adult male criminal outcomes. *Archives of General Psychiatry* 56: 215–19.

Brock, D.B., Guralnik, J.M., and Brody, J.A. (1990). Demography and epidemiology of aging in the United States. In *Handbook of the Biology of Aging*, E.L. Schneider and J.W. Rose (eds.). San Diego: Academic Press, Inc., pp. 3–23.

Brody, J. (1983). Limited importance of cancer and competing risk theories of aging. *Journal of Clinical Experimental Gerontology* 5: 141–54.

Broe, G.A. and Creasey, H. (1995). Brain aging and neurodegenerative diseases: a minor public health issue of the twenty-first century. *Perspectives in Human Biology*, 1: 53–8.

Brown, D.T., Samuels, D.C., Michael, E.M., Turnbull, D.M., and Chinnery, P.F. (2001). Random genetic drift determines the level of mutant mtDNA in human primary oocytes. *American Journal of Human Genetics* 68: 533–6.

Brownlee, M., Cerami, A., and Vlassara, H. (1988). Advanced glycosylation end products in tissue and the biochemical basis of diabetic complications. *Seminars in Medicine of the Beth Israel Hospital, Boston* 318: 1315–21.

Buck, S., Nicholson, M., Dudas, S. *et al.* (1993). Larval regulation of adult longevity in a genetically-selected long-lived strain of *Drosophila*. *Heredity* 71: 23–32.

Buckley, C.E. and Roseman, J.M. (1976). Immunity and survival. *Journal of the American Geriatric Society* 24: 241–8.

Burnet, F.M. (1970). An immunological approach to aging. *Lancet* 2: 358–60.

Burnet, F.M. (1970). *Immunological Surveillance.* New York: Pergamon Press.

Campbell, D.M., Hall, M.H., Barker, D.J.P., Cross, J., Shiell, A.W., and Godfrey, K.M. (1996). Diet in pregnancy and the offspring's blood pressure 40 years later. *British Journal of Obstetrics and Gynecology* 103: 273–80.

Carey, J.R. (1997). What demographers can learn from fruit fly actuarial models and biology. *Demography* 34: 17–30.

Carmichael, C.M. and McGue, M. (1995). A cross-sectional examination of height, weight, and body mass index in adult twins. *Journals of Gerontology, Series A, Biological and Medical Sciences* 50: B237–44.

Caro, T., Sellen, D., Parrish, A. *et al.* (1995). Termination of reproduction in non-human and human female primates. *International Journal of Primatology* 16: 205–20.

Carey, J.R., Liedo, P., Müeler, H.-G., Love, B., Harshman, L., and Partridge, L. (2001). Female sensitivity to diet and irradiation treatments underlies sex-mortality differentials in the Mediterranean fruit fly. *Gerontology: Biological Sciences* 56A: B89–95.

Caspari, R. and Wolpoff, M. (2003). *Race and Human Evolution.* Cambridge, M.A.: Westview Press.

Cassel, J. (1976). The contribution of the social environment to host resistance. *American Journal of Epidemiology* 104: 107–23.

Cassel, C.K. (2001). How increased life expectancy and medical advances are changing geriatric care. *Geriatrics* 56: 35–9.

Cauley, J.A., Eichmer, J.E., Kamboh, M.I., Ferrell, R.E., and Kuller, L.H. (1993). Apolipoprotein E allele frequencies in younger (age 42–50) vs. older (65–90) women. *Genetics and Epidemiology* 10: 27–34.

The *C. elegans* Sequencing Consortium. (1998). Genome sequence of the nematode *C. elegans:* a platform for investigating biology. *Science* 282: 2012–18.

Cerami, A. (1985). Hypothesis: glucose as a mediator of aging. *Journal of the American Geriatrics Society* 33: 626–34.

Cerami, A. (1986). Aging of proteins and nucleic acids: what is the role of glucose? *Trends in Biochemical* Sciences, 11: 311–14.

Cerami, A., Vlassara, H., and Brownlee, M. (1987). Glucose and aging. *Scientific American* 256: 90–6.

Chagnon, N.A. (1968). *Yanomamo: the Fierce People.* New York: Rinehard and Winston.

Chandra, R. (1992). Nutrition and immunity in the elderly. *Nutrition Reviews* 50: 367–71.

Chang-Claude, J., Frentzel-Beyrne, R., and U. Eliber (1992). Mortality pattern of German vegetarians after 11 years of follow-up. *Epidemiology* 3: 395–401.

Charlesworth, B. (1994). *Evolution in Age-Structured Populations*, 2nd edition. Cambridge: Cambridge University Press.

Charlesworth, B. (2001). Patterns of age-specific means and genetic variances of mortality rates predicted by the mutation-accumulation theory of aging. *Journal of Theoretical Biology* 210: 47–65.

Charnov, E.L. (1993). *Life History Invariants: Some Explanations of Symmetry in Evolutionary Ecology.* New York: Oxford University Press.

Chartier-Harlin, M.C., Parfitt, M., Legrain, S. *et al.* (1994). Apolipoprotein E, epsilon 4 allele as a major risk factor for sporadic early and late onset forms of Alzheimer's disease: analysis of the 19q13.2 chromosomal region. *Human Molecular Genetics* 3: 569–74.

Chen, F.-C. and Lei, W.-H. (2001). Genomic divergences between humans and other hominoids and the effective population size of the common ancestor of humans and chimpanzees. *American Journal of Human Genetics* 68: 444–56.

Clare, M.J. and Luckinball, L.S. (1985). The effects of gene–environment interaction on the expression of longevity. *Heredity* 55: 19–29.

Clark, W.R. (1999). *A Means to an End: The Biological Basis of Aging and Death.* New York: Oxford University Press.

Cohen, M.N. (1989). *Health and the Rise of Civilization.* New Haven, C.T.: Yale University Press.

Comfort, A. (1979). *Aging: The Biology of Senescence, 3rd edition.* New York: Elsevier North Holland Inc.

Comuzzie, A.G., Blangero, J., Mahaney, M.C. *et al.* (1995). Major gene with sex-specific effects influences fat mass in Mexican Americans. *Genetic Epidemiology* 12: 475–88.

Coon, C.S. (1965). *The Living Races of Man.* New York: Alfred A. Knopf.

Cooper C., Fall, C., Egger, P., Eastell, R., and Barker, D. (1996). Growth in infancy and bone mass in later life. *Annals of the Rheumatic Diseases* 55: 1–6.

Corr, J.A. (2000). *The Effects of Aging on Social Behavior in Male and Female Rhesus Macaques of Cayo Santiago.* Ph.D. dissertation, The Ohio State University.

Corder, E.H., Lannfelt, L., Viitanen, M. *et al.* (1996). Apoliopoprotein E genotype determines survival in the oldest old (85 years or older) who have good cognition. *Archives of Neurology* 53: 418–22.

Cotton, P. (1994). Alzheimer's/Apo E link grows stronger. *JAMA* 272: 1483.

Cowgill, D.O. (1986). *Aging Around the World*. Belmont, C.A.: Wadsworth.

Crews, D.E. (1988). Multiple causes of death and the epidemiological transition in American Samoa. *Social Biology* 35: 198–213.

Crews, D.E. (1990a). Anthropological issues in biological gerontology. In: RL Rubinstein (ed.) *Anthropology and Aging: Comprehensive Reviews*. Dordecht, The Netherlands: D. Reidel Publishing Co., pp.11–38.

Crew, D.E. (1990b). Multiple causes of death, chronic diseases, and aging. *Collegium Anthropologicum* 14: 197–204.

Crews, D.E. (1993a). Biological Anthropology and Aging: Some Current Directions in Aging Research. *Annual Review of Anthropology*, 22: 395–420.

Crews, D.E. (1993b). Cultural lags in social perceptions of the aged. *Generations* 17: 29–33.

Crews, D. (1993c). The organizational concept and vertebrates with out sex chromosomes. *Brain, Behavior, and Evolution* 42: 202–14.

Crews, D.E. (1997). Aging and gerontology: a paradigm of transdisciplinary research. *Collegium Anthropologicum* 21: 83–92.

Crews, D.E. and Bindon, J.R. (1991). Ethnicity as a taxonomic tool in biomedical and biosocial research. *Ethnicity and Disease* 1: 42–9.

Crews D.E. and Gerber, L.M. (1994). Chronic degenerative diseases and aging. In *Biological Anthropology and Aging: Perspectives on Human Variation Over the Lifespan*. D.E. Crews and R.M. Garruto (eds.). New York, Oxford University Press, pp. 174–208.

Crews D.E. and Gerber, L.M. (1999). Evolutionary perspectives on chronic degenerative diseases. In *Evolutionary Medicine*, W. Trevathan, J. McKenna, and N. Smith (eds.). New York: Oxford University Press, pp. 443–69.

Crews D.E. and Garruto, R.M. (1994). *Biological Anthropology and Aging: Perspectives on Human Variation Over the Life Span*. Oxford: Oxford University Press.

Crews, D.E. and Harper, G.J. (1998). Renin, ANP, ACE polymorphisms, Blood pressure, and Age in American Samoans. *American Journal of Human Biology* 10: 439–49.

Crews, D.E. and Harper, G.J. (1997). Congregate care retirement communities, a case study of current directions. *Gerontologist* 37: 179.

Crews, D.E. and James, G.D. (1991). Human evolution and the genetic epidemiology of chronic degenerative diseases. In *Applications of Biological Anthropology of Human Affairs*, G.W. Lasker and N. Mascie-Taylor (eds.). Cambridge: Cambridge University Press.

Crews, D.E. and Losh, S.R. (1994). Structural modeling of blood pressure in Samoans. *Collegium Anthropologicum* 18: 101–13.

Crews, D.E. and MacKeen, P.C. (1982). Mortality related to cardiovascular disease and diabetes mellitus in a modernizing population. *Social Science and Medicine* 16:175–81.

Crews D.E. and Williams, S. R. (1999). Molecular aspects of blood pressure regulation. *Human Biology* 71: 475–503.

Crews, D.E., Murrell, S.A. and Lang, C.A. (1985). High mental health occurs in physically healthy elderly women. *The Gerontologist* 26: 165.

Cristofalo, V.J., Allen, R.G., Pignolo, R.J., Martin, B.G., and Beck, J.C. (1998). Relationship between donor age and the replicative lifespan of human cells in culture: a reevaluation. *Proceeding of the National Academy of Sciences USA* 95: 10614–19.

Cristofalo, V.G., Tresini, M., Francis, M.K., and Volker, C. (1999). Biological theories of senescence. In *Handbook of Theories of Aging*, V.L. Bengston and K.W. Shaie (eds.). New York: Springer Publishing Company, pp. 98–112.

Crognier, E., Baali, A., Hilali, M-K. (2001). Do helpers at the nest increase their parents' reproductive success? *American Journal of Human Biololgy* 13: 365–73.

Crose, R. (1997). *Why Women Live Longer Than Men and What Men Can Learn From Them*. San Francisco: Jossey-Bass Publishers.

Cummings, S.R. and Nevitt, N.C. (1989). A hypothesis: the cause of hip fractures. *Journal of Gerontology* 44: M107–11.

Curtin, P.D. (1992). The slavery hypothesis for hypertension among African-Americans: the historical evidence. *American Journal of Public Health* 82: 1681–6.

Curtsinger, J.W., Fukui, H.H., Townsend, D.R., and Vaupel, J.W. (1992). Demography of genotypes: failure of the limited life-span paradigm in *Drosophila melanogaster*. *Science* 258: 461–3.

Cutler, R.G. (1975). Evolution of human longevity and the genetic complexity governing aging rate. *Proceedings of the National Academy of Sciences USA* 72: 4664–8.

Cutler, R.G. (1976). Evolution of longevity in primates. *Journal of Human Evolution* 5: 169–202.

Cutler, R.G. (1980). Evolution of human longevity. In *Aging, Cancer, and Cell Membranes*, C. Borek, C.M. Fenoglis, and D.W. King (eds.). New York: Thieme-Straton, Inc., pp. 43–79.

Cutler, R.G. (1991). Antioxidants and aging. *American Journal of Clinical Nutrition* 53: 373S–9S.

Darwin, C.R. (1859). *On the Origin of Species by Means Of Natural Selection, or the Preservation of Favoured Races In the Struggle For Life*, 2nd edition. London: Murray.

Davies, K.J.A. (1987). Protein damage and degradation by oxygen radicals, I. General aspects. *Journal of Biological Chemistry* 262: 9895–901.

Davignon, J., Gregg, R.E., and Sing, C.F. (1988). Apolipoprotein E polymorphism and atherosclerosis. *Atherosclerosis* 8: 1–21.

Davison, A.N. (1987). Pathophysiology of the aging brain. *Gerontology* 33: 129–35.

Dawkins, R. (1976). *The Selfish Gene*. New York: Oxford Univeristy Press.

De Benedictus, Ross, G., Carrieri, G. *et al.* (2000a). Mitochondrial DNA inherited variants are associated with successful aging and longevity in humans. *FASEB Journal* 13: 1532–1536.

De Benedictus, Ross, G., Carrieri, G. *et al.* (2000b). Inherited variability of the mitochondrial genome and successful aging in humans. *Annals of the New York Academy of Sciences* 908: 208–218.

Deevey, E.S. (1950). The probability of death. *Scientific American* 18: 58–30.

DeFronzo, R.A. and Ferrannini, E. (1991). Insulin resistance: a multifaceted syndrome responsible for NIDDM, obesity, hypertension, dyslipidemia, and atherosclerotic cardiovascular disease. *Diabetes Care* 14: 173–94.

De Lange, T. (1998). Telomeres and senescence: ending the debate. *Science* 249: 334–5.

De Grey, A.D.N.J. (1997). A proposed refinement of the mitochondrial free radical theory of aging. *Bioessays* 19: 161–6.

De Haan, J.B., Cristano, F., Iannello, R.C., and Kola, I. (1995). Cu/Zn–superoxide dismutase and glutathione peroxidase during aging. *Biochemistry and Molecular Biology International* 35: 1281–97.

de la Torre, M.R., Casado, A., Lopez-Fernandez, M.E. *et al.* (1996). Human aging brain disorders: role of antioxidant enzymes. *Neurochemical Research* 21: 885–8.

Demographic Yearbook (1996). New York: United Nations Department of Economic and Social Affairs.

Demographic Yearbook (1998). New York: United Nations Department of Economic and Social Affairs.

Dennison, E., Fall, C., Cooper, C., and Barker, D. (1997). Prenatal factors influencing long-term outcome. *Hormone Research* 48(Suppl.1): 25–6.

DeRousseau, C. (1990). *Primate Life History and Evolution*. New York: Wiley-Liss.

Dietz, J.M. and Baker, A.J. (1993). Polygyny and female reproductive success in golden lion tamarins, (*Leotopithecus rosalia*). *Animal Behavior* 46: 1067–78.

Dietz, J.M. and Baker, A.J. (1994). Seasonal variation in reproduction, juvenile growth, and adult body mass in golden lion tamarins. *American Journal Primatology* 34: 115–132.

Doblhammer, G. and Kytir, J. (2001). Compression or expansion of morbidity? Trend in healthy-life expectancy in the elderly Austrian population between 1978 and 1998. *Social Science and Medicine* 52: 385–91.

Dressler, W.W. (1991). Social class, skin color, and arterial blood pressure in two societies. *Ethnicity and Disease* 1: 60–77.

Dressler, W.W. (1996). Hypertension in the African community: social, cultural, and psychological factors. *Seminars in Nephrology* 16: 71–82.

Dressler, W.W., Bindon, J.R., and Neggers, Y.H. (1998). Culture, socioeconomic status and coronary heart disease risk factors in an African American community. *Journal of Behavioral Medicine* 21: 527–45.

Dressler, W.W., Grell, G.A., and Viteri, F.E. (1982). Intracultural diversity and the sociocultural correlates of blood pressure: a Jamaican example. *Medical Anthroplogy Quarterly* 9: 291–313.

Drinkwater, B. (1994). Does physical activity play a role in preventing osteoporosis? *Research Quarterly for Exercise and Sports* 65: 197–206.

Driscoll, M. (1995). Genes controlling programmed cell death: relation to mechanisms of cell senescence and aging? In *Molecular Aspects of Aging*, K. Esser and G.M. Martin (eds.). New York: John Wiley and Sons, pp. 45–60.

Duara, R., London, E.D., and Rapoport, S.I. (1985). Changes in structure and energy metabolisms of the aging brain. In *Handbook of the Biology of Aging*, 2nd edition, E.L. Schnieder and C.E. Finch (eds.). New York: Van Nostrand Reinhold, pp. 595–616.

Dudas, S.P. and Arking, R. (1995). A coordinate upregulation of antioxidant gene activities is associated with the delayed onset of senescence in a long-lived strain of *Drosophila*. *Journal of Gerontology* 50A: B117–27.

Dufour, D.L. (1992). Nutritional ecology in the tropical rainforests of Amazonia. *American Journal of Human Biology* 4: 197–207.

Dunbar, R.I.M. (1988). *Primate Social Systems, Studies in Behavioral Adaptation*. London: Croom Helm.

Durham, W.H. (1991). *Coevolution: Genes, Minds, and Human Diversity*. Palo Alto, C.A.: Stanford University Press.

Easterbrook, P.J., Berlin, J.A., Goplan, R., and Matthews, D.R. (1991). Publication bias in clinical research. *Lancet* 1: 867–72.

Eaton, S.B. and Konner, M.J. (1985). Paleolithic nutrition: a consideration of its nature and current implications. *New England Journal of Medicine* 312: 283–9.

Eaton, S.B. and Nelson, D.A. (1991). Calcium in evolutionary perspective. *American Journal of Clinical Nutrition* 54 (suppl.): 2815–75.

Eaton, S.B., Konner, M., and Shostak, M. (1988). Stone agers in the fast lane: chronic degenerative diseases in evolutionary perspective. *American Journal of Medicine*, 84: 739–749.

Elson, J.L., Samuels, D.C., Turnbull, D.M., and Chinnery, P.F. (2001). Random intercellular drift explains the clonal expansion of mitochondrial DNA mutations with age. *American Journal of Human Genetics* 68: 802–6.

Erdos, E.S. and Skidgel, R.A. (1987). The angiotensin I-converting enzyme. *Laboratory Investigation* 56: 345–348.

Erhardt, C.L. and Berlin, J.E. (eds.) (1974). *Mortality and Morbidity in the United States*. Cambridge, M.A.: Harvard University Press.

Esser K. and Martin, G.E. (eds.). (1995). *Molecular Aspects of Aging*. New York: John Wiley & Sons.

Fall, C.H.D., Barker, D.J.P., Osmond, C., Winter, P.D., Clark, P.M.S., and Hales, C.N.s (1992). Relation of infant feeding to adult serum cholesterol concentration and death from ischaemic heart disease. *British Medical Journal* 304: 801–5.

Federal Interagency Forum on Aging Related Statistics (FIFARS). (2000). *Older Americans 2000: Key Indicators of Well-Being*. Washington, D.C.: U.S. Government Printing Office.

Fenner, F. (1970). The effects of changing social organization on the infectious diseases of man. In *The Impact of Civilization on the Biology of Man*, S.V. Boyden (ed.). Toronto: Toronto University Press, pp. 48–68.

Fergusson, D.M. (1999). Prenatal smoking and antisocial behavior. *Archives of General Psychiatry* 56: 223–4.

Fergusson, D.M., Woodward, L.J., and Harwood, L.J. (1998). Maternal smoking during pregnancy and psychiatric adjustment in late adolescence. *Archives of General Psychiatry* 55: 721–7.

Ferrannini, E.S., Haffner, S.M., Mitchell, B.D., and Stern, S.P. (1991). Hyperinsulinemia: the key feature of a cardiovascular and metabolic syndrome. *Diabetologia* 34: 416–22.

Finch, C.E. (1990). *Longevity, Senescence, and the Genome*. Chicago: Chicago University Press.

Finch, C.E. (1994). *Longevity, Senescence and the Genome*. 2nd edition. Chicago: Chicago University Press.

Finch, C.E. and Hayflick, L. (eds.) (1977). *Handbook of the Biology of Aging*. New York: Van Nostrand Reinhold Company.

Finch, C.E. and Rose, M.R. (1995). Hormones and the physiological architecture of life history evolution. *The Quarterly Review of Biology* 70: 1–52.

Finch, C.E. and Seeman, T.E. (1999). Stress theories of aging. In *Handbook of Theories of Aging*, V.L. Bengston and K.W. Schaie (eds.). New York: Springer Publishing Co., pp. 81–97.

Finch, C.E. and Tanzi, R.E. (1997). Advances in aging research: genetics of aging. *Science* 278: 407–11.

Fisher, R.A. (1930). *The Genetical Theory of Natural Selection*. Oxford: Clarendon Press.

Fitton, L.J. (1999). *Is Acculturation Healthy? Biological, Cultural, and Environmental Change Among the Cofan of Ecuador*. Ph.D. Dissertation, The Ohio State University.

Fozard, J.L., Metter, E.J., and Brant, L.J. (1990). Next steps in describing aging and disease in longitudinal studies. *Journal of Gerontology* 45: pp. 116–27.

Fridovich, I. (1978). The biology of oxygen radicals. *Science* 201: 875–880.

Fridovich, I. (1995). Superoxide radical and superoxide dismutases. *Annals of Developmental Biochemistry* 64: 97–112.

Fried, L.P., Tangen, C.M., Walston, J. *et al.* for the Collaborative Health Study Collaborative Research Group. (2001). Fraility in older adults: evidence for a phenotype. *Journal of Gerontology* 56A: M146–56.

Friedlaender, J.S. (1987). *The Solomon Islands Project*. New York: Oxford University Press.

Friedman, D.B. and Johnson, T.E. (1988). A mutation in the *age-1* gene in *Caenorhabditis elegans* lengthens life and reduces hermaphrodite fertility. *Genetics* 118: 75–86.

Fries, J.F. (1980). Aging, natural death, and the compression of morbidity. *New England Journal of Medicine* 303: 130–5.

Fries, J.F. (1983). The compression of morbidity. *Milbank Memorial Fund Quarterly* 61: 397–419.

Fries, J.F. (1984). The compression of morbidity: miscellaneous comments about a theme. *The Gerontologist* 24: 354–9.

Fries, J.F. (1988). Aging, illness, and health policy: implications of the compression of morbidity. *Perspectives in Biology and Medicine* 31: 407–28.

Fries, J.F. (2001). Aging, cumulative disability, and the compression of morbidity. *Comprehensive Therapy* 27: 322–9.

Fries, J.F. and Carpo, L.M. (1981). *Vitality and Aging: Implications of the Rectangular Curve*. San Francisco: W.H. Freeman and Company.

Frisancho, A.R. (1993). *Human Adaptation and Accommodation*. Ann Arbor: University of Michigan Press.

Frolkis, V.V. (1993). Stress-age syndrome. *Mechanisms of Aging and Development* 69: 93–107.

Garn, S. (1994). Fat, lipid, and blood pressure changes in adult years. In *Biological Anthropology and Aging: Perspectives on Human Variation over the Life Span*, D.E. Crews and R.M. Garruto (eds.). New York: Oxford University Press, pp. 301–20.

Gavrilov, L.A. and Gavrilova, N.S. (1986). *The Biology of Life Span: A Quantitative Approach.* New York: Harwood Academic Publishers.

Gavrilov, L.A. and Gavrilova, N.S. (1991). *The Biology of Life Span: A Quantitative Approach*, V.P. Skulachev (ed.), Trans. J. Payne and L. Payne, New York: Harwood Academic.

Gavrilov, L.A. and Gavrilova, N.S. (2001). The reliability theory of aging and longevity. *Journal of Theoretical Biology* 213: 527–45.

Geodakian, V.A. (1982). Sexual dimorphism and the evolution of duration of ontogenesis and its stages. In *Evolution and Environment*, V.J.A. Novák and J. Mlikovsky (eds.). Prague: CSAV.

Gerber, L.M. and Crews, D.E. (1999). Evolutionary perspective on chronic degenerative diseases. In *Evolutionary Medicine*, W. Trevathan, J. McKenna, and N. Smith (eds.). New York: Oxford University Press, pp. 443–69.

Gibbons, A. (1990). Gerontology research comes of age. *Science* 250: 622–5.

Goodall, J. (1986). *The Chimpanzees of Gombe: Patterns of Behavior.* Cambridge, M.A.: Belknop Press of Harvard University Press.

Goodman, A. and Leatherman, T. (1999). *Building a Biocultural Synthesis.* Ann Arbor, M.I.: University of Michigan Press.

Goodman, L.A. (1969). The analysis of population growth when the birth and death rates depend upon several factors. *Biometrics* 25: 659–81.

Gould, S.J. (1996). The *Mismeasure of Man: Revised and Expanded.* New York: W.W. Norton and Co.

Greenberg, L.J. and Yunis, E.J. (1978). In *Birth Defects: Original Article Series, Genetic control of autoimmune disease and immune responsiveness and the relationship to aging*, vol. XIV, no.1, pp. 249–60.

De Grey, A.D.N.J. (1997). A proposed refinement of the mitochondrial free radical theory of aging. *BioEssays* 19: 161–5.

Grigsby, J.S. (1991). Paths for future population aging. *Gerontologist* 31: 195–203.

Grossman, C.J. (1985). Interactions between the gonadal steroids and the immune system. *Science* 227: 257–61.

Greksa, L.P., Hilde, S., and Caceres, E. (1994). Total lung capacity in young highlanders of aumara ancestry. *American Journal of Physical Anthropology* 94: 477–86.

Guralnik, J.M., Simonsick, E.M., Ferrucci, L. *et al.* (1994). A short physical performance battery assessing lower extremity function: association with self-reported disability and prediction of mortality and nursing home admission. *Journal of Gerontology* 49: M85–94.

Guralnik, J.M., Yanagishita, M., and Schneider, E.L. (1988). Projecting the older population of the United States: lessons from the past and prospects for the future. *Milbank Memorial Fund Quarterly* 66: 622–5.

Hales, C.N. (1997). Fetal and infant origins of adult disease. *Journal of Clinical Pathology* 50: 359–60.

Hales, C.N. and Barker, D.J.P. (1992). Type 2 (non-insulin dependent) diabetes mellitus: the thrifty phenotype hypothesis. *Diabetologia* 35: 595–601.

Hamilton, W.D. (1966). The moulding of senescence by natural selection. *Journal of Theoretical Biology*, 12: 12–45.

Harley, C.B., Futcher, A.B. and Greider, C.W. (1990). Telomeres shorten during aging of human fibroblasts. *Nature* 345: 458–60.

Harman, D. (1956). Aging: a theory based on free radical and radiation chemistry. *Journal of Gerontology* 11: 298–300.

Harman, D. (1981). The aging process. *Proceeding of the National Academy of Sciences USA* 178: 7124–8.

Harman, D. (1984). Free radicals and the origination, evolution, and present status of the free radical theory in aging. In *Free Radicals in Molecular Biology, Aging, and Disease*, D. Armstrong, R.S. Sohal, R.G. Cutler, and T.F. Slater (eds.). New York: Raven Press, pp. 1–12.

Harman, D. (1988). Free radicals in aging. *Molecular and Cellular Biochemistry* 84: 155–61.

Harman, D. (1999). Aging: prospects for further increases in the functional lifespan. *Age* 17: 119–46.

Harper, G.J. (1998). *Stress and Adaptation Among Elders in Life-Care Communities*, Ph.D. Dissertation, The Ohio State University.

Harper, G.J. and Crews, D.E. (2000). Aging, senescence, and human variation. In *Human Biology*, S. Stinson, R. Huss-Ashmore, and D. O'Rourke (eds.). New York: Wiley-Liss, pp. 465–505.

Harper, G.J., Crews, D.E. and Wood, J.W. (1996). Lack of age-related blood pressure increase in the Gainj, Papua New Guinea: another low blood pressure population. *American Journal of Human Biology* 6: 122–3 (abstract).

Harrington, C.R. and Wischik, C.M. (1995). Pathogenic mechanisms in Alzheimer syndromes and related disorders. In *Molecular Aspects of Aging*, K. Esser and G.M. Martin (eds.). New York: John Wiley and Sons, pp. 227–40.

Harris, M. (1983). Classification and diagnostic criteria for diabetes mellitus. *Tohoku Journal of Experimental Medicine* 141: 21–8.

Harrison, G.A. (1973). The effects of modern living. *Journal of Biosocial Science* 5: 217–28.

Hart, R.W. and Tuturro, A. (1983). Theories of Aging. In *Review of Biological Research in Aging*, Vol. 1. New York: Alan R. Liss Inc., pp. 5–18.

Harvey, P.H. and Bennet, P.M. (1983). Brain size, energetics, ecology and life history patterns. *Nature* 306: 314–15.

Harvey, P.H., Martin, R.D., and Clutton-Brock, T.H. (1987). Life histories in comparative perspective. In *Primate Societies,* B.B. Smuts *et al.* (eds.). Chicago: University of Chicago Press, pp. 181–96.

Hastie, N.D., Dempster, M., Dunlop, M.G., Thompson, A.M., Green, D.K., and Allshire, R.C. (1990). Telomere reduction in human colorectal carcinoma and with ageing. *Nature* 346: 866–8.

Hauspie, R.C. and Susanne, C. (1998). Genetics of child growth. In *The Cambridge Encyclopedia of Human Growth and Development*, S.J. Ulijaszek, F.E. Johnston, and M.A. Preece (eds.). Cambridge: Cambridge University Press, pp. 124–8.

Hawkes, K., O'Connell, J.F., and Blurton Jones, N.G. (1997). Hadza women's time allocation, offspring provisioning, and the evolution of long postmenopausal life span. *Current Anthropology* 48: 551–77.

Hawkes, K., O'Connell, J.F., Blurton Jones, N.G., Alvarez, H., and Charnov, E.L. (1998). Grandmothering, menopause, and the evolution of human life histories. *Proceedings of the National Academy of Science USA* 95: 1336–9.

Hayflick, L. (1965). The limited *in vitro* lifetime of human diploid cell strains. *Experimental Cell Research* 37: 614–36.

Hayflick, L. (1987). Origins of longevity. In *Modern Biological Theories of Aging*, H.R. Warner *et al.* (eds.). New York: Raven Press, pp. 21–34.

Hayflick, L. (1988). Why do we live so long? *Geriatrics* 43: 77–83.

Hayflick, L. and Moorhead, P.S. (1961). The serial cultivation of human diploid cell strains. *Experimental Cell Research* 25: 585–621.

Hazzard, W.R. (1986). Biological basis of the sex differential in longevity. *Journal of the American Geriatric Society* 34: 455–71.

Hendricks, J. and Achenbaum, A. (1999). Historical development of theories of aging. In: *Handbook of Theories of Aging,* V.L. Bengtson and K.W. Schaie (eds.). New York: Springer, pp. 21–39.

Hernández, M. (1998). Hormonal regulation of fetal growth. In *The Cambridge Encyclopedia of Human Growth and Development*, S.J. Ulijaszek, F.E. Johnston, M.A. Preece (eds.). Cambridge: Cambridge University Press, pp. 151–3.

Hill, K. and Hurtado, A.M. (1991). The evolution of premature reproductive senescence in human females: an evaluation of the "Grandmother Hypothesis". *Human Nature* 2: 313–50.

Hill, K. and Hurtado, A.M. (1999). Packer and colleagues' model of menopause for humans. *Human Nature* 10: 199–204.

Hindmarsh, P. (1998). Endocrinological regulation of post-natal growth. In *The Cambridge Encyclopedia of Human Growth and Development*, S.J. Ulijaszek, F.E Johnston, M.A. Preece (eds.). New York: Cambridge University Press, pp. 182–3.

Hirokawa, K. (1992). Understanding the mechanism of the age-related decline in immune function. *Nutrition Reviews* 50: 361–6.

Hodge, A.M., Dowse, G.K., Gareeboo, H., Tuomilehto, J., Alberti K.G.M.M., J., and Zimmer, P.Z. (1996). Incidence, increasing prevalence, and prediction of change in obesity and fat distribution over 5 years in the rapidly developing population of Mauritius. *International Journal of Obesity* 20: 137–46.

Hoffman, W.R. (1983). On the presumed coevolution of brain size and evolution. *Journal of Human Evolution* 13: 371–6.

Hoffman, W.R. (1984). Energy metabolism, brain size, and longevity in mammals. *The Quarterly Review of Biology* 58: 495–512.

Holden, C. (1987). Why do women live longer than men? *Science* 238: 158–60.

Holmes, L.D. and Rhodes, E.C. (1983). Aging and change in Samoa. In *Growing Old in Different Societies: Cross-Cultural Perspectives,* J. Sokolovsky (ed.). Belmont, C.A.: Wadsworth, pp. 119–29.

Holt, I.J., Harding, A.E., and Morgan Hughes, J.A. (1988). Deletions of muscle mitochondrial DNA in patients with mitochondrial myopathies. *Nature* 331: 717–19.

Hong, Y., de Faire, U., Heller, D.A., McClearn, G.E. and Pedersen, N. (1994). Genetic and environmental influences on blood pressure in elderly twins. *Hypertension* 24: 663–70.

Howell, T.H. (1963). Multiple pathology in nonagenarians. *Geriatrics* 18: 899–902.

Hunt, K.D. (1994). The evolution of human bipedality: ecology and functional morphology. *Journal of Human Evolution* 26: 183–202.

Idler, E.L. and Kasl, S. (1991). Health perspectives and survival: do global evaluations of health status really predict mortality? *Journal of Gerontology* 46: S55–65.

Imahori, K. (1992). How I understand aging. *Nutrition Reviews* 50(12): 351–352.

Ingram, D.K., Cutler, R.G., Weindruch, R. *et al.* (1990). Dietary restriction and aging: the initiation of a primate study. *Journal of Gerontology* 45: B148–63.

Ivanova, R., Henon, N., Lepage, V., Charron, D., Vincent, E., and Schachter, F. (1998). HLA-Dr alleles display sex-dependent effects on survival and discriminate between individual and familial longevity. *Human Molecular Genetics* 7: 187–94.

James, G.D. and Baker, P.T. (1990). Human population biology and hypertension: evolutionary and ecological aspects of blood pressure. In *Hypertension: Pathophysiology, Diagnosis, and Management*, J.H. Laragh, B.M. Brenner (eds). New York: Raven Press, pp. 137–145.

James, G.D., Crews, D.E., and Pearson, J. (1989). Catecholamine and stress. In *Human Population Biology: A Transdisciplinary Science*, M.A. Little and J.D. Haas (eds.), New York: Oxford University Press, pp. 280–95.

Jialal, I. and Grundy, S.M. (1992). Influence of antioxidant vitamins on LDL oxidation. *Annals of the New York Academy of Sciences* 669: 237–48.

Johnson, R.J. and Wolinsky, F.D. (1994). Gender, race, and health: the structure of health status among older adults. *The Gerontologist* 34: 24–35.

Johnson, T.E. (1990). Increased life-span of age-1 mutants in *Caenorhabditis elegans* and lower Gompertz rate of aging. *Science* 249: 908–12.

Johnson, T.E. *et al.* (1995). Group report: research on diverse model systems and the genetic basis of aging and longevity. In *Molecular Aspects of Aging*, K. Esser and G.M. Martin (eds.). New York: John Wiley and Sons, pp. 5–15.

Johnston, F.E. (1999). Screening for growth failure in developing countries. In *The Cambridge Encyclopedia of Human Growth and Development*, S.J. Ulijaszek, F.E. Johnston and M.A. Preece (eds.). Cambridge: Cambridge University Press.

Joint National Committee V (JNC V) (1993). The Fifth Report of the Joint National Committee on Detection, Evaluation, and Treatment of High Blood Pressure. *Archives of Internal Medicine* 153: 154–83.

Judge, D.S. and Carey, J.R. (2000). Postreproductive life predicted by primate patterns. *Journal of Gerontology* 55A: B201–9.

Jun, T., Ke-yan, F., and Catalano, M. (1996). Increased superoxide anion production in humans: a possible mechanism for the pathogenesis of hypertension. *Journal of Human Hypertension* 10: 305–9.

Kaplan, G.A. (1991). Epidemiologic observations on the compression of mortality: evidence from the Alameda County Study. *Journal of Aging Health* 3: 155–71.

Katz, S.A. and Armstrong, D.A. (1994). Cousin marriage and the X-chromosome: evolution of longevity. In *Biological Anthropology and Aging: Perspectives on Human Variation Over the Lifespan*, D.E. Crews and R.M. Garruto (eds.). New York: Oxford University Press, pp. 101–26.

Katz, S.A., Ford, A.B., Moskowitz, R.W., Jackson, B.A., and Jaffee, M.W. (1963). Studies of illness in the aged. The index of ADL: a standardized measure of biological and psychosocial function. *Journal of the American Medical Association* 185: 94–101.

Katzmarzyk, P.T. and Leonard, W.R. (1998). Climatic influences on human body size and proportions: ecological adaptations and secular trends. *American Journal of Physical Anthropology* 106: 483–503.

Kemnitz, J.W., Weindruch, R., Roecker, E.B., Crawford, K., Kaufman, P.L., and Ershler, W.B. (1993). Dietary restriction of adult male rhesus monkeys: design, methodology and preliminary findings from the first year of study. *Journal of Gerontology* 45: B17–26.

Kerber, R.A., O'Brian, E., Smith, K.R., and Cawthan, R.M. (2001). Familial excess longevity in Utah genealogies. *Journal of Gerontology: Biological Sciences* 56A: B130–9.

Keyfitz, N. (1968). *Introduction to the Mathematics of Population*. Reading, M.A.: Addison-Wesley.

Khaw, K.T., Wareham, N., Luben, R. *et al.* (2001). Glycated hemoglobin, diabetes, and mortality in the Norfolk cohort of the European Prospective Investigation of Cancer and Nutrition (EPIC-Norfolk). *British Medical Journal* 322: 15–18.

Khazaeli A.A., Xiu, L., and Curtsinger, J.W. (1996). Effect of adult cohort density on age-specific mortality in *Drosophila*: a density supplementation experiment. *Genetica* 98: 21–31.

Kiecolt-Glasser, J.K., Dura, J.R., Speicher, C.E., Trask, O.J., and Glasser, R. (1991). Spousal caregivers of dementia victims: longitudinal changes in immunity and health. *Psychosomatic Medicine* 53: 345–62.

Kiecolt-Glasser, J.K., Fisher, C.D., Ogrocki, P., Stout, J.C., Spencer, C.E., and Glasser, R. (1987). Marital quality, marital disruption, and immune function. *Psychosomatic Medicine* 49: 13–32.

Kinsella, K. and R. Suzman. (1992). Demographic dimensions of population aging in developing countries. *American Journal of Human Biology* 4: 3–8.

Kirkwood, TBL. (1977). Evolution of aging. *Nature* 270: 301–4.

Kirkwood, TBL. (1981). Repair and its evolution: survival versus reproduction. In *Physiological Ecology: An Evolutionary Approach to Resource Use*, C.R. Townsend and P. Calow (eds.). Oxford: Blackwell.

Kirkwood, TBL. (1990). The disposable soma theory of aging. In *Genetic Effects on Aging II*, D.E. Harrison (ed.). New Jersey: Telford Press, pp. 9–19.

Kirkwood, TBL. (1995). What can evolution theory tell us about the mechanisms of aging? In *Molecular Aspects of Aging*, K. Esser and G.M. Martin (eds.). New York: John Wiley and Sons.

Kirkwood, TBL. (2000). Molecular gerontology – bridging the simple and the complex. *Molecular and Cellular Gerontology* 908: 14–20.

Kirkwood, TBL. and Austad, S.N. (2000). Why do we age? *Nature* 408: 233–8.

Kirkwood, TBL. and Cremer, T. (1982). Cytogerontology since 1881: a reappraisal of August Weismann and a review of modern progress. *Human Genetics* 60: 101–21.

Kirkwood, T.B. and Kowald, A. (1997). Network theory of aging. *Experimental Gerontology* 32: 395–9.

Knopman, D., Boland, L.L., Mosley, T. *et al.* (2001). Cardiovascular risk factors and cognitive decline in middle-aged adults. *Neurology* 56: 42–8.

Kondo, O., Dode, Y., Akazawa, T., and Muhesen, S. (2000). Estimation of stature from the skeletal reconstruction of an immature Neanderthal from Dederiyeh Cave, Syria. *Journal of Human Evolution*, 38: 457–73.

Korpelainen, H. (2000). Variation in the heritability and evolvability of human lifespan. *Naturwissenschaften* 87: 566–8.

Kosciolek, B.A. and Rowley, P.T. (1998). Human lymphocyte telomerase is genetically regulated. *Genes, Chromosomes, and Cancer* 21: 124–30.

Kowald, A. and Kirkwood, TBL. (1994). Towards a network theory of aging: a model combining the free radical theory and the protein error theory. *Journal of Theoretical Biology* 168: 75–94.

Krall, E. (1997). Bone mineral density and biochemical markers of bone turnover in healthy elderly men and women. *Journal of Gerontology* 52: M61–7.

Kreisberg, R.A. (1987). Aging, glucose, metabolism, and diabetes: current concepts. *Geriatrics*, 42: 67–72.

Kurtz, T.W. and Spence, M.A. (1993). Genetics of essential hypertension. *American Journal of Medicine* 94: 77–84.

Kusnetsova, S.M. (1987). Polymorphism of heterochromatin areas on chromosomes 1, 9, 16, and T in long lived subjects and persons of different ages in two regions of the Soviet Union. *Archives of Gerontology and Geriatrics* 6: 177–86.

Lagaay, A.M., D'Amaro, J., Ligthart, G.J., Th. Schreuder, G.M., van Rood, J.J., and Hijmans, W. (1991). Longevity and heredity in humans: association with the human leucocyte antigen phenotype. *Annals of the New York Academy of Sciences*, 621: 78–89.

Lakatta, E.G. (1990). Heart and circulation. In *Handbook of the Biology of Aging*, E.L. Schneider and J.W. Rowe (eds.). San Diego: Academic Press, Inc., pp. 181–218.

Lakowski, B. and Hekimi, S. (1996). Determination of life-span in *Caenorhabditis elegans* by four clock genes. *Science* 272: 1010–13.

Lancaster, J.B. and King, B.J. (1985). An evolutionary perspective on menopause. In *In Her Prime*, J.K. Brown and V. Kerns (eds.). South Hadley, M.A.: Bergin and Garvey.

Lane, M.A., Ingram, D.K., and Roth, G.S. (1997). Beyond the rodent model: calorie restriction in rhesus monkeys. *Age* 20: 45–56.

Lapidot, M.B. (1987). Does the brain age uniformly? Evidence from effects of smooth pursuit eye movements on verbal and visual tasks. *Journal of Gerontology* 42: 329–31.

Lasker, G.W., Crews, D.E. (1996). Behavioral influences on the evolution of human genetic evolution. *Molecular Phylogenetics and Evolution* 5: 232–40.

Laveist, T.A., Nickerson, K.J. and Bowie, J.V. (2000). Attitudes about racism, medical mistrust and satisfaction with care among African American and white cardiac patients. *Medical Care Research and Review* 57: 146–61.

Law, C.M. (1996). Fetal and infant influences on non-insulin dependent diabetes mellitus (NIDDM). *Diabetic Medicine* 13: 549–52.

Law, C.M., de Swiet, M., and Osmond, C. (1993). Initiation of hypertension in utero and its amplification throughout life. *British Medical Journal* 306: 24–7.

Lee, P.C. (1998). Primate life-history. In *The Cambridge Encyclopedia of Human*

Growth and Development, S.J. Ulijaszek, F.E. Johnston, and M.A. Preece (eds.). Cambridge: Cambridge University Press, pp. 102–3.

Leidy, L.E. (1999). Menopause in evolutionary perspective. In *Evolutionary Medicine*, W.R. Trevathan, E.O. Smith and J.J. McKenna (eds). New York: Oxford University Press, pp. 407–27.

Leidy-Sievert, L. (2001). Menopause as a measure of population health: an overview. *American Journal of Human Biology* 13: 429–33.

Leigh, S.R. (1996). Evolution of human growth spurts. *American Journal of Physical Anthropology* 101: 455–74.

Le Sourd, B. (1995). Protein undernutrition as the major cause of decreased immune function in the elderly: clinical and functional implications. *Nutrition Reviews* 53: 586–91.

LeSourd, B. (1997). Nutrition and immunity in the elderly: modification of immune responses with nutritional treatments. *American Journal of Clinical Nutrition* 66: 478S–84.

LeSourd, B., Mazori, L., and Ferry, M. (1998). The role of nutrition in immunity in the aged. *Nutrition Reviews* 56: S113–25.

Levi, L. and Anderson, L. (1975). *Psychological Stress*. New York: Spectrum Publications.

Lezhava, T. (2001). Chromosome and aging: genetic conception of aging. *Biogerontology* 2: 253–60.

Lifton, R.P. (1996). Molecular aspects of human blood pressure variation. *Science* 272: 676–80.

Little, M.A. and Haas, J.D. (eds.). (1989). *Human Population Biology: A Transdisciplinary Science*. New York: Oxford University Press.

Livshits, G., Karasik, D., Pavlovsky, O., and Kobyliansky, E. (1999). Segregation analysis reveals a major gene effect in compact and cancellous bone mineral density in 2 populations. *Human Biology* 71: 155–72.

Loth, S.R., and Isçan, M.V. (1994). Morphological indicators of skeletal aging: implications for paleodemography and paleogerontology. In *Biological Anthropology and Aging: Perspectives on Human Variation Over the Life Span*, D.E. Crews and R.M. Garruto (eds.). Oxford: Oxford University Press, pp. 394–425.

Luciani, F., Turchetti, G., Franceschi, C. and Valensin, S. (2001). A mathematical model for the immunosenescence. *S. Rivista Di Biologia* 94: 305–18.

Luckinbill, L.S., Arking, R., Clare, M.J., Cirocco, W.C., and Buck, S.A. (1984). Selection for delayed senescence in *Drosophila melangaster*. *Evolution* 38: 996–1003.

Ly D.H., Lockhart, D.J., Lerner, R.A., Schultz, P.G. (2000). Mitotic misregulation and human aging. *Science* 287: 2486–92.

Lynn J. (1994). Bad to the bone. *Prevention* 46: 73.

Macdonald, D. (1989). *The Encyclopedia of Mammals*, 2nd edition. Oxford: Equinox.

Macintyre, S., Hunt, K., and Sweeting, H. (1996). Gender differences in health: are things really as simple as they seem? *Social Science and Medicine* 42: 617–24.

Macurová, H., Ivanyi, P., Sajdlova, H., and Trojan, J. (1975). HLA antigens in aged persons. *Tissue Antigens* 6: 269–71.

Malino, R.M. and Bouchard, C. (1991). *Growth, Maturation, and Physical Activity*. Champaign, I.L: Human Kinetics Books.

Mancilha-Carvalho, J.J. (1985). *Estudo da presso arterial de Indios Yanami.* Rio de Janiero, Brazil: Fese de Doutorado.

Mancilha-Carvalho, J.J. and Crews, D.E. (1990). Lipid profiles of Yanomamo Indians. *Preventative Medicine* 19: 66–75.

Mangel, M. (2001). Complex adaptive system, aging and longevity. *Journal of Theoretical Biology* 213: 559–71.

Mann, A., Lampl, M., and Monge, J.M. (1996). The evolution of childhood: dental evidence for the appearance of human maturation patterns. *American Journal of Physical Anthropology* 21(Suppl.): 156 (abstract).

Manton, K.G. (1986a). Cause specific mortality patterns among the oldest old: multiple cause of death trends 1968 to 1980. *Journal of Gerontology* 41: 282–9.

Manton, K.G. (1986b). Past and future life expectancy increases at later ages: their implications for the linkage of chronic morbidity, disability, and mortality. *Journal of Gerontology* 41: 672–81.

Manton, K.G. and Gu, X. (2001). Changes in the prevalence of chronic disability in the United States black and nonblack population above age 65 from 1982 to 1999. *Proceedings of the National Academy of Sciences of the United States* 98: 6354–9.

Manton, K.G. and Stallard, E. (1984). *Recent Trends in Mortality Analysis.* Academic Press: New York.

Manton, K.G., Corder, L., Stallard, E. (1997). Chronic disability trends in elderly United States populations: 1982–1994. *Proceedings of the National Academy of Sciences of the United States* 94: 2593–8.

Manton, K.G., Stallard, E., Tolley, H.D. (1991). Limits to human life expectancy. *Population and Development Review* 17: 603–37.

Manton, K.G., Woodbury, M., and Stallard, E. 1995. Sex differences in human mortality and aging at late ages: the effects of mortality selection and state dynamics. *The Gerontologist* 35: 597–608.

Marks, J. (1994). *Human Biodiversity: Genes, Race, and History.* New York: Aldine de Gruyter.

Marmot M.G., Shipley, M.J., Rose, G. (1984). Inequalities in sex-specific explanations of a general pattern? *Lancet* 1984(i): 1003–6.

Martin, G.M. (1978). Genetic syndromes in man with potential relevance to the pathobiology of aging. *Birth Defects Original Article Series* 14: 5–39.

Martin, G.M. and Turker, M.S. (1994). Genetics of human disease, longevity, and aging. In *Principles of Geriatric Medicine and Gerontology*, W.R. Hazzard, E.L. Bierman, J.P. Blass *et al.* (eds.). New York: McGraw-Hill.

Martin, G.M. *et al.* (1995). Group report: do common underlying mechanisms of aging contribute to the pathogenesis of major geriatric disorders? In *Molecular Aspects of Aging*, K. Esser and B.M. Martin (eds.). New York: John Wiley & Sons.

Martin, G.M., Austad, S.N., and Johnson, T.E. (1996). Genetic analysis of ageing: role of oxidative damage and environmental stresses. *Nature Genetics* 13: 25–34.

Marx, J. (2000). Chipping away at the causes of aging. *Science* 287: 2390.

Mascie-Taylor, C.G.N. (1991). Nutritional status: its measurement and relation to health. In *Applications of Biological Anthropology to Human Affairs* C.G.N. Masice-Taylor and G.W. Lasker (eds). Cambridge: Cambridge University Press, pp. 55–82.

Masoro, E.J. (1995). McCay's Hypothesis: Undernutrition and Longevity. *Proceedings of the Nutrition Society* 54: 657–664.

Masoro, E.J. (1996). The biological mechanism of aging: is it still an enigma? *Age* 19: 141–5.

Mattock, C., Marmot, M., and Stern, G. (1988). Could Parkinson's disease follow intrauterine influenza? A speculative hypothesis. *Journal of Neurology, Neurosurgery, and Psychiatry* 51: 753–6.

Mayer, P.J. (1982). Evolutionary advantage of the menopause. *Human Ecology* 10: 477–93.

Mayer, P.J. (1991). Inheritance of longevity evinces no secular trend among members of six New England families born in 1650–1874. *American Journal of Human Biology* 3: 49–58.

Mayer, P.J. (1994). Human immune system aging: approaches, examples, and ideas. In *Biological Anthropology and Aging: Perspectives on Human Variation over the Life Span*, D.E. Crews and R.M. Garruto (eds.). Oxford: Oxford University Press, pp. 182–213.

McCay, C.M., Crowell, M.F., and Maynard, L.A. (1935). The Effect of Retarded Growth on the Length of Life Span and upon Ultimate Body Size. *Journal of Nutrition* 10: 255–63.

McEwen, B.S. (1998). Protective and damaging effects of stress mediators. *New England Journal of Medicine* 338: 171–9.

McEwen, B.S. (1999). Allostasis and allostatic load: implications for neuropsychopharmacology. *Neuropsycopharmacology* 22: 108–24.

McEwen, B.S. and Seeman, T. (1999). Protecting and damaging effects of mediators of stress: elaborating and testing the concepts of allostasis and allostatic load. *Annals of the New York Academy of Science* 896: 30–47.

McEwen, B.S. and Steeler, E. (1993). Stress and the individual: mechanisms leading to disease. *Archives of Internal Medicine* 153: 2093–101.

McGarvey, S.T., Levinson, P.D., Bausserman, L., Galanis, D.L., and Hornick, C.A. (1993). Population change in adult obesity and blood lipids in American Samoa from 1976–1978 to 1990. *American Journal of Human Biology* 5: 17–30.

McIntosh, W.A., Kaplan, H.B., Kubena, K.S., and Landman, W.A. (1993). Life events, social support, and immune response in elderly individuals. *International Journal of Aging and Human Development* 37: 23–36.

McLong, P.A. (2000). *Aging and Glycemia in a Nonwesternized Context: Rural Maya Females in Yucatan, Mexico*, Ph.D. Dissertation, Southern Illinois University, Carbondale, I.L.

Mead, J.F. (1976). Free radical mechanisms of lipid damage and consequences for cellular membranes. In *Free Radicals in Biology*, vol. 1, W.A. Pryor (ed.). New York: Academic Press, pp. 51–68.

Medawar, P.B. (1946). Old age and natural death. *Mod. Quart.* 1: 30–56.

Medawar, P.B. (1952). *An Unsolved Problem of Biology*. London: H.K. Lewis.

Menzel, H-J., Kladetzky, R-G., and Assmann, G. (1983). Apoliopoprotein E polymorphism and coronary artery disease. *Arteriosclerosis* 3: 310–15.

Miller, R.A. (1995). Geroncology: the study of aging as the study of cancer. In *Molecular Aspects of Aging*, K. Esser and G.M. Martin (eds.). New York: John Wiley and Sons, pp. 265–80.

Minc-Golomb, D., Knobler, H., and Groner, Y. (1991). Gene dosage of CuZnSOD and Down's Syndrome: diminished prostaglandin synthesis in human trisomy 21, transfected cells and transgenic mice. *EMBO. J.* 10: 2119–2124.

Mitrushina, M., Fogel, T., D'Elia, L., Uchiyama, C., and Satz, P. (1995). Performance on motor tasks as an indication of increased behavioral asymmetry with advancing age. *Neuropsychologia* 33: 359–64.

Mitrushina, M., Uchiyama, C., and Satz, P. (1995). Heterogeneity of cognitive profiles in normal aging: implications for early manifestations of Alzheimer's disease. *Journal of Clinical and Experimental Neuropsychology Research* 17: 374–82.

Miyamoto, H., Manabe, N., Akiyama, Y., Mitani, Y., Sugimoto, M., and Salo, E. (1994a). Quantitative studies on spermatogenesis in the senescence accelerated mouse. In *The SAM Model of Senescence*, T. Takeda (ed.). Elsevier Science B.V., Amsterdam, pp. 275–8.

Miyamoto, H., Manabe, N., Watanabe, T. *et al.* (1994b). Female reproductive characteristics of the senescence accelerated mouse. In *The SAM Model of Senescence*, T. Takeda (ed.). Elsevier Science B.V., Amsterdam, pp. 279–90.

Molnar, S. (2000). *Human Variation: Races, Types, and Ethnic Groups*, 5th edition. Englewood Cliffs, Prentice Hall.

Monnier, V.M., Sell, D.R., Nagara, R.H., and Miyata, S. (1991). Mechanisms of protection against damage mediated by the maillard reactions in aging. *Gerontology* 37: 152–65.

Montagu, M.F.A. (1962). The concept of race. *American Anthropologist* 64: 919–28.

Morrison, N.A., Tokita, A., Kelly, P.J., Crofts, L., Nguyen, T.V., Sambrook, N., and Eisman, J.A. (1994). Prediction of bone density from vitamin D reception alleles. *Nature*, 367: 284–87.

Mossey, J.M. and Shapiro, E. (1982). Self-related health: a predictor of mortality among the elderly. *American Journal of Public Health* 72: 800–8.

Mueller, W.H. (1998). Genetic and environmental influences on fetal growth. In *The Cambridge Encyclopedia of Human Growth and Development*, S.J. Ulijaszek, F.E. Johnston, and M.A. Preece (eds.). Cambridge: Cambridge University Press, pp. 133–56.

Muldoon, M.F., Manuck, S.B., and Matthews, K.A. (1990). Lowering cholesterol of concentrations and mortality: a quantitative review of primary prevention trials. *British Medical Journal* 301: 309–14.

Mulrow, C.D., Gerety, M.B., Cornell, J.E., Lawrence, V.A., and Katen, D.N. (1994). The relationship of disease and function and perceived health in very frail elders. *Journal of the American Geriatrics Society* 42: 374–80.

Multhaup, G., Ruppert, T., Schlicksupp, A. *et al.* (1997). Reactive oxygen species and Alzheimer's disease. *Biochemical Pharmacology* 54: 533–9.

Murphy, S.L. 2000. *Deaths: Final Report for 1998. National Vital Statistics Reports 48.* Hyattsville, M.D.: NCHS.

Myers G.C. and Manton, K.G. (1984). Compression of mortality: myth or reality? *The Gerontologist* 24: 346–53.

Namba, Y., Tomonaga, M., Kawasaki, H., Otomo, E., and Ikeda, K. (1991). Apolipoprotein E immunoreactivity in cerebral amyloid deposits and neurofibrillary tangles in Alzheimer's disease and kuryplaque amyloid in Creutzfeldt-Jakob disease. *Brain Research* 541: 163–6.

Nathanielsz, P.W. (1996). The timing of birth. *American Scientist* 84: 562–9.

National Center for Health Statistics. (1987). *Aging in the eighties, functional limitations of individuals age 65 years and over*. Advance Data from Vital Health Statistics (No. 133. DHHS Publ. No (PHS) 87–1250). Hyattsville, M.D.: Public Health Services.

National Cholesterol Education Program Expert Panel (NCEP) (1993). Expert Panel on Detection and Treatment of High Blood Cholesterol in Adults. Summary of the second report of the National Cholesterol Education Program (NCEP) Expert Panel on Detection, Evaluation, and Treatment of High Blood Cholesterol in Adults (Adult Treatment Panel II). *Journal of the American Medical Association* 269: 3015–23.

Neel, J.V. (1962). Diabetes mellitus: a "thrifty" genotype rendered detrimental by "progress." *American Journal of Human Genetics* 14: 353–62.

Neel, J.V. (1982). The thrifty genotype revisited. In *The Genetics of Diabetes Mellitus*, J. Kobberling and R. Tattersall (eds.). London: Academic Press.

Newcomb, T.G. and Leob, L.A. (1998). Oxidative DNA damage and mutagenesis. In *DNA Damage and Repair*, J. Nickoloff and M. Hoekstra (eds.). New Jersey: Human Press.

Nimchimsky, E.A., Gilssen, E., Allman, J.M., Perl, D.P. Erwin, J.M., and Hof, P.R. (1999). A neuronal morphologic type unique to humans and great apes. *Proceedings of the National Academy of Science* 96: 5268–73.

O'Hara, R., Yesavage, J.A., Kraemer, H.C. *et al.* (1998). The APOE 4 allele is associated with decline on delayed recall performance in community-dwelling elderly. *Journal of the American Geriatrics Society* 46: 1493–8.

O'Connor, K., Holman, D.J., and Wood, J.W. (1998). Declining fecundity and ovarian aging in natural fertility populations. *Maturitas* 21: 103–13; 30: 127–36.

Okabe T., Hamaguchi, K., Inafuku, T., and Hara, M. (1996). Aging and superoxide dismutase activity in cerebrospinal fluid. *Journal of Neurological Science* 141: 100–4.

Olofsson, P., Schwalb, O., Chakraborty, R., and Kimmel, M. (2001). An application of a general branching process in the study of the genetics of aging. *Journal of Theoretical Biology* 213: 547–57.

Oliver, D.L. (1961). *The Pacific Islanders*. Garden City, N.Y.: Doubleday and Company. Inc.

Oliver, M.F. (1992). Doubts about preventing coronary heart disease. Multiple interventions in middle aged men may do more harm than good. *British Medical Journal* 304: 393–4.

Olovnikov, A.M. (1996). Telomeres, telomerase, and aging: origin of the theory. *Experimental Gerontology* 4: 443–8.

Olshansky, S.J. and Carnes, B.A. (1994). Demographic perspectives of human senescence. *Population and Development Review* 20: 57–80.

Olshansky, S.J., Carnes, B.A., and Cassel, C. (1990). In search of Methuselah: estimating the upper limits to human longevity. *Science* 250: 634–40.

Olshansky, S.J., Carnes, B.A., and Grahn, D. (1998). Confronting the boundaries of human longevity. *American Scientist* 86: 52–61.

Omran, A.R. (1971). The epidemiologic transition: a theory of the epidemiology of population change. *Milbank Memorial Fund Quarterly* 49: 509–38.

Omran, A.R. (1983). The epidemiological transition theory. A preliminary update. *Journal of Tropical Pediatrics* 29: 305–16.

Orgel, L.E. (1970). The maintenance of the accuracy of protein synthesis and its relevance to aging. *Proceedings of the National Academy of Sciences USA* 49: 517–21.

Orgel, L.E. (1973). Aging of clones of mammalian cells. *Nature* 243: 441–5.

Orr, W.C. and Sohal, R.S. (1994). Extension of life span by overexpression of superoxide dismutase and catalase in *Drosophila melanogaster*. *Science* 263: 1128–30.

Ossa, K. and Crews, D.E. (in press). Biological and genetic theories of the process of aging throughout life. In *Age-Between Nature and Culture*, C. Sauvain-Dugergil and N. Mascie-Taylor (eds.). International Union for the Scientific Study of Population.

Packer, C., Tatar, M., and Collin, A. (1998). Reproductive cessation in female mammals. *Nature* 392: 807–10.

Paffenbarger, R.S., Hyde, R.T., Wing, A.L., and Hsieh, C. (1986). Physical activity, all-cause mortality, and longevity of college alumni. *New England Journal of Medicine* 314: 605–13.

Page, *et al.* (1974).

Pamuk, E., Makue, D., Heck, K., Reuben, C., and Lochner, K. (1998). *Socioeconomic Status and Health Chartbook*. Hyattsville, M.D.: National Center for Health Statistics.

Papasteriades, C., Boki, K., Pappa, H., Hedonopoulos, S., Papasteriadis, E., and Economidou, J. (1997). HLA phenotypes in healthy aged subjects. *Gerontology* 43: 176–81.

Partridge, L. and Fowler, K. (1992). Direct and correlated responses to selection on age at reproduction in *Drosophila melangaster*. *Evolution* 46: 76–91.

Pavelka, M.S.M. and Fedigan, L.M. (1991). Menopause: a comparative life history perspective. *Yearbook of Physical Anthropology*. 34: 13–38.

Paz, *et al.* (1995). Chapter 11. In *Molecular Aspects of Aging* K. Esser and G.M. Martin (eds.). New York: John Willey and Sons.

Pearl, R. (1922). *The Biology of Death*. Philadelphia: J.B. Lippincott Co.

Pearl, R. (1928). *The Rate of Living Theory*. New York: A.A. Knopf.

Pearl, R. (1931). Studies on human longevity III: the inheritance of longevity. *Human Biology* 3: 245–53.

Pearl, R. and Pearl, H.D. (1934). *The Ancestry of the Long-Lived*. Baltimore: Johns Hopkins University Press.

Pearson, J.D. and Crews, D.E. (1998). Apoptosis: programmed cell death. In *The Cambridge Encyclopedia of Human Growth and Development*, S.J. Ulijaszek, F.E Johnston, M.A. Preece (eds.). Cambridge: Cambridge University Press, p. 431.

Peccei, J.S. (1995a). A hypothesis for the origin and evolution of menopause. *Mauritas* 21: 83–9.

Peccei, J.S. (1995b). The origin and evolution of menopause: the altriciality-lifespan hypothesis. *Etiology and Sociobiology* 16: 425–49.

Peccei, J.S. (2001a). A critique of the grandmother hypotheses: old and new. *American Journal of Human Biology* 13(4): 434–452.

Peccei, J.S. (2001b). Menopause: Adaptation or Epiphenomenon? *Evolutionary Anthropology*. 10 (2001): 43–57.

Perls, T.T. (1995). The oldest old. *Scientific American* 272: 70–5.

Perls, T.T. (2001). Genetic and phenotypic markers among centenarians. *Journal of Gerontology* 56A: M67–70.

Perls, T.P. and Silver, M.H. with Lanerman, J.F. (1995). *Living to 100: Lessons in Living to your Maximum Potential at any Age.* New York : Basic Books.

Pfeiffer, S. (1982). The evolution of human longevity: distinctive mechanisms? *Canadian Journal on Aging* 9: 95–103.

Plassman, B.L., Welsh, K.A., Helms, M., Brandt, J., Page, W.F., and Breitner, J.C. (1995). Intelligence and education as predictors of cognitive state in late life: a 50-year follow-up. *Neurology* 45: 1446–50.

Pletcher, S.D. and Curtsinger, J.W. (1998). Age-specific properties of spontaneous mutations affecting mortality in *Drosophila melangaster. Genetics* 148: 287–303.

Prothero, J. and Jürgens. K.D. (1987). Scaling of maximum life span in mammals: a review. *Basic Life Sciences* 42: 49–74.

Proust, J., Moulias, R., Fumero, F., Bekkhoucha, F., Busson, M., and Schmid, M. (1982). HLA and longevity. *Tissue Antigens* 19: 168–73.

Pryor, W.A. (1984). Free radicals in autoxidation and in aging. In *Free Radicals in Molecular Biology, Aging, and Disease*, D. Armstrong, R.S. Sohal. R.G. Cutler, and T.F. Slater (eds.). New York: Raven Press, pp. 13–42.

Puca, A., Shea, M., Bowen, J. *et al.* (2000). Familial clustering for extreme longevity and patterns of inheritance. *Journal of the American Geriatrics Society* 48: 1483–5.

Raeven, G.M. and Raeven, E.P. (1985). Age, glucose intolerance, and non-insulin-dependent diabetes mellitus. *Journal of the Geriatric Society* 33: 286–90.

Raeven, G.M. (1988). Banting Lecture. Role of insulin resistance in human disease. *Diabetes* 37: 1595–607.

Rakowski, W., and Pearlman, D.N. (1995). Demographic aspects of aging: current and future trends. In *Care of the Elderly: Clinical Aspects of Aging*, W. Reichel (ed.). Baltimore: Williams and Wilkins, pp. 488–495.

Reff, M.E. (1985). RNA and protein metabolism. In *Handbook of the Biology of Aging*, 2nd edition, C.E. Finch and E.L. Schneider (eds.). New York: Van Nostrand Reinhold CO.: pp. 225–54.

Resnick, D. (1985). Costs of reproduction: an evaluation of the empirical evidence. *Oikos* 44: 257–67.

Richter, C. (1988). Do mitochondrial DNA fragments promote cancer and aging? *FEBS Letters* 241: 1–5.

Richter, C. (1995). Oxidative damage to mitochondrial DNA and its relationship to aging. In *Molecular Aspects of Aging*, K. Esser and G.M. Martin, (eds.). New York: John Wiley and Sons, pp. 99–108.

Richter, C., Park, J.W., and Ames, B.N. (1988). Normal oxidative damage to mitochondrial nuclear DNA is extensive. *Proceedings of the National Academy of Sciences of the United States* 85: 6465–67.

Rogers, R. (1992). Living and dying in the U.S.A.: sociodemographic determinants of death among blacks and whites. *Demography*, 29: 287–303.

Rose, M.R. (1984). Laboratory evolution of postponed senescence in *Drosophila melangaster. Evolution* 38: 1004–10.

Rose, M.R. (1991). *The Evolutionary Biology of Aging.* Oxford: Oxford University Press.

Rosen, D.R., Siddique, T., and Patterson, D. (1993). Mutations on Cu/Zn superoxide dismutase gene are associated with familial amyotrophic lateral sclerosis. *Nature* 362: 59–62.

Rosenwaike, L. (1985). A demographic portrait of the oldest old. *Milbank Memorial Fund Quarterly* 63: 187–205.

Roth, G.S., Ingram, D.K., and Lane, M.A. (1995). Slowing aging by caloric restriction. *Nature Medicine* 1: 414–15.

Rowland, R. and Roberts, J. (1982). Blood pressure levels and hypertension in persons ages 6–74 years: United States, 1976–1980. In *Vital Health Statistics*, no. 84. Department of Health and Human Services Publication 82–1250. Washington D.C.: Government Printing Office.

Sacher, G.A. (1975). Maturation and longevity in relation to cranial capacity in hominid evolution. In *Primate Functional Morphology and Evolution*, R. Tuttle (ed.). The Hague: Mouton, pp. 417–41.

Sacher, G.A. (1977). Life table modification and life prolongation. In *Handbook of the Biology of Aging*, L. Hayflick and C.E. Finch (eds.). New York: Academic Press, pp. 582–638.

Sacher, G.A. (1980). Mammalian life histories: their evolution and molecular genetic mechanism. In *Aging, Cancer, and Cell Membranes*, C. Borek, C.M. Fenoglis and D.W. King (eds.). New York: Thieme-Straton Inc.

Santos, R.V. and Coimbra, C.A.E. (1996). Socioeconomic differentiation and body morphology in the Surui of Southwestern Amazonia. *Current Anthropology* 37: 851–6.

Sapolsky, R.M. (1990). The adrenocortical axis. In *Handbook of the Biology of Aging*, E.L. Schneider and J.W. Rowe (eds.). San Diego: Academic Press, Inc., pp. 330–48.

Sapolsky, R.M. (1992). *Stress, the Aging Brain, and the Mechanism of Neuron Death.* Cambridge, M.A.: NIT Press.

Satta, Y., O'Huigin, C., Takahata, N., and Klein, J. (1994). Intensity of natural selection at the major histocompatibility complex loci. *Proceedings of the National Academy of Sciences of the United States* 91: 7184–8.

Schächter, F., Faire–Delanef, L., Guenot, F. *et al.* (1994). Genetic association with human longevity at the APOE and ACE loci. *Nature Genetics* 6: 29–32.

Schell, L.M. (1991). Pollution and human growth: lead, noise, polychlorobiphenyl compounds and toxic wastes. In *Applications of Biological Anthropology in Human Affairs*, C.G.N. Mascie-Taylor and G.W. Lasker (eds.). New York Cambridge University Press, pp. 83–116.

Schlenker, E. (1998). *Nutrition in Aging.* Dubuque, I.A.: WCB/McGraw-Hill.

Schoenfeld, D.E., Malmrose, L.C., Blazer, D.G., Gold, D.T., and Seeman, T.E. (1994). Self-rated health and mortality in the high-functioning elderly – a closer look at healthy individuals: MacArthur field study of successful aging. *Journal of Gerontology* 49: M109–15.

Schulz-Aellen, M.F. (1997). *Aging and Human Longevity.* Boston: Birkhauser.

Schuster, H., Wienker, T.F., Stremmler, U., Noll, B., Steinmetz, A., and Luft, F.C. (1995). An angiotensin-converting enzyme gene variant is associated with acute myocardial infarction in women but not in men. *American Journal of Cardiology* 76: 601–3.

Schrimshaw, N.S. and Young, V.R. (1989). Adaptation to low protein and energy intakes. *Human Organization* 48: 20–30.

Seeman, T.E., Singer, B.H., Rowe, J.W., Horowitz, R.I., and McEwen, B.S. (1997). Price of adaptation – allostatic load and its consequences. *Archives of Internal Medicine* 157: 2259–68.

Seltzer, C.C. (1966). Some re-evaluations of the build and blood pressure study, 1959, as related to ponderal index, somatotype and mortality. *New England Journal of Medicine* 274: 254–9.

Selye, H. (1956). *The Stress of Life*. New York: McGraw-Hill.

Service, P.M. (1987). Physiological mechanisms of increased stress resistance in *Drosophila melanogaster* selected for postponed senescence. *Physiological Zoology* 60: 321–6.

Service, P.M. (2000a). The genetic structure of female life history in *D. melanogaster*: comparisons among populations. *Genetic Research* 75: 153–66.

Service, P.M. (2000b). Heterogeneity in individual mortality risk and its importance for evolutionary studies of senescence. *The American Naturalist* 156: 1–13.

Service, P.M., Hutchinson, E.W., Mackinley, M.D., and Rose, M.R. (1985). Resistance to environmental stress in *Drosophila melangaster* selected for postponed senescence. *Physiological Zoology* 58: 380–9.

Service, P.M., Micheli, C.A., and McGill. (1998). Experimental evolution of senescence: an analysis using a "heterogeneity" mortality model. *Evolution* 52: 1844–50.

Sevanian, A. (1985). Mechanisms and consequences of lipid peroxidation in biological systems. *Annual Review of Nutrition* 5: 365–90.

Severson, L.D. (2001). *HLA, microsatellite, and viral DNA sequences in American Samoans and Yanomami Indians*. Ph.D. dissertation, The Ohio State University.

Severson, L.D., Crews, D.E., and Lang, R.W. (1997). Application of SSP/ARMS to HLA class 1 loci in Samoans. In: *Proceedings of the 12th International Histocompatibility Workshop and Conference*, vol. 11 (ed. D. Charron). Paris: EDK.

Shay, J.W. (1997). Molecular pathogenesis of aging and cancer: are telomeres and telomerase the connection? *Journal of Clinical Pathology* 50: 799–800.

Shay, J.W. and Werbin, H. (1992). New evidence for the insertion of mitochondrial DNA into the human genome: significance for cancer and aging. *Mutation Research* 275: 227–35.

Shay, J.W. *et al.* (1995). Chapter 13. In *Molecular Aspects of Aging*, K. Esser and G.M. Martin (eds.). New York: John Wiley and Sons.

Shay, J.W. (1998). Molecular pathogenesis of aging and cancer: are telomeres and telomerase: the connection. *Journal of Clinical Pathology* 50: 799–800.

Shea, B. (1998). Growth in non-human primates. In *The Cambridge Encyclopedia of Human Growth and Development*, S.J. Ulijaszek, F.E. Johnston, and M.A. Preece (eds.). Cambridge: Cambridge University Press, pp. 100–3.

Shimokata, H., Muller, D.C., Fleg, J.L., Sorkin, J., Ziemba, A.W., and Andres, R. (1991). Age as an independent determinant of glucose tolerance. *Diabetes Care* 40: 44–51.

Shippen, D.E. (1993). Telomeres and telomerase. *Current Opinions in Genetics of Development* 3: 759–63.

Shock, N.W. (1984). *Normal Human Aging: The Baltimore Longitudinal Study of Aging*. U.S. Department of Health and Human Services: NIH Publication No. 84–2450.

Shock, N.W. (1985). Longitudinal studies of aging in humans. In *Handbook of the Biology of Aging*, 2nd edition, C.E. Finch and E.L. Schneider (eds.). New York: Van Nostrans Reinhold. pp. 721–43.

Silva, H.P. (2001). *Growth, development, nutrition, and health in Caboclo populations from the Brazilian Amazon*, PhD dissertation. The Ohio State University.

Silva, H.P., Crews, D.E. (in preparation). Race and ethnicity in biomedical and biosocial research: where are we and where do we go from here?

Silva, H.P., Crews, D.E., and Neves, W.A. (1995). Subsistence patterns and blood pressure variation in two rural "Caboclo" communities on Marajo Island, Para, Brazil. *American Journal of Human Biology* 7: 535–42.

Sing, C.F.E., Boerwinkle, E.A., and Moll, P.P. (1985). Strategies for elucidating the phenotypic and genetic heterogeneity of a chronic disease with a complex etiology. In *Diseases of Complex Etiology in Small Populations*, R. Chakraborty and E.J.E. Szathmary (eds.). New York: Alan R. Liss.

Slagboom, P.E., Droog, S., and Boomsma, D.I. (1994). Genetic determination of telomere size in humans: a twin study of three age groups. *American Journal of Human Genetics* 55: 876–82.

Smith, B.H. (1986). Dental development in *Australopithecus* and early *Homo*. *Nature* 323: 327–30.

Smith, B.H. (1991). Dental development and the evolution of life history in *Hominidae*. *American Journal of Physical Anthropology* 86: 157–74.

Smith, B.H. (1992). Life history and the evolution of human maturation. *Evolutionary Anthropology* 1: 134–42.

Smith, B.H. (1993). The physiological age of KNM-WT 1500. In The *Nariokotome Homo erectus Skeleton*, A. Walker and R. Leakey (eds.). New York: Harvard University Press, pp. 195–220.

Smith, B.H. and Tompkins, R.L. (1995). Toward a life history of the *Hominidae*. *Annual Review of Anthropology* 24: 257–79.

Smith, D. (1993). *Human Longevity*. New York: Oxford University Press.

Snieder, H., van Doornen, L.J., and Boomsma, D.I. (1997). The age dependency of gene expression for plasm lipids, lipoproteins, and apolipoproteins. *American Journal of Human Genetics* 60: 638–50.

Snowdon, D. (1990). Early natural menopause and the duration of postmenopausal life: findings for a mathematical model of life expectancy. *Journal of the American Geriatric Society* 38: 402–408.

Snowdon, D.A., Lane, R.A., Beeson, W.L. *et al.* (1989). Is early natural menopause a biological marker of health and aging? *American Journal of Public Health* 79: 709–14.

Sohal, R.S. and Orr, W.C. (1995). Is oxidative stress a causal factor in aging? In *Molecular Aspects of Aging*, K. Esser and G.M. Martin (eds.). New York: John Wiley and Sons, pp. 109–38.

Sohal, R.S. and Weindruch, R. (1996). Oxidative stress, caloric restriction, and aging. *Science* 273: 59–63.

Sorlie, P.D., Backlund, E., and Keller, J.B. (1995). U.S. mortality by economic, demographic, and social characteristics: the national longitudinal mortality study. *American Journal Public Health* 85: 949–56.

Soubrier, F. and Bonnardeaux, A. (1994). Role and identification of the genes involved in human hypertension. *Journal of Hypertension* 8: 579–85.

Sozou, P.D. and Krikwood, TBL. (2001). A stochastic model of cell replicative senescence based on telomere shortening, oxidative stress, and somatic mutations in nuclear and mitochondrial DNA. *American Journal of Human Biology* 213: 573–86.

Stadtman, E.R. (1992). Protein oxidation and aging. *Science* 257: 1220–3.

Stadtman, E.R. (1995). The status of oxidatively modified proteins as a marker of aging. In *Molecular Aspects of Aging*, K. Esser and G.M. Martin (eds.). New York: John Wiley and Sons, pp. 129–144.

Sternberg, E.M. (1997). Perspectives series: cytokines and the brain. Neural-immune interactions in health and disease. *Journal of Clinical Investigation* 100: 2641–7.

Sterns, S.C. (1992). *The Evolution of Life Histories*. New York: Oxford University Press.

Sterns, S.C. and Hoekstra, R.F. (2000). *Evolution: An Introduction*. New York: Oxford University Press.

Steward, J.H. (1977). *Evolution and Ecology: Essays on Social Transformation*, J.C. Steward and R.F. Murphy, (eds.). University of Illinois Press.

Stini, W.A. (1971). Evolutionary implications of changing nutritional patterns in human populations. *American Anthropologist* 73: 1019–30.

Stini, W.A. (1972). Malnutrition, body size, and proportion. *Ecology of Food and Nutrition* 1: 1–6.

Stini, W.A. (1990). Changing patterns of morbidity and mortality and the challenge to health care delivery systems of the future. *Collegium Anthropologium* 14: 189–95.

Stini, W.A. (1991). Nutrition and aging: intraindividual variation. In *Biological Anthropology and Aging: Perspectives on Human Variation over the Life Span*, D.E. Crews and R.M. Garruto (eds.). Oxford: Oxford University Press, pp. 232–71.

Stini, W.A. (in press). Sex differences in bone loss: an evolutionary perspective on a clinical problem. *Collegium Anthropologicum*.

Stinson, S. (1985). Sex differences in environmental sensitivities during growth and development. *Yearbook of Physical Anthropology* 28: 123–47.

Strehler, B. (1982). *Time, Cells, and Aging*. New York: Academic Press.

Strittmatter, J.J. and Roses, A.D. (1995). Apolipoprotein E and Alzheimer's disease. *Proceedings of the National Academy of Science USA* 92: 4725–7.

Strong, M.J. and Garruto, R.M. (1994). Neuronal aging and age-related disorders. In *Biological Anthropology and Aging: Perspectives on Human Variation Over the Life Span*, D.E. Crews and R.M. Garruto (eds.). Oxford: Oxford University Press, pp. 214–31.

Sugiyama, Y. (1994). Age-specific birth rate and lifetime reproductive success of chimpanzees at Bossou, Guinea. *American Journal of Primatology* 32: 311–18.

Sullivan, T.V. (2002). *The Growth Status of Guatamallian Street Children*. PhD. thesis, Suny-Buffalo.

Susser, M., Watson, W., and Hopper, K. (1985). *Sociology in Medicine*. New York: Oxford University Press.

Swartz, H.M. and Mäder, K. (1995). Free radicals in aging: theories, facts, and artifacts. In *Molecular Aspects of Aging*, K. Esser and B.M. Martin (eds.). New York: John Wiley and Sons.

Swedlund, A.C. *et al.* (1976). Population studies in the Connecticut Valley: a prospectus. *Journal of Human Ecology*, 5: 75.

Swedlund A.C., Meindl, R.S., and Gradie, M.I. (1980). Family reconstitution in the Connecticut valley: progress on record linkage and the mortality survey. In *Genealogical Demography*, B. Dyke, W. Morrill (eds.). *New York: Academic Press*. pp. 139–55.

Swedlund, A.C., Meindl, R.S., Nydon., J., and Gradie, M.I. (1983). Family patterns in longevity and longevity patterns of the family. *Human Biology* 55: 115–29.

Takata, H., Suzuki, M., Ishi, T., Seiguchi, S., and Iri, H. (1987). Influence of major histocompatibility complex region genes on human longevity among Okinawan-Japanese centenarians and nonagenarians. *Lancet* 2: 824–6.

Takeda, T., Hosokawa, M., and Higuchi, K. (1991). Senescence-accelerated mouse (SAM). A novel murine model of aging. 39: 911–19.

Takeda, T., Hosokawa, M., and Higuchi, K. (1994). Senescence-accelerated mouse (SAM). A novel murine model of aging. In *The SAM model of Senescence*, T. Takeda (ed.). Amsterdam: Elsevier Science, pp. 15–22.

Tanaka, M., Gong, J-S., Zang, J., Yaneda, M., and Yagi, K. (1998). Mitochondrial genotype associated with longevity. *Lancet* 351: 185–6.

Tanner, J.M. (1981). *A History of the Study of Human Growth.* Cambridge: Cambridge University Press.

Tanner, C.M., Offman, R., Goldman, S.M. *et al.* (1999). Parkinson's disease in twins: an etiologic study. *Journal of the American Medical Association* 281: 341–6.

Tanner, J.M. (1990). *Fetus Into Man: Physical Growth from Conception to Maturity* (revised and enlarged). Cambridge MA: Harvard Press.

Templeton, A.R. (2002). Out of Africa again and again. *Nature* 416: 45–51.

Thomas, P.D., Goodwin, J.M., and Goodwin, J.S. (1985). Effect of social support on stress-related changes in cholesterol level, uric acid level, and immune function in an elderly sample. *American Journal of Psychiatry* 142: 735–7.

Thompson, J.S, Wekstein, D.R., Rhoades, J.L. *et al.* (1984). The immune Status of healthy centenarians. *Journal of the American Geriatrics Society* 32: 274–81.

Thompson, J.L. and Nelson, A.J. (2000). The place of Neanderthals in the evolution of hominids patterns of growth and development. *Journal of Human Evolution* 38: 475–95.

Tietlebaum, M.S. (1975). Relevance of demographic transition theory to developing populations. *Science* 188: 420–5.

Tissenbaum, H.A. and Ruvkun, G. (1998). An insulin-like signaling pathway effects both longevity and reproduction in *Caenorhabditis elegans*. *Genetics* 148: 703–17.

Torrey, E. F., Miller, J., Rawlings, R., and Yolken, R. (2000). Seasonal birth patterns of neurological disorders. *Neuroepidemiology* 19: 177–85.

Toupance, B., Gordelle, B., Gouyon, P-H., and Schacter, F. (1998). A model for antagonistic pleiotropic gene action for mortality and advanced age. *American Journal of Human Genetics* 62: 1525–34.

Tucker, K. (1995). Micronutrient status and aging. *Nutritional Reviews* 53: 59–15.

Tuomo, P., Sulkava, R., Haltia, M. *et al.* (1996). Apolipoprotein E, dementia, and cortical deposition of β-amyloid protein. *New England Journal of Medicine* 333: 1242–7.

Turke, P.W. (1997). Hypothesis: menopause discourages infanticide and encourages continued investment by agnates. *Evolution and Human Behavior* 18: 3–13.

Turner, T. and Weiss, M. (1994). The genetic of longevity in humans: models, biomarkers and population data. In *Biological Anthropology and Aging: Perspectives on Human Variation Over the Life Span*, D.E. Crews and R.M. Garruto (eds.). New York: Oxford University Press.

Ulijaszek, S. (1998a). Comparative growth and development of mammals. In *The Cambridge Encyclopedia of Human Growth and Development*, S.J. Ulijaszek, F.E. Johnston, and M.A. Preece, (eds.). Cambridge: Cambridge University Press, pp. 96–9.

Ulijaszek, S. (1998b). The genetics of growth. In *The Cambridge Encyclopedia of Human Growth and Development*, S.J. Ulijaszek, F.E. Johnston, and M.A. Preece (eds.). Cambridge: Cambridge University Press, pp. 121–3.

Ulijaszek S.J., Johnston, F.E., and Preece M.A. (eds.) (1999). *The Cambridge Encyclopedia of Human Growth and Development.* Cambridge: Cambridge University Press.

United Nations. (1996). *Demographic Yearbook.* New York: United Nations.

Vaag, A., Alford, F., and Becknielsen, H. (1996). Intracellular glucose and fat metabolism in identical twins discordant for non-insulin-dependent diabetes mellitus (NIDDM): aquired verses genetic metobolic defects? *Diabetic Medicine* 13: 806–15.

van Bockxmeer, F.M. (1994). *ApoE* and *ACE* genes: impact on human longevity. *Nature Genet.* 6: 4–5.

VanVoorhies, W.A. (1992). Production of sperm reduces nematode life span. *Nature* 360: 456–8.

Vaupel, J.W. (1988). Inherited frailty and longevity. *Demography* 25: 277–87.

Vaupel, J.W., Carey, J.R., Christensen, K. *et al.* (1998). Biodemographic trajectories of longevity. *Science* 280: 855–60.

Vaupel, J.W., Manton, K.G., and Stallard, E. (1979). The impact of heterogeneity in individual frailty on the dynamics of mortality. *Demography* 16: 439.

Verbrugge, L.M. (1984). Longer life but worsening health: trends in health and mortality of middle-aged and older persons. *Millbank Memorial Fund. Quartely* 62: 475–519.

Verbrugge, L.M. (1989). The dynamics of population aging and health. In *Aging and Health*, S.J. Lewis and P. Baran (eds.). Chelsea: Lewis Publishers, Inc., pp. 23–40.

Vijg, J. and Wei, J.Y. (1995). Understanding the biology of aging: the key to prevention and therapy. *Journal of the American Geriatric Society* 43: 426–34.

Waldron, I. (1983). The role of genetic and biological factors in sex differences in mortality. In *Sex Differentials in Mortality: Trends, Determinants, and Consequences*, A.D. Lopez and L.T. Ruzicka (eds.). Canberra: Australian National University Press.

Walford, R.L. (1968). *The 120-Year Diet.* New York: Simon and Schuster.

Walford, R.L. (1969). *The Immunological Theory of Aging.* Copenhagen: Munksgaard Kobenhaven.

Walford, R. (1984). *Maximum Life Span.* New York: Avon Printing.

Wallace, D.C. (1992a). Diseases of the mitochondrial DNA. *Annual Review of Biochemistry* 61: 1175–212.

Wallace, D.C. (1992b). Mitochondrial genetics: a paradigm for aging and degenerative diseases? *Science* 256: 628–32.

Wallace, D.C. (2001). A mitochondrial paradigm for degenerative diseases and aging. *Novartis Foundation Symposium* 235: 247–63.

Wallace, D.C., Bohr, V.A., Cortopassi, G. *et al.* (1995). Group report: The role of bioenergetic and mitochondrial DNA mutations in aging and age-related diseases. In *Molecular Aspects of Aging*, K. Esser and G.M. Martin (eds.). New York: John Wiley and Sons, pp. 199–225.

Wallace, D.C., Singh, G., Lott, M.T. *et al.* (1988). Mitochondrial DNA mutation associated with Lieber's hereditary optic neuropathy. *Science* 242: 1427–30.

Warner, H.R. (1994). Superoxide dismutase, aging and degenerative disease. *Free Radical Biology and Medicine* 17: 249–58.

Webster, N. (1983). *Webster's New Universal Unabridged Dictionary*, Deluxe Second Edition. New York: Simon and Schuster.

Weder, A.B. and Schork, N.J. (1994). Adaptation, allometry, and hypertension. *Hypertension* 24: 145–56.

Wei, Y.H. (1998). Oxidative stress and mitochondrial DNA mutations in human aging. *Publication of the Society for Experimental Biology and Medicine* 217: 53–63.

Weindrunch, R. (1989). Dietary restriction, tumors, and aging in rodents. *Journal of Gerontology* 44: 67–71.

Weindrunch, R. and Walford, R.L. (1988). *The Retardation of Aging and Disease by Dietary Restriction*. Springfield, IL: Charles C. Thomas.

Weindrunch, R., Marriott, B.M., Conway, J. *et al.* (1995). Measures of body size and growth in rhesus and squirrel monkeys subject to long-term dietary restriction. *American Journal of Primatology* 35: 207–28.

Weismann, A. (1889). The Duration of Life (a paper presented in 1881). In *Essays upon Heredity and Kindred Biological Problems*, E.B. Poulton, S. Schonland, and A.E. Shipley (eds.). Oxford: Clarendon Press, pp. 1–66.

Weiss, K.M. (1981). Evolutionary perspectives on aging. In *Other Ways of Growing Old: Anthropological Perspectives*, P.T. Amoss and S. Harrell (eds.). Stanford, C.A.: Stanford University Press, pp. 25–58.

Weiss, K.M. (1984). On the number of members of genus *Homo* who have ever lived, and some evolutionary interpretations. *Human Biology* 56: 637–49.

Weiss, K.M. (1989a). Are Known Chronic Diseases Related to the Human Lifespan and its Evolution? *American Journal of Human Biology* 1: 307–19.

Weiss, K.M. (1989b). A survey of biodemography. *J Quantitative Anthropology* 1: 79–151.

Weiss, K.M. (1990). The biodemography of variation in human frailty. *Demography* 27: 185–206.

Weiss, K.M., Ferrell, R.E., and Hanis, C.L. (1984). A New World "Syndrome" of metabolic diseases with a genetic and evolutionary basis. *Yearbook of Physical Anthropology* 27: 153.

Westendorp, R.J.P. and Kirkwood, TBL. (1998). Human longevity at the cost of reproductive success. *Nature* 396: 743–6.

Whitbourne, S.K. (1985). *The Aging Body: Physiological Changes and Psychological Consequences*. New York: Springer-Verlag.

Wilding, G. (1995). Endocrine control of prostate cancer. *Cancer Surveys* 23: 43–62.

Williams, D.R. (1998). African American health: the role of the social environment. *Journal of Urban Health Bulletin of the New York Academy of Medicine* 75: 300–21.

Williams, G.C. (1957). Pleiotropy, natural selection, and the evolution of senescence. *Evolution* 11: 398–411.

Wilmoth, J.R., Deegan, L.J., Lunström, H., Horiuchi, S. (2000). Increase of maximum life-span in Sweden, 1861–1999. *Science* 289: 2366–8.

Wilson, E.O. and Bossert, W.H. (1971). *A Primer of Population Biology*. Stamford, C.T.: Sinauer Associates.

Wilson, T.W. and Grim, C.E. (1991). Biohistory of slavery and blood pressure differences in Blacks today: a hypothesis. *Hypertension* 17: I122–8.

Wolpoff, M. (1998). *Paleoanthropology*. McGraw-Hill.

Wood, J.W., Bentley, G.R., and Weiss, K.M. (1994). Human population biology and the evolution of human aging. In *Biological Anthropology and Aging: Perspectives on Human Variation Over the Life Span*, D.E. Crews and R.M. Garruto (eds.). New York: Oxford University Press.

World Health Organization (2002). Diabetes Mellitus Fact Sheet. http://www.who.int/inffs/en/fact138.html

Wright, R.O. (1997). *Life and Death in the United States*. Jefferson, N.C.: McFarland and Company.

Yarnell, J.W.G., St. Leger, A.S., Balfour, I.C., and Russell, R.B. (1979). The distribution, age effects, and disease associations of HLA antigens and other blood group markers in a random sample of an elderly population. *Journal Chronic Disease* 32: 555–61.

Yegorov, Y. *et al.* (2001). Senescent accelerated mouse (SAM): a model that binds in vivo and in vitro aging. *Journal of Anti-Aging Medicine* 4: 39–47.

Yeung, A.C. (1997). Estrogen for men: reversal of cardiovascular misfortune. *American College of Cardiology* 29: 1445–6.

Yu, B.P. (1993). Oxidative damage by free radicals and lipid peroxidation in aging. In: *Free Radicals in Aging*, B.P. Yu (ed.). Boca Raton, FL: CRC Press.

Yunis, E.J. and Salazar, M. (1993). Genetics of life span in mice. *Genetica* 91: 211–23.

Zeman, F.D. (1962). Pathologic anatomy of old age. *Archives of Pathology* 73: 126–45.

Zhao, Z., Weiss, K., and Stock, D.W. (1998). Growth factor and development. In *The Cambridge Encyclopedia of Human Growth and Development*, S.J. Uljaszek, F.E Johnston, M.A. Preece (eds.). New York: Cambridge University Press, pp.151–3.

Ziegler, R. and Scheidt-Nave, C. (1995). Pathophysiology of osteoporosis: unresolved problems and new insights. *Journal of Nutrition* 125: 2033S–7S.

Zwaan, B., Bijlmsa, R., and Hoekstra, R.F. (1995). Direct selection on life span in *Drosophila melanogaster*. *Evolution* 49: 649–59.

Index